CALCULATION
and
CALCULATORS

CALCULATION
and
CALCULATORS

THOMAS J. McHALE
PAUL T. WITZKE

Milwaukee Area Technical College, Milwaukee, Wisconsin

ADDISON-WESLEY PUBLISHING COMPANY
Reading, Massachusetts · Menlo Park, California
London · Amsterdam · Don Mills, Ontario · Sydney

OTHER BOOKS OF INTEREST

ARITHMETIC MODULE SERIES

Programmed Modules

Module 1: Whole Numbers
Module 2: Fractions
Module 3: Decimal Numbers
Module 4: Percent, Ratio, Proportion
Module 5: Rounding, Estimation, Squares and Square Roots,
Formula Evaluation, Exponents and Scientific Notation

Nonprogrammed Text

A single text with the content of all five modules in
shorter, nonprogrammed form.

INTRODUCTORY ALGEBRA - Programmed
INTRODUCTORY ALGEBRA

Reproduced by Addison-Wesley from camera-ready copy supplied by the
authors.

ISBN 0-201-04771-3

29 30 31 32 33 34 35-CRW-010099

PREFACE

<u>CALCULATION</u> <u>AND</u> <u>CALCULATORS</u> is designed to teach a wide range of calculation skills using a "scientific" or "slide-rule" calculator with an algebraic-entry method. Besides elementary operations, skills involving scientific notation, formula evaluation, trigonometric ratios, powers, roots, and both common and natural logarithms are included. The calculation skills are presented in the context of teaching the related mathematical concepts and principles. Where necessary, separate instruction is given for calculators with parentheses symbols and calculators without parentheses symbols. The content, which is ideally suited for students in science, technology, pre-engineering, and mathematics itself, is presented in a programmed format with assignment self-tests with answers within each chapter and supplementary problems for each assignment at the end of each chapter. Answers for the supplementary problems are given in the back of the text.

The text is accompanied by the book <u>TESTS</u> <u>FOR</u> <u>CALCULATION</u> <u>AND</u> <u>CALCULATORS</u> which contains a diagnostic test, thirty assignment tests, ten chapter tests, five multi-chapter tests (every two chapters), and a comprehensive test. Three parallel forms are provided for chapter tests, multi-chapter tests, and the comprehensive test. A full set of answer keys for the tests is included. Because of the large number of tests provided, various options are possible in using them. For example, an instructor can use the chapter tests alone, the multi-chapter tests alone, or some combination of the two types.

> Note: The test book is provided <u>only</u> to teachers. Copies of the tests
> for student use must be made by some copying process.

The following features make the instruction effective and efficient for students:

1. The instruction, which is based on a task analysis, contains examples of all types of problems that appear in the tests.

2. The full and flexible set of tests can be used as a teaching tool to identify learning difficulties which can be remedied by tutoring or class discussion.

3. Because of the programmed format and the full and flexible set of tests provided, the text is ideally suited for individualized instruction.

Various options are possible in sequencing the chapters in the text. The first four chapters should be covered in order. Chapters 1, 3, and 4 are prerequisites for Chapters 5 through 9 which can be covered in any order, with the exception that Chapter 6 should follow Chapter 5. Chapter 10 can be covered any time after Chapter 3. It is probably wise to cover Chapter 9 after the first eight chapters are completed in order to avoid unproductive discussion about the number of significant digits that should be reported in answers.

The authors wish to thank Ms. Arleen D'Amore and Ms. Marylou Hansen who typed and proofread the camera-ready copy, Ms. Peggy McHale who prepared the drawings, Mr. Gail W. Davis who did the final proofreading, and Mr. Allan A. Christenson who prepared the Index.

HOW TO USE THE TEXT AND TESTS

This text and the tests available in TESTS FOR CALCULATION AND CALCULATORS can be used in various instructional strategies ranging from paced instruction with all students taking the same content to totally individualized instruction. The general procedure for using the materials is outlined below.

1. The diagnostic test can be administered either to simply get a measure of the entry skills of the students or as a basis for prescribing an individualized program.

2. Each chapter is covered in a number of assignments (see below). After the students have completed each assignment and the assignment self-test (in the text), the assignment test (from the test book) can be administered, corrected, and used as a basis for tutoring. The assignment tests are simply a teaching tool and need not be graded. The supplementary problems at the end of the chapter can be used at the instructor's discretion for students who need further practice.

   Note: Instead of using the assignment self-tests (in the text) as an integral part of the assignments, they can be used at the completion of a chapter as a chapter review exercise.

3. After the appropriate assignments are completed, either a chapter test or a multi-chapter test can be administered. Ordinarily, these tests should be graded. Parallel forms are provided to facilitate the test administration, including the retesting of students who do not achieve a satisfactory score.

4. After all desired chapters are completed in the manner above, the comprehensive test can be administered. Since the comprehensive test is a parallel form of the diagnostic test, the difference score can be used as a measure of each student's improvement.

ASSIGNMENTS FOR CALCULATION AND CALCULATORS

Ch. 1:	#1 (pp. 1-15)	Ch. 4:	#10 (pp. 114-128)	Ch. 7:	#20 (pp. 266-277)
	#2 (pp. 15-26)		#11 (pp. 129-142)		#21 (pp. 278-292)
	#3 (pp. 27-36)		#12 (pp. 143-153)		#22 (pp. 293-302)
			#13 (pp. 153-172)		
Ch. 2:	#4 (pp. 40-53)			Ch. 8:	#23 (pp. 306-317)
	#5 (pp. 54-62)	Ch. 5:	#14 (pp. 176-193)		#24 (pp. 317-328)
	#6 (pp. 63-73)		#15 (pp. 194-207)		#25 (pp. 329-343)
			#16 (pp. 208-223)		
Ch. 3:	#7 (pp. 76-89)			Ch. 9:	#26 (pp. 346-359)
	#8 (pp. 89-100)	Ch. 6:	#17 (pp. 226-240)		#27 (pp. 360-373)
	#9 (pp. 101-111)		#18 (pp. 240-252)		#28 (pp. 374-384)
			#19 (pp. 252-262)		
				Ch. 10:	#29 (pp. 387-398)
					#30 (pp. 399-409)

C O N T E N T S

Chapter 1

ELEMENTARY CALCULATOR OPERATIONS

In this chapter, we will discuss the method for performing the four fundamental operations (adding, subtracting, multiplying, and dividing) and three other operations (squaring, square roots, and reciprocals) on a calculator. The operations are extended to signed numbers. A review of the decimal number system and a review of the rules for rounding are included.

> Note: The instruction in this book is designed for calculators with an "algebraic entry method" as opposed to an "RPN entry method".

1-1   WHOLE NUMBERS

In this section, we will discuss place-names in whole numbers, the two expanded forms for whole numbers, and place-value in whole numbers.

---

1.   Our number system is called the "decimal" number system because it uses <u>ten</u> symbols called "digits". The ten digits and their word names are:

0 zero	2 two	4 four	6 six	8 eight
1 one	3 three	5 five	7 seven	9 nine

Most whole numbers contain two or more digits written in definite places. The names of the first ten places in whole numbers are given in the diagram below.

In 6,058,197,342: the "4" is in the <u>tens</u> place.

   a) the "7" is in the _____ place.

   b) the "8" is in the _____ place.

   c) the "6" is in the _____ place.

2.   Name the place in which the "3" appears in each number below.

    a) 4,381 _____   d) 73,689 _____

    b) 930,245 _____   e) 317,690,458 _____

    c) 375,612 _____   f) 635,088,144 _____

a) thousands

b) millions

c) billions

---

3.   Any whole number with two or more digits can be expanded in <u>place-name</u> form. For example:

$$62 = 6 \text{ tens} + 2 \text{ ones}$$
$$495 = 4 \text{ hundreds} + 9 \text{ tens} + 5 \text{ ones}$$
$$3,018 = 3 \text{ thousands} + 0 \text{ hundreds} + 1 \text{ ten} + 8 \text{ ones}$$

Write the number corresponding to each place-name expansion.

    a) 3 tens + 9 ones = _____

    b) 1 hundred + 7 tens + 0 ones = _____

    c) 8 thousands + 4 hundreds + 2 tens + 2 ones = _____

a) hundreds

b) ten-thousands

c) hundred-thousands

d) thousands

e) hundred-millions

f) ten-millions

---

4.   The terms in the place-name expansion of whole numbers can also be written as <u>place-numbers</u>. For example:

$$5 \text{ ones} = 5$$
$$2 \text{ tens} = 20$$
$$9 \text{ hundreds} = 900$$
$$7 \text{ thousands} = 7,000$$

Therefore, any whole number with two or more digits can also be expanded in place-number form. That is:

$$38 = 30 + 8$$
$$149 = 100 + 40 + 9$$
$$6,572 = 6,000 + 500 + 70 + 2$$

Write the number corresponding to each place-number expansion.

    a) 70 + 3 = _____

    b) 200 + 10 + 6 = _____

    c) 9,000 + 400 + 50 + 8 = _____

a) 39

b) 170

c) 8,422

---

a) 73

b) 216

c) 9,458

5. The meaning of a digit in a number depends on its _place_ in the number. That is:

    In 215, the "5" means 5 ones or 5.
    In 856, the "5" means 5 tens or 50.

    In 504, the "5" means _____ or _____ .

---

6. Each place-name term below was converted to a place-number by writing the non-zero digit in the named place with 0's in all places to its right. For example:

        8 ten-thousands =   80,000

     2 hundred-thousands =   200,000

          4 millions = 4,000,000

Convert each place-name term to a place-number.

    a) 5 ten-millions = _____   c) 7 billions = _____

    b) 6 hundred-millions = _____   d) 1 ten-thousand = _____

5 hundreds or 500

---

7. In 8,147,926,305 : the "7" means 7 millions or 7,000,000

    a) the "4" means _____ or _____

    b) the "9" means _____ or _____

    c) the "8" means _____ or _____

a)   50,000,000

b)   600,000,000

c) 7,000,000,000

d)   10,000

---

8. When "0" appears in any place in a number, that place contributes "0" to the value of the number. For example:

    In 508 , the "0" means 0 tens or 0.
   In 40,125 , the "0" means 0 thousands or 0.

    In 2,088,644 , the "0" means _____ or _____ .

a) 4 ten-millions
   or 40,000,000

b) 9 hundred-thousands
   or 900,000

c) 8 billions
   or 8,000,000,000

---

0 hundred-thousands or 0

1-2   WORD NAMES FOR WHOLE NUMBERS

In this section, we will discuss the word names for whole numbers.

---

9.   Some two-digit numbers have a one-word name. They are given in the tables below.

10	ten	15	fifteen
11	eleven	16	sixteen
12	twelve	17	seventeen
13	thirteen	18	eighteen
14	fourteen	19	nineteen

20	twenty	60	sixty
30	thirty	70	seventy
40	forty	80	eighty
50	fifty	90	ninety

The word names for all other two-digit numbers contain a hyphen ("-").
For example:

34 is called "thirty-four".
81 is called "eighty-one".

Write the number corresponding to each word name.

a) ninety-eight = _____   c) seventy-five = _____

b) twenty-three = _____   d) forty-seven = _____

---

10.   The word names for three-digit numbers contain the word "hundred".
For example:

200 is called "two hundred".
603 is called "six hundred three".
470 is called "four hundred seventy".
861 is called "eight hundred sixty-one".

Write the number corresponding to each word name.

a) five hundred = _____   c) one hundred nine = _____

b) seven hundred twenty = _____   d) nine hundred forty-four = _____

a) 98   c) 75

b) 23   d) 47

---

11.   The word name for any whole number with four, five, or six digits contains the word "thousand". For example:

3,000 = three thousand

7,164 = seven thousand, one hundred sixty-four

58,400 = fifty-eight thousand, four hundred

950,015 = nine hundred fifty thousand, fifteen

a) 500   c) 109

b) 720   d) 944

Continued on following page.

11.   Continued

Write the number corresponding to each word name.

    a) five thousand, two hundred = _____

    b) ten thousand, fifty = _____

    c) seventy-three thousand, one hundred ninety = _____

    d) two hundred thirty thousand, five hundred eighty-eight = _____

---

12.   The word name for any whole number with seven, eight, or nine digits contains the word "million". For example:

        5,000,000 = five million

        1,750,000 = one million, seven hundred fifty thousand

      84,200,000 = eighty-four million, two hundred thousand

    316,000,000 = three hundred sixteen million

Write the number corresponding to each word name.

    a) two million, five hundred thousand = _____

    b) eleven million, seventy thousand = _____

    c) forty million, two hundred thirty thousand = _____

    d) eight hundred ninety-nine million = _____

a)	5,200
b)	10,050
c)	73,190
d)	230,588

---

13.   The numbers below contain more than nine digits. Their word names contain the word "billion".

        9,000,000,000 = nine billion

    20,800,000,000 = twenty billion, eight hundred million

Write the number corresponding to each word name.

    a) thirty-five billion, six hundred million = _____

    b) one hundred fifty billion, seven hundred ninety million =

        _____

a)	2,500,000
b)	11,070,000
c)	40,230,000
d)	899,000,000

a)	35,600,000,000
b)	150,790,000,000

14.   We have seen that the word "thousand" can be used to name a four-digit whole number. For example:

$$2,000 = \text{two thousand}$$

$$1,700 = \text{one thousand, seven hundred}$$

$$6,390 = \text{six thousand, three hundred ninety}$$

We can also name four-digit numbers by using the word "hundred". For example:

$$2,000 = \text{twenty hundred}$$

$$1,700 = \text{seventeen hundred}$$

$$6,390 = \text{sixty-three hundred ninety}$$

Write the number corresponding to each word name.

a) thirty hundred = _____   c) sixty-eight hundred fifty = _____

b) eleven hundred = _____   d) ninety-one hundred twelve = _____

a) 3,000   b) 1,100   c) 6,850   d) 9,112

1-3   DECIMAL NUMBERS

In this section, we will discuss place-names in decimal numbers, the two expanded forms of decimal numbers, and place-value in decimal numbers.

---

15.   A fraction is called a "decimal" fraction if its denominator is a number like 10, 100, 1,000 , and so on. The word names for some decimal fractions are given below.

$$\frac{1}{10} = \text{one tenth}$$

$$\frac{7}{100} = \text{seven hundredths}$$

$$\frac{80}{1,000} = \text{eighty thousandths}$$

$$\frac{613}{1,000,000} = \text{six hundred thirteen millionths}$$

Write the decimal fraction corresponding to each word name.

a) nine tenths = _____   c) fifty-one thousandths = _____

b) fifteen hundredths = _____   d) one hundred-thousandth = _____

a) $\frac{9}{10}$ b) $\frac{15}{100}$ c) $\frac{51}{1,000}$ d) $\frac{1}{100,000}$

16. By "decimal places" in a decimal number, we mean the number of digits
     (including 0's) to the right of the decimal point. For example:

                  .5   has one decimal place
                  .307   has three decimal places
                  .0098 has four decimal places

     Identify the number of decimal places in each number below.

     a) .007 _____   b) .59 _____   c) .000101 _____

---

17. Any decimal fraction can be converted to a decimal number. The decimal
     number contains the digit or digits of the numerator and as many decimal
     places as there are 0's in the denominator. For example:

     $\frac{3}{10}$ = .3   (Since 10 has one "0", .3 has one decimal place.)

     $\frac{509}{1,000}$ = .509 (Since 1,000 has three 0's, .509 has three decimal places.)

     Sometimes we have to add one or more 0's in front of the digit or digits of
     the numerator to get enough decimal places. For example:

     $\frac{5}{100}$ = .05

     (We need two decimal places since 100 has two 0's.)

     $\frac{903}{1,000,000}$ = .000903

     (We need six decimal places since 1,000,000 has six 0's.)

     Convert each decimal fraction to a decimal number.

     a) $\frac{1}{10}$ = _____   d) $\frac{25}{100}$ = _____

     b) $\frac{84}{1,000}$ = _____   e) $\frac{647}{10,000}$ = _____

     c) $\frac{7}{100,000}$ = _____   f) $\frac{309}{1,000,000}$ = _____

a) three

b) two

c) six

---

18. When the numerator ends with one or more 0's, the decimal number
     ends with one or more 0's. For example:

     $\frac{90}{1,000}$ = .090 $\frac{2,600}{1,000,000}$ = .002600

     Convert each decimal fraction to a decimal number.

     a) $\frac{70}{100}$ = _____   b) $\frac{120}{10,000}$ = _____   c) $\frac{800}{100,000}$ = _____

a) .1

b) .084

c) .00007

d) .25

e) .0647

f) .000309

a) .70   b) .0120   c) .00800

19. When a decimal fraction is converted to a decimal number, the name of the place of the last digit is the same as the name of the denominator. That is:

$$\frac{5}{10} = .5 \qquad \text{(The "5" is in the \underline{tenths} place.)}$$

$$\frac{17}{100} = .17 \qquad \text{(The "7" is in the "\underline{hundredths}" place.)}$$

$$\frac{83}{1,000} = .083 \qquad \text{(The "3" is in the "\underline{thousandths}" place.)}$$

Following the pattern above, we named the first six places to the right of the decimal point in the diagram below.

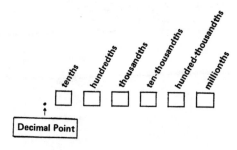

Therefore, in .371928 : the "3" is in the <u>tenths</u> place.
the "1" is in the <u>thousandths</u> place.

the "2" is in the _____ place.

---

20. Name the place in which the "7" appears in each decimal number below.

a) .75 _____   d) .071 _____

b) .607 _____   e) .00473 _____

c) .000187 _____   f) .000279 _____

hundred-thousandths

---

21. Any mixed number containing a decimal fraction can be converted to a decimal number larger than "1". To do so, we simply substitute a decimal part for the fraction part. For example:

$$25\frac{1}{10} = 25.1 \qquad 98\frac{47}{100} = 98.47 \qquad 1\frac{9}{1,000} = 1.009$$

Convert each mixed number to a decimal number.

a) $47\frac{3}{10} =$ _____   b) $7\frac{9}{100} =$ _____   c) $3\frac{159}{1,000} =$ _____

a) tenths

b) thousandths

c) millionths

d) hundredths

e) ten-thousandths

f) hundred-thousandths

---

a) 47.3

b) 7.09

c) 3.159

22. Name the place in which the "3" appears in each number.

a) 37.8 _____   c) 93.5 _____

b) 5.63 _____   d) 4.31 _____

---

23. Any decimal number can be expanded in place-name form. For example:

$$2.87 = 2 \text{ ones} + 8 \text{ tenths} + 7 \text{ hundredths}$$
$$.049 = 0 \text{ tenths} + 4 \text{ hundredths} + 9 \text{ thousandths}$$

Write the number corresponding to each place-name expansion.

a) 5 tens + 0 ones + 1 tenth = _____

b) 8 ones + 3 tenths + 6 hundredths = _____

c) 2 tenths + 1 hundredth + 7 thousandths = _____

a) tens

b) hundredths

c) ones

d) tenths

---

24. The place-name terms for digits to the right of the decimal point can also be converted to <u>place-numbers</u>. That is:

$$5 \text{ tenths} = .5$$
$$8 \text{ hundredths} = .08$$
$$3 \text{ thousandths} = .003$$
$$1 \text{ ten-thousandth} = .0001$$

Therefore any decimal number can also be expanded in <u>place-number</u> form. That is:

$$3.72 = 3 + .7 + .02$$
$$.485 = .4 + .08 + .005$$

Write the number corresponding to each place-number expansion.

a) 40 + 7 + .1 = _____   c) .6 + .05 = _____

b) 9 + .4 + .08 = _____   d) .2 + .01 + .009 = _____

a) 50.1

b) 8.36

c) .217

---

25. The meaning of a digit to the right of the decimal point also depends on its <u>place</u> in the number. That is:

In .607 , the "6" means 6 tenths or .6 .
In .368 , the "6" means 6 hundredths or .06 .

In .716 , the "6" means _____ or _____ .

a) 47.1   c) .65

b) 9.48   d) .219

---

26. Each place-name term below can be converted to a place-number by writing the digit in the named place. That is:

$$3 \text{ ten-thousandths} = .0003$$

a) 9 hundred-thousandths = _____

b)   4 millionths = _____

6 thousandths or .006

27.   In .000647, the "6" means 6 ten-thousandths or .0006 .

    a) In .000169, the "6" means _____ or _____ .

    b) In .000536, the "6" means _____ or _____ .

      a) .00009

      b) .000004

---

28.   When writing a decimal number smaller than "1", a "0" is sometimes written in the <u>ones</u> place. For example:

          0.25 is written instead of .25 .

          0.008 is written instead of .008 .

Of course, writing a "0" in the <u>ones</u> place does not change the value of the number. That is:

    0.3 = .3   0.019 = .019   0.000668 = _____

      a) 6 hundred-thousandths
         or .00006

      b) 6 millionths
         or .000006

      .000668

1-4   WORD NAMES FOR DECIMAL NUMBERS

In this section, we will discuss the word names for decimal numbers.

---

29.   To name a decimal number smaller than "1", we simply name it as if it were a whole number <u>and</u> <u>then</u> <u>add</u> <u>the</u> <u>name</u> <u>of</u> <u>the</u> <u>place</u> <u>of</u> <u>the</u> <u>last</u> <u>digit</u>. Some examples are given below.

          .4 = four <u>tenths</u>

         .71 = seventy-one <u>hundredths</u>

       .003 = three <u>thousandths</u>

    .000205 = two hundred five <u>millionths</u>

Write the number corresponding to each word name.

    a) twelve hundredths = _____

    b) nine millionths = _____

    c) twenty-four thousandths = _____

    d) four hundred three ten-thousandths = _____

---

30.   Write the number corresponding to each word name.

    a) eighty-one hundred-thousandths = _____

    b) two hundred sixty-six ten-thousandths = _____

    c) four hundred seventy-five millionths = _____

      a) .12

      b) .000009

      c) .024

      d) .0403

31. Decimal numbers ending with one or more 0's are named in the usual way. For example:

    .20 = twenty hundredths
    .0750 = seven hundred fifty ten-thousandths

    Write the decimal number corresponding to each word name.

    a) forty thousandths = _____

    b) six hundred thirty hundred-thousandths = _____

    c) one hundred millionths = _____

    a) .00081
    b) .0266
    c) .000475

32. When naming a decimal number larger than "1", the name of the whole number part and the name of the decimal part are connected by the word "and". For example:

    12.3 = twelve and three tenths
    4.07 = four and seven hundredths

    Write the number corresponding to each word name.

    a) twenty and five tenths = _____

    b) nine and fifteen hundredths = _____

    c) one and fifty-seven thousandths = _____

    d) eighteen and ten thousandths = _____

    a) .040
    b) .00630
    c) .000100

33. To name a decimal number smaller than "1", we sometimes use the word "point" for the decimal point and then name the digits from left to right. For example:

    .56 is named "point five six".
    .037 is named "point zero three seven".

    Write the number corresponding to each word name.

    a) point zero seven = _____   b) point zero two nine = _____

    a) 20.5
    b) 9.15
    c) 1.057
    d) 18.010

    a) .07
    b) .029

34.   The "point" method can also be used to name decimal numbers larger than "1". When doing so:

1) either we name all the digits from left to right and use the word "point" for the decimal point. For example:

9.1 is named "nine point one".
47.66 is named "four seven point six six".

2) or we use the ordinary name for the whole number part followed by the word "point" and the names of the digits to the right of the decimal point. For example:

64.8 is named "sixty-four point eight".
20.03 is named "twenty point zero three".

Write the number corresponding to each word name.

a) three zero point seven five = _____

b) fifteen point zero nine = _____

c) one hundred thirty-eight point three = _____

a) 30.75   b) 15.09   c) 138.3

1-5   COMPARING NUMBERS

In this section, we will discuss methods for comparing the size of numbers.

35.   To compare the size of whole numbers, we can use the rules below.

1) If they contain a different number of digits, the one with more digits is larger. For example:

2,375 is larger than 986,
because 2,375 contains 2 thousands.

2) If they contain the same number of digits, we compare their digits from left to right until we find a place where their digits differ. For example:

685 is larger than 499,
because 6 hundreds is larger than 4 hundreds.

5,098,621 is larger than 5,093,888 ,
because 8 thousands is larger than 3 thousands.

Identify the larger number in each pair.

a) 1,099 or 998   c) 8,011 or 7,999

b) 25,647 or 203,900   d) 3,750,000 or 3,850,000

a) 1,099   b) 203,900   c) 8,011   d) 3,850,000

36.   To compare decimal numbers with the same number of decimal places, we simply ignore the decimal points and compare them as if they were whole numbers. For example:

      75.3 is larger than 18.7   (because 753 is larger than 187)
      5.71 is larger than 5.26   (because 571 is larger than 526)
      .0046 is larger than .0033   (because 46 is larger than 33)

Identify the larger number in each pair.

    a) 5.75 and 8.12   c) .00269 and .00274

    b) 10.7 and 10.4   d) .00015 and .00009

---

37.   We get an equivalent or "equal" number if we add one or more final 0's to a decimal number. That is:

        .47 = .470 = .4700 = .47000
        8.1 = 8.10 = 8.100 = 8.1000

Therefore, to compare decimal numbers with a different number of decimal places, we add a final "0" or 0's until both have the same number of decimal places. Two examples are shown.

    To compare .57 and .6, we added <u>one</u> "0" to .6 below.

      .57   .57   Since .60 is larger than .57 ,
      .6   .60   .6 is larger than .57 .
             ↑

    To compare 4.298 and 4.3, we added <u>two</u> 0's to 4.3 below.

     4.298   4.298   Since 4.300 is larger than 4.298 ,
     4.3   4.300   4.3 is larger than 4.298 .
           ↑↑

Identify the larger number in each pair.

    a) .23 and .2   c) .007 and .0071

    b) 5.103 and 5.1   d) .0001 and .000066

---

38.   Though it is not always written, any whole number has a decimal point after the "ones" digit. For example:

      9 = 9.   45 = 45.

We get an equivalent number if we add one or more 0's after the decimal point in a whole number. For example:

      9 = 9.0 = 9.00 = 9.000
      45 = 45.0 = 45.00 = 45.000

Therefore to compare a whole number and a decimal number, we add one or more 0's after the decimal point until both have the same number of decimal places. Two examples are shown on the following page.

Continued on following page.

Answer column:

37.
a) 8.12   c) .00274
b) 10.7   d) .00015

38.
a) .23   c) .0071
b) 5.103   d) .0001

38.   Continued

To compare 6.75 and 6 below, we added two 0's to 6.

6.75	6.75	Since 6.75 is larger than 6.00,
6	6.00	6.75 is larger than 6 .
	↑↑	

To compare .538 and 4 below, we added three 0's to 4.

.538	.538	Since 4.000 is larger than .538,
4	4.000	4 is larger than .538.
	↑↑↑	

Identify the larger number in each pair.

a) 12 or 12.46   b) 18.7 or 14   c) .99 or 9

---

39.   Identify the larger number in each pair.

a) 10,001 and 9,999   c) 6,125,780,000 and 6,125,690,000

b) 28.62 and 29.01   d) 14.7 and 14.2

a) 12.46

b) 18.7

c) 9

---

40.   Identify the larger number in each pair.

a) .0175 and .0099   c) 15.1 and 15.09

b) .008 and .01   d) 1.701 and 1.7

a) 10,001

b) 29.01

c) 6,125,780,000

d) 14.7

---

41.   Identify the larger number in each pair.

a) 16 and 25.01   c) 2 and .98

b) 99.9 and 99   d) .001 and .01

a) .0175   c) 15.1

b) .01   d) 1.701

---

a) 25.01   b) 99.9   c) 2   d) .01

SELF-TEST 1 (pages 1-15)

Name the place in which the "7" appears in each number.

1. 872,530,000	2. 638.74	3. 0.002176
_____	_____	_____

In 459.36 :   4. What digit is in the hundredths place? _____

5. What digit is in the hundreds place? _____

Continued on following page.

Self-Test 1 (Continued)

Write the number corresponding to each expanded form.

    6. 8 hundreds + 0 tens + 3 ones = _____

    7. 5 ones + 2 tenths + 6 hundredths = _____

    8. 0 tenths + 0 hundredths + 9 thousandths = _____

    9. 20 + 8 + .4 + .03 + .005 = _____

Convert each of the following to a decimal number.

    10. $\dfrac{29}{10,000}$ = _____   11. $14\dfrac{53}{1,000}$ = _____

Write the number corresponding to each word name.

    12. eighty thousand, one hundred sixty-five = _____

    13. three million, forty-two thousand = _____

    14. four hundred seven billion, eight hundred million = _____

    15. sixty-five hundred twenty = _____

    16. thirty-one millionths = _____

    17. seventeen and fifty-two hundredths = _____

    18. two hundred ten point zero eight one = _____

Identify the <u>larger</u> number in each pair.

| 19. 27,301 and 27,096 | 20. 13.05 and 13.2 | 21. .0009 and .005 |

ANSWERS:

1. ten-millions	6. 803	12. 80,165	19. 27,301
2. tenths	7. 5.26	13. 3,042,000	20. 13.2
3. hundred-thousandths	8. .009	14. 407,800,000,000	21. .005
4. 6	9. 28.435	15. 6,520	
5. 4	10. .0029	16. .000031	
	11. 14.053	17. 17.52	
		18. 210.081	

1-6   ADDING, SUBTRACTING, MULTIPLYING, DIVIDING

In this section, we will discuss the procedure for performing the four fundamental operations on a calculator. The procedure for converting fractions to decimal numbers is also discussed.

42.   Turn on your calculator. The display shows "0".

To enter a whole number, press the digits in order <u>from</u> <u>left</u> <u>to</u> <u>right</u>.
Try this one.

1) Enter 2,740. The display shows "2740".

2) Press the "clear" key $\boxed{\text{C}}$ to clear. The display shows "0".

To enter a decimal number, press the digits in order <u>from</u> <u>left</u> <u>to</u> <u>right</u>.
Press the decimal-point key $\boxed{\,\cdot\,}$ when needed. Try these.

1) Enter 15.96 . The display shows "15.96".

2) Press $\boxed{\text{C}}$ to clear. The display shows "0".

3) Enter .0037 . The display shows "0.0037".

4) Press $\boxed{\text{C}}$ to clear. The display shows "0".

---

43.   When explaining how to use the calculator for operations, we will use the
<u>Enter</u>-<u>Press</u>-<u>Display</u> format.

<u>Enter</u> - gives the number to be entered.

<u>Press</u> - gives the operation key or function key to be used.

<u>Display</u> - gives what is seen on the display after the operation key
or function key is pressed.

To perform 8 + 7 = 15 on a calculator, follow these steps.

Enter	Press	Display
8	$\boxed{+}$	8.
7	$\boxed{=}$	15.

Note: When $\boxed{=}$ is pressed, the calculator is automatically cleared
for the next operation.

---

44.   We performed 3,658 + 949 = 4,607 at the right.
To perform it on a calculator, follow these steps.

$$\begin{array}{r} 3,658 \\ +\ 949 \\ \hline 4,607 \end{array}$$

Enter	Press	Display
3,658	$\boxed{+}$	3658.
949	$\boxed{=}$	4607.

---

45.   We performed 9.75 + 17.6 + 6.34 = 33.69 at the right.
<u>Notice</u> <u>how</u> <u>we</u> <u>lined</u> <u>up</u> <u>the</u> <u>decimal</u> points. To perform
it on a calculator, follow these steps.

$$\begin{array}{r} 9.75 \\ 17.6 \\ +\ 6.34 \\ \hline 33.69 \end{array}$$

Enter	Press	Display
9.75	$\boxed{+}$	9.75
17.6	$\boxed{+}$	27.35
6.34	$\boxed{=}$	33.69

46.   Use a calculator for these.

   a) $367,000 + 9,088,000 =$ _____

   b) $.0075 + .00106 + .00099 =$ _____

---

47.   We performed $1,403 - 927 = 476$ at the right.
       To perform it on a calculator, follow these steps.

Enter	Press	Display
1,403	$-$	1403.
927	$=$	476.

$$\begin{array}{r} 1,403 \\ -\ 927 \\ \hline 476 \end{array}$$

a) 9,455,000

b) 0.00955
    (or .00955)

---

48.   We performed $7.31 - 2.99 = 4.32$ at the right.
       Notice how we lined up the decimal points. To
       perform it on a calculator, follow these steps.

Enter	Press	Display
7.31	$-$	7.31
2.99	$=$	4.32

$$\begin{array}{r} 7.31 \\ -\ 2.99 \\ \hline 4.32 \end{array}$$

---

49.   Use a calculator for these.

   a) $574,200 - 85,500 =$ _____

   b) $.007 - .0019 =$ _____

---

50.   We performed $74 \times 508 = 37,592$ at the right.
       To perform it on a calculator, follow these steps.

Enter	Press	Display
74	$\times$	74.
508	$=$	37592.

$$\begin{array}{r} 5\ 0\ 8 \\ \times\ 7\ 4 \\ \hline 2\ 0\ 3\ 2 \\ 3\ 5\ 5\ 6 \\ \hline 3\ 7,5\ 9\ 2 \end{array}$$

a) 488,700

b) .0051

---

51.   We performed $3.76 \times 12.4 = 46.624$ at the right.
       Notice that the number of decimal places in the
       product equals the sum of the number of decimal
       places in the factors. To perform it on a
       calculator, follow these steps.

Enter	Press	Display
3.76	$\times$	3.76
12.4	$=$	46.624

$$\begin{array}{r} 3.7\ 6 \quad \text{(2 places)} \\ \times\ 1\ 2.4 \quad \text{(1 place)} \\ \hline 1\ 5\ 0\ 4 \\ 7\ 5\ 2 \\ 3\ 7\ 6 \\ \hline 4\ 6.6\ 2\ 4 \quad \text{(3 places)} \end{array}$$

52.   To perform 1.5 x 2.8 x 3.2 = 13.44 on a calculator, follow these steps.

Enter	Press	Display
1.5	x	1.5
2.8	x	4.2
3.2	=	13.44

---

53.   Use a calculator for these.

a) 45 x 184 = _____

b) .034 x 8.5 = _____

c) 3.25 x .0406 x 7,800 = _____

---

54.   We performed 1,215 ÷ 45 = 27 at the right. To perform it on a calculator, follow these steps.

$$\begin{array}{r} 27 \\ 45\overline{)1,215} \\ -90 \\ \hline 315 \\ -315 \\ \hline \end{array}$$

Enter	Press	Display
1,215	÷	1215.
45	=	27.

a) 8,280

b) .289

c) 1,029.21

---

55.   We performed 106.95 ÷ 15 = 7.13 at the right. Notice that the decimal point in the quotient is directly above the decimal point in 106.95 . To perform it on a calculator, follow these steps.

$$\begin{array}{r} 7.1\,3 \\ 15\overline{)1\,0\,6.9\,5} \\ -1\,0\,5 \\ \hline 1\,9 \\ -1\,5 \\ \hline 4\,5 \\ -4\,5 \\ \hline \end{array}$$

Enter	Press	Display
106.95	÷	106.95
15	=	7.13

---

56.   We performed 121 ÷ 44 = 2.75 at the right. Notice that we had to add two 0's after the decimal point in 121 . To perform it on a calculator, follow these steps.

$$\begin{array}{r} 2.7\,5 \\ 44\overline{)1\,2\,1.0\,0} \\ -8\,8 \\ \hline 3\,3\,0 \\ -3\,0\,8 \\ \hline 2\,2\,0 \\ -2\,2\,0 \\ \hline \end{array}$$

Enter	Press	Display
121	÷	121.
44	=	2.75

57. We performed $.885 \div 1.25 = .708$ at the right.
     Note:

$$1.25\overline{)\begin{array}{r} .708 \\ .88500 \\ \underline{-875} \\ 100 \\ \underline{-0} \\ 1000 \\ \underline{-1000} \end{array}}$$

1) We shifted the decimal point 2 places
    to the right in both 1.25 and .885 .

2) We placed the decimal point in the quotient
    directly above the decimal point in 88.5 .

To perform it on a calculator, follow these steps.

Enter	Press	Display
.885	$\boxed{\div}$	0.885
1.25	$\boxed{=}$	0.708

58. Use a calculator for these:

   a) $25{,}056 \div 348 = $ _____

   b) $52.7 \div .00425 = $ _____

59. Divisions are frequently written as fractions. For example:

$$3 \div 4 = \frac{3}{4} \qquad 97 \div 20 = \frac{97}{20} \qquad 39.6 \div 2.4 = \frac{39.6}{2.4}$$

Therefore, to convert a fraction to a decimal number, we divide the
numerator by the denominator. Use a calculator for these conversions.

   a) $\dfrac{3}{4} = 3 \div 4 = $ _____

   b) $\dfrac{97}{20} = 97 \div 20 = $ _____

   c) $\dfrac{39.6}{2.4} = 39.6 \div 2.4 = $ _____

a) 72

b) 12,400

a) .75   b) 4.85   c) 16.5

1-7   SQUARES, SQUARE ROOTS, RECIPROCALS

In this section, we will show how a calculator can be used to square a number, find the square root of a
number, and find the reciprocal of a number.

---

60. To <u>square</u> a <u>number</u>, we multiply the number by itself. For example:

   The "square of 7" is 49, since 7 x 7 = 49 .

   The "square of 2.5" is 6.25, since 2.5 x 2.5 = 6.25 .

Continued on following page.

60.   Continued

To indicate that a number should be squared, we write a small "2"
(called an <u>exponent</u>) slightly above and to the right of the number.
Therefore:

$$7^2 = 7 \times 7 = 49$$

$$2.5^2 = 2.5 \times 2.5 = 6.25$$

To square a number on a calculator, we use the $\boxed{x^2}$ key. Following the
steps below, find 7^2 and 2.5^2 .

Enter	Press	Display
7	$\boxed{x^2}$	49.
2.5	$\boxed{x^2}$	6.25

---

61.   Use the calculator for these:

a) $14^2 = $ _____

b) $256^2 = $ _____

c) $11.8^2 = $ _____

d) $.52^2 = $ _____

---

62.   The <u>square root</u> of a number is that number whose square equals the
original number. For example:

The <u>square root of 81</u> is 9, since $9^2 = 81$.

The <u>square root of 17.64</u> is 4.2, since $4.2^2 = 17.64$.

Instead of saying "the square root of", we use the symbol $\sqrt{}$
called the <u>square root</u> <u>radical</u>. For example:

$$\sqrt{81} = 9 \qquad\qquad \sqrt{17.64} = 4.2$$

To find the square root of a number on a calculator, we use the $\boxed{\sqrt{x}}$
key. Following the steps below, find $\sqrt{81}$ and $\sqrt{17.64}$.

Enter	Press	Display
81	$\boxed{\sqrt{x}}$	9.
17.64	$\boxed{\sqrt{x}}$	4.2

a) 196

b) 65,536

c) 139.24

d) 0.2704

---

63.   Use the calculator for these.

a) $\sqrt{169} = $ _____

b) $\sqrt{348,100} = $ _____

c) $\sqrt{129.96} = $ _____

d) $\sqrt{.000729} = $ _____

a) 13   c) 11.4

b) 590   d) .027

64. Two numbers are a <u>pair of reciprocals</u> if their product is "1". For example:

  2 and $\frac{1}{2}$ are a pair of reciprocals, since $2\left(\frac{1}{2}\right) = \frac{2}{2} = 1$

  .8 and $\frac{1}{.8}$ are a pair of reciprocals, since $.8\left(\frac{1}{.8}\right) = \frac{.8}{.8} = 1$

To find the reciprocal of a number, we divide "1" by that number. That is:

  The reciprocal of 2 is $\frac{1}{2} = 1 \div 2 = .5$

  The reciprocal of .8 is $\frac{1}{.8} = 1 \div .8 = 1.25$

To find the reciprocals above on a calculator, we use the $\boxed{1/x}$ key. Following the steps below, find the reciprocals of 2 and .8 .

Enter	Press	Display
2	$\boxed{1/x}$	0.5
.8	$\boxed{1/x}$	1.25

---

65. Use a calculator to find the reciprocal of each number.

  a) 40 _____ b) .25 _____ c) 320 _____

---

66. Any fraction whose numerator is "1" is the reciprocal of its denominator. That is:

  $\frac{1}{4}$ is the reciprocal of 4.

  $\frac{1}{125}$ is the reciprocal of 125.

Therefore, to convert fractions like those above to decimal numbers, we can simply find the reciprocals of their denominators. We did so below.

Enter	Press	Display
4	$\boxed{1/x}$	0.25
125	$\boxed{1/x}$	0.008

  Therefore: $\frac{1}{4} = .25$ $\frac{1}{125} = $ _____

a) .025

b) 4

c) .003125

---

67. Using the reciprocal key, convert each fraction to a decimal number.

  a) $\frac{1}{8} = $ _____   b) $\frac{1}{64} = $ _____   c) $\frac{1}{1.25} = $ _____

.008

---

a) .125   b) .015625   c) .8

68.   When "0" is a factor in a multiplication, the product is always "0".
For example:

$$0 \times 7 = 0 \qquad 0 \times 5.8 = 0 \qquad 0 \times .009 = 0$$

Since we cannot get "1" as the product when multiplying by "0", the number
"0" has no reciprocal. Therefore, FINDING THE RECIPROCAL OF "0"
IS AN IMPOSSIBLE OPERATION.

To indicate an IMPOSSIBLE OPERATION, a calculator can show a flashing
display, a series of dots, the word "ERROR", or something else. Try to
find the reciprocal of "0" on your calculator.

Note: Press $\boxed{C}$ to return the display to "0".

69.   Any fraction (or division) can be converted to a multiplication by multiplying
the numerator by the reciprocal of the denominator. For example:

$$\frac{12}{3} = 12(\text{the reciprocal of 3}) = 12\left(\frac{1}{3}\right)$$

When a division by "0" is converted to a multiplication, one of the factors
is "the reciprocal of 0". For example:

$$\frac{6}{0} = 6(\text{the reciprocal of 0})$$

Since "0" has no reciprocal, DIVISION BY "0" IS AN IMPOSSIBLE
OPERATION. Try to divide 6 by "0" on your calculator.

Note: Press $\boxed{C}$ to return the display to "0".

1-8   OPERATIONS INVOLVING NEGATIVE NUMBERS

In this section, we will show how a calculator can be used for operations involving negative numbers.

70.   For practical purposes, the absolute value of a signed number is its
numerical value with the sign ignored. For example:

The absolute value of +12 is 12.
The absolute value of -15 is 15.

To enter a negative number on a calculator, we enter its absolute value and
then use the "change sign" key $\boxed{+/-}$. Following the steps below, enter -17
and -6.45 .

Enter	Press	Display
17	$\boxed{+/-}$	-17.
	Note: Press $\boxed{C}$ to clear.	
6.45	$\boxed{+/-}$	-6.45
	Note: Press $\boxed{C}$ to clear.	

71.   The rule for adding two negative numbers is:

> 1. The <u>absolute value of the sum</u> is the sum of their absolute values.
>
> 2. The <u>sign of the sum</u> is negative.

We used the above rule for the addition below.

(-4) + (-7) :   The absolute value of the sum is 11 (from 4 + 7).
                   The sign of the sum is negative.
                   Therefore: (-4) + (-7) = -11.

To perform the addition above on a calculator, follow these steps.

Enter	Press	Display
4	+/- +	-4.
7	+/- =	-11.

---

72.   Use a calculator for these.

a) (-489) + (-276) = _____   b) (-3.45) + (-7.66) = _____

---

73.   The rule for adding a positive and a negative number is:

> 1. The <u>absolute value of the sum</u> is obtained by subtracting the smaller absolute value from the larger absolute value.
>
> 2. The <u>sign of the sum</u> is the same as the sign of the number with the larger absolute value.

We used the above rule for the additions below.

10 + (-7):   The absolute value of the sum is 3 (from 10 - 7).
                The sign of the sum is positive because 10 is positive.
                Therefore: 10 + (-7) = 3.

(-8) + 2:   The absolute value of the sum is 6 (from 8 - 2).
                The sign of the sum is negative because -8 is negative.
                Therefore: (-8) + 2 = -6.

To perform the same additions on a calculator, follow these steps:

Enter	Press	Display
10	+	10.
7	+/- =	3.

Enter	Press	Display
8	+/- +	-8.
2	=	-6.

a) -765   b) -11.11

74.   Use a calculator for these. Be sure to notice whether the sum is negative or not.

   a) (−58) + 82 = _____   b) 2.85 + (−7.03) = _____

75.   The <u>opposite</u> of a signed number is a number with the same absolute value but the opposite sign. For example:

               The opposite of   5 is −5.
               The opposite of −9 is   9.

To perform a subtraction involving a negative number, we convert the subtraction to an addition by <u>ADDING THE OPPOSITE OF THE SECOND NUMBER</u> and then perform the addition. For example:

      6 − (−8) =   6 + (the opposite of −8) =   6 + 8   = 14

   (−9) − 7 = (−9) + (the opposite of 7) = (−9) + (−7) = −16

To perform the same subtractions on a calculator, follow these steps.

Enter	Press	Display
6	−	6.
8	+/− =	14.

Enter	Press	Display
9	+/− −	−9.
7	=	−16.

a) 24   b) −4.18

76.   Use a calculator for these. Be sure to notice whether the answer is negative or not.

   a) 87 − (−66) = _____   c) (−2.4) − (−7.4) = _____

   b) (−510) − 430 = _____   d) .785 − .966 = _____

77.   In a multiplication, if one factor is positive and the other factor is negative, the product is negative. For example:

         (7)(−6) = −42   (−8)(9) = −72

To perform the same multiplications on a calculator, follow these steps.

Enter	Press	Display
7	x	7.
6	+/− =	−42.

Enter	Press	Display
8	+/− x	−8.
9	=	−72.

a) 153   c) 5

b) −940   d) −.181

78.　If both factors in a multiplication are negative, the product is positive. For example:

$$(-6)(-4) = 24$$

To perform the above multiplication on a calculator, follow these steps.

Enter	Press	Display
6	+/-　　x	-6.
4	+/-　　=	24.

79.　Use a calculator for these.

a)　$(47)(-56) = $ _____

c)　$(-1.88)(20) = $ _____

b)　$(-8)(-7.5) = $ _____

d)　$(-2.4)(-8.5) = $ _____

80.　In a division, if one term is positive and the other term is negative, the quotient is negative. For example:

$$\frac{-45}{5} = -9 \qquad \frac{60}{-3} = -20$$

To perform the same divisions on a calculator, follow these steps.

Enter	Press	Display
45	+/-　　÷	-45.
5	=	-9.

Enter	Press	Display
60	÷	60.
3	+/-　　=	-20.

a)　-2,632

b)　60

c)　-37.6

d)　20.4

81.　In a division, if both terms are negative, the quotient is positive. For example:

$$\frac{-75}{-25} = 3$$

To perform the above division on a calculator, follow these steps.

Enter	Press	Display
75	+/-　　÷	-75.
25	+/-　　=	3.

82.　Use a calculator for these.

a)　$\dfrac{396}{-12} = $ _____

b)　$\dfrac{-42}{168} = $ _____

c)　$\dfrac{-7.48}{-1.36} = $ _____

83.   To square a negative number, we multiply the number by itself.
Therefore, the square of a negative number is positive. For example:

$$(-7)^2 = (-7)(-7) = 49$$

$$(-1.2)^2 = (-1.2)(-1.2) = 1.44$$

To find $(-7)^2$ and $(-1.2)^2$ on a calculator, follow these steps.

Enter	Press	Display
7	$\boxed{+/-}$ $\boxed{x^2}$	49.
1.2	$\boxed{+/-}$ $\boxed{x^2}$	1.44

a) -33

b) -.25

c) 5.5

84.   Since the square of any number (positive or negative) is <u>positive</u>, there is
no ordinary number that is the square root of a negative number. For
example:

$$\sqrt{-16} \neq +4, \quad \text{since} \quad (+4)^2 = +16$$

$$\sqrt{-16} \neq -4, \quad \text{since} \quad (-4)^2 = +16$$

<u>FINDING THE SQUARE ROOT OF A NEGATIVE NUMBER IS AN
IMPOSSIBLE OPERATION</u>. Try $\sqrt{-16}$ on your calculator.

<u>Note</u>:   Press $\boxed{C}$ to return the display to "0".

SELF-TEST 2 (pages 15-26)

Use a calculator for the following problems.

1. 258.7 + 463.9 = _____   2. .0087 + .353 + .049 = _____   3. 5.04 - 1.37 = _____

4. 2,960 x 170 = _____   5. 3.8 x 1.4 x 2.5 = _____

6. 73.1 ÷ 17 = _____   7. 3,750 ÷ .0625 = _____   8. $\dfrac{8.4}{150}$ = _____

9. Convert $\dfrac{5}{16}$ to a decimal number.   _____

10. $(31.6)^2$ = _____   11. $(.29)^2$ = _____   12. $\sqrt{5,476}$ = _____   13. $\sqrt{86.49}$ = _____

14. Find the reciprocal of 80.   _____

15. (-796) + (-548) = _____   16. (-8.1) - 3.7 = _____   17. 38 - (-45) = _____

18. (6.4)(-8.3) = _____   19. $\dfrac{-39}{650}$ = _____   20. $\dfrac{-5.85}{-4.68}$ = _____   21. $(-24)^2$ = _____

ANSWERS:

1. 722.6	6. 4.3	12. 74	18. -53.12
2. .4107	7. 60,000	13. 9.3	19. -.06
3. 3.67	8. .056	14. .0125	20. 1.25
4. 503,200	9. .3125	15. -1,344	21. 576
5. 13.3	10. 998.56	16. -11.8	
	11. .0841	17. 83	

1-9   ROUNDING WHOLE NUMBERS

Rounding is frequently used to report calculator answers. In this section, we will discuss the rules for rounding whole numbers to a specific place.

85.   To round numbers between 70 and 80 to the nearest ten, we can examine the number line below.

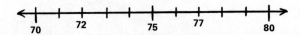

72 is closer to 70. We round 72 down to 70.
77 is closer to 80. We round 77 up to 80.
75 is halfway between 70 and 80. We agree to round 75 up to 80.

To round numbers between 420 and 430 to the nearest ten, we can examine the number line below.

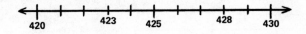

a) 423 rounds to _____   b) 428 rounds to _____   c) 425 rounds to _____

86.   To round numbers between 300 and 400 to the nearest hundred, we can examine the number line below.

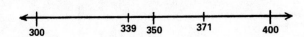

339 is closer to 300. We round 339 down to 300.
371 is closer to 400. We round 371 up to 400.
350 is halfway between 300 and 400. We agree to round 350 up to 400.

To round numbers between 8,600 and 8,700 to the nearest hundred, we can examine the number line below.

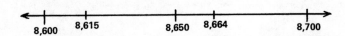

a) 8,615 rounds to _____   c) 8,650 rounds to _____

b) 8,664 rounds to _____

a) 420

b) 430

c) 430

a) 8,600

b) 8,700

c) 8,700

87.   To round numbers between 8,000 and 9,000 to the nearest thousand, we can examine the number line below.

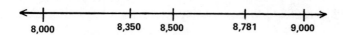

8,350 is closer to 8,000. We round 8,350 down to 8,000.
8,781 is closer to 9,000. We round 8,781 up to 9,000.
8,500 is halfway between 8,000 and 9,000. We agree to
   round 8,500 up to 9,000.

To round numbers between 11,000 and 12,000 to the nearest thousand, we can examine the number line below.

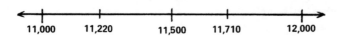

a) 11,220 rounds to _____   c) 11,500 rounds to _____

b) 11,710 rounds to _____

---

88.   Instead of using the number line to round whole numbers to a specific place, we ordinarily use the rules below.

> 1) If the digit in the place immediately to the right of the specific place is a 0, 1, 2, 3, or 4, we round down. That is, we:
>
>   a) leave the digit in the specific place unchanged, and
>   b) replace all digits to the right of the specific place with 0's.
>
> 2) If the digit in the place immediately to the right of the specific place is a 5, 6, 7, 8, or 9, we round up. That is, we:
>
>   a) add "1" to the digit in the specific place, and
>   b) replace all digits to the right of the specific place with 0's.

Some examples of using the rules to round to tens, hundreds, and thousands are shown below. Notice that we drew an arrow over the digit in the place rounded to.

      ↓
Rounding 274 to tens, we round down to 270.
   (Because of the 4 in the "ones" place, we left the 7 unchanged.)

     ↓
Rounding 5,365 to hundreds, we round up to 5,400.
   (Because of the 6 in the "tens" place, we added "1" to the 3 and got 4.)

     ↓
Rounding 68,500 to thousands, we round up to 69,000.
   (Because of the 5 in the "hundreds" place, we added "1" to the 8 and got 9.)

Continued on following page.

a) 11,000

b) 12,000

c) 12,000

88.   Continued

Using the rules, complete these.

a) Round 891 to <u>tens</u>. _____   c) Round 7,655 to <u>hundreds</u>. _____

b) Round 159 to <u>tens</u>. _____   d) Round 3,499 to <u>thousands</u>. _____

---

89.   The same rules are used to round to places to the left of the "thousands" place. For example:

Rounding 471,299 to <u>ten-thousands</u>, we round down to 470,000.
   (Because of the "1" in the "thousands" place,
   we left the 7 unchanged.)

Rounding 16,588,100 to <u>millions</u>, we round up to 17,000,000.
   (Because of the 5 in the "hundred-thousands" place,
   we added "1" to the 6 and got 7.)

Using the rules, complete these.

a) Round 635,250 to <u>ten-thousands</u>. _____

b) Round 408,999 to <u>hundred-thousands</u>. _____

c) Round 98,199,600 to <u>millions</u>. _____

d) Round 2,774,000,000 to <u>billions</u>. _____

a) 890	c) 7,700
b) 160	d) 3,000

---

90.   When rounding down, we can get a "0" in the place rounded to. For example:

Rounding 203 to <u>tens</u>, we get 200.

Rounding 700,295 to <u>thousands</u>, we get 700,000.

Complete: a) Round 6,044,000 to <u>hundred-thousands</u>. _____

           b) Round 800,357,000 to <u>millions</u>. _____

a) 640,000
b) 400,000
c) 98,000,000
d) 3,000,000,000

---

91.   When rounding up, we can also get a "0" in the place rounded to. For example:

Rounding 397 to <u>tens</u>, we get 400.

Rounding 9,957,600 to <u>hundred-thousands</u>, we get 10,000,000.

Complete: a) Round 89,525 to <u>thousands</u>. _____

           b) Round 499,677,000 to <u>millions</u>. _____

a) 6,000,000
b) 800,000,000

a) 90,000   b) 500,000,000

92.   Complete these. Begin by drawing an arrow over the place rounded to.

    a) Round 20,511 to <u>hundreds</u>. _____

    b) Round 10,550,000 to <u>millions</u>. _____

    c) Round 300,250 to <u>thousands</u>. _____

    d) Round 19,755,000,000 to <u>billions</u>. _____

---

a) 20,500   b) 11,000,000   c) 300,000   d) 20,000,000,000

1-10   ROUNDING DECIMAL NUMBERS

In this section, we will discuss the rules for rounding decimal numbers to places to the left and right of the decimal point.

---

93.   To round decimal numbers to a place to the left of the "ones" place, we examine the digit in the place immediately to its right. Some examples are shown. <u>Notice that we dropped all digits to the right of the decimal point</u>.

    Rounding to <u>hundreds</u>, 768.92 rounds up to 800.

    Rounding to   tens   43.256 rounds down to 40.

Complete: a) Round 4,720.75 to <u>hundreds</u>. _____

              b) Round 98.13 to <u>tens</u>. _____

---

a) 4,700

b) 100

---

94.   To round numbers between 4 and 5 to the nearest whole number (or nearest "one"), we can examine the number line below.

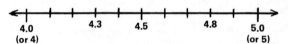

4.3 is closer to 4. We round 4.3 down to 4.
4.8 is closer to 5. We round 4.8 up to 5.
4.5 is halfway between 4 and 5. We agree to round 4.5 up to 5.

To round numbers between 11 and 12 to the nearest whole number (or nearest "one"), we can examine the number line below.

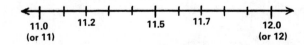

a) 11.2 rounds to _____   b) 11.7 rounds to _____   c) 11.5 rounds to _____

---

a) 11   b) 12   c) 12

95. Instead of using the number line to round to the nearest whole number (or nearest "one"), <u>we</u> <u>can</u> <u>examine</u> <u>the</u> <u>digit</u> <u>in</u> <u>the</u> "<u>tenths</u>" <u>place</u>. Some examples are shown. Notice again that we dropped all digits to the right of the decimal point.

    ↓
37.2 rounds down to 37   (Because of the "2" in the "tenths" place, we left the "7" unchanged.)

    ↓
94.87 rounds up to 95   (Because of the "8" in the "tenths" place, we added "1" to the "4" and got "5".)

Round to the nearest whole number.

a) 125.7 _____

b) 17.03 _____

c) 210.44 _____

d) 199.6 _____

---

96. To round numbers between 6.7 and 6.8 to the nearest tenth, we can examine the number line below.

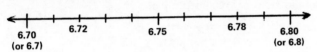

6.72 is closer to 6.7. We round 6.72 down to 6.7 .
6.78 is closer to 6.8. We round 6.78 up to 6.8 .
6.75 is halfway between 6.7 and 6.8. We agree to round 6.75 up to 6.8

To round numbers between 1.2 and 1.3 to the nearest tenth, we can examine the number line below.

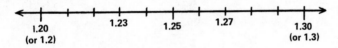

a) 1.23 rounds to _____   b) 1.27 rounds to _____   c) 1.25 rounds to _____

a) 126   c) 210

b) 17   d) 200

---

97. To round numbers between 4.28 and 4.29 to the nearest hundredth, we can examine the number line below.

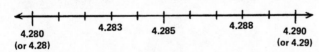

4.283 is closer to 4.28. We round 4.283 down to 4.28 .
4.288 is closer to 4.29. We round 4.288 up to 4.29 .
4.285 is halfway between 4.28 and 4.29. We agree to round 4.285 up to 4.29 .

a) 1.2

b) 1.3

c) 1.3

Continued on following page.

97.   Continued

To round numbers between .31 and .32 to the nearest hundredth, we can examine the number line below.

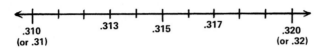

.310   .313   .315   .317   .320
(or .31)   (or .32)

a) .313 rounds to _____   b) .317 rounds to _____   c) .315 rounds to _____

---

98.   Instead of using the number line, we ordinarily use the rules below to round decimal numbers to a place to the right of the decimal point.

> 1) If the digit <u>in the place</u> <u>immediately to the right</u> of the specific place is a 0, 1, 2, 3, or 4, we round down. That is, we:
>
>   a) leave the digit in the specific place unchanged, and
>   b) drop all digits to the right of the specific place.
>
> 2) If the digit <u>in the place</u> <u>immediately to the right</u> of the specific place is a 5, 6, 7, 8, or 9, we round up. That is, we:
>
>   a) add "1" to the digit in the specific place, and
>   b) drop all digits to the right of the specific place.

Some examples of using the rules to round to tenths, hundredths and thousandths are shown below.

Rounding to <u>tenths</u>, 8.6↓35 rounds down to 8.6 .
(Because of the "3" in the "hundredths" place,
we left the "6" unchanged.)

Rounding to <u>hundredths</u>, .03↓59 rounds up to .04 .
(Because of the "5" in the "thousandths" place,
we added "1" to the "3" and got "4".)

Rounding to <u>thousandths</u>, .024↓69 rounds up to .025 .
(Because of the "6" in the "ten-thousandths" place,
we added "1" to the "4" and got "5".)

Using the rules, complete these:

a) Round 17.3↓6 to <u>tenths</u>. _____

b) Round .1↓285 to <u>tenths</u>. _____

c) Round 6.18↓05 to <u>hundredths</u>. _____

d) Round .028↓51 to <u>thousandths</u>. _____

a) .31

b) .32

c) .32

---

a) 17.4   b) .1   c) 6.18   d) .029

99. The same rules are used to round to places to the right of the "thousandths" place. For example:

Rounding .007794 to <u>ten-thousandths</u>, we round up to .0078 .
(Because of the 9 in the "hundred-thousandths" place, we added "1" to the 7 and got 8.)

Rounding .0003425 to <u>hundred-thousandths</u>, we round down to .00034 .
(Because of the 2 in the "millionths" place, we left the 4 unchanged.)

Complete: a) Round .015836 to <u>ten-thousandths</u>. _____

b) Round .00004175 to <u>millionths</u>. _____

---

100. When rounding to a place to the right of the decimal point, we do not leave 0's in places to its right. For example:

Rounding 17.825 to <u>tenths</u>, we get 17.8 (not 17.800).

Rounding .02795 to <u>thousandths</u>, we get .028 (not .02800).

Complete. Begin by writing an arrow over the place rounded to.

a) Round 4.66531 to <u>hundredths</u>. _____

b) Round .472085 to <u>thousandths</u>. _____

c) Round .00366991 to <u>hundred-thousandths</u>. _____

a) .0158

b) .000042

---

101. When rounding to a place to the right of the decimal point:

We can get a "0" in that place when rounding down. For example:

Rounding to <u>tenths</u>, 25.03 rounds to 25.0 .

Rounding to <u>thousandths</u>, .70045 rounds to .700 .

We can get a "0" in that place when rounding up. For example:

Rounding to <u>hundredths</u>, 7.195 rounds to 7.20 .

Rounding to <u>ten-thousandths</u>, .003975 rounds to .0040 .

Complete: a) Round 1.9036 to <u>hundredths</u>. _____

b) Round .0087963 to <u>hundred-thousandths</u>. _____

a) 4.67

b) .472

c) .00367

---

a) 1.90

b) .00880

102.   When we get a "0" to the right of the decimal point in the place rounded to,
that "0" is reported so that we know what place was rounded to. For
example:

Rounding to <u>tenths</u>, 16.035 rounds to 16.0 (not 16).

Rounding to <u>thousandths</u>, .19967 rounds to .200 (not .2).

Complete: a) Round 5.0043 to <u>hundredths</u>. _____

b) Round .036955 to <u>ten-thousandths</u>. _____

103.   Rounding to a definite number of decimal places is the same as rounding
to a specific place to the right of the decimal point. For example:

"Round to <u>one</u> decimal place" means "round to <u>tenths</u>".

"Round to <u>two</u> decimal places" means "round to <u>hundredths</u>".

"Round to <u>three</u> decimal places" means "round to <u>thousandths</u>".

Round each number to <u>one</u> decimal place (or "tenths").

a) 6.405 _____   b) 17.97 _____   c) 29.999 _____

a)	5.00
b)	.0370

104.   Round to <u>two</u> decimal places (or "hundredths").

a) 3.6937 _____   b) 1.89526 _____   c) 9.9975 _____

a)	6.4
b)	18.0
c)	30.0

105.   Round to <u>three</u> decimal places (or "thousandths").

a) .41025 _____   b) .91866 _____   c) .049853 _____

a)	3.69
b)	1.90
c)	10.00

106.   a) Round .067149 to <u>four</u> decimal places. _____

b) Round .0085044 to <u>five</u> decimal places. _____

c) Round .0001996743 to <u>six</u> decimal places. _____

a)	.410
b)	.919
c)	.050

a) .0671   b) .00850   c) .000200

1-11   ROUNDING CALCULATOR ANSWERS

Calculator answers frequently contain more digits than are needed. In such cases they are rounded. We will discuss calculations of that type in this section.

---

107.   Calculator answers can contain as many as ten digits. For example:

$$.987 \times 5.48 = 5.40876$$
$$44.7 \div 13.3 = 3.360902256$$
$$\sqrt{17} = 4.123105626$$

Answers like those above are usually rounded to less decimal places. Rounding each to <u>hundredths</u>, we get:

a) .987 x 5.48 = _____

b) 44.7 ÷ 13.3 = _____

c) $\sqrt{17}$ = _____

---

108.   Round each answer to <u>tenths</u>.

a) 17.5 + 3.76 + 1.205 = _____   c) 69.7 ÷ 2.6 = _____

b) 47.9 - 8.666 = _____   d) 8.77 x 5.33 = _____

a) 5.41
b) 3.36
c) 4.12

---

109.   Convert each fraction to a decimal number. Round to <u>two</u> decimal places.

a) $\frac{2}{3}$ = _____   b) $\frac{1}{7}$ = _____   c) $\frac{17.8}{3.70}$ = _____

a) 22.5   c) 26.8
b) 39.2   d) 46.7

---

110.   Round each answer to <u>thousandths</u>.

a) 1.17 x 3.89 x .204 = _____

b) The reciprocal of 13 is _____

c) $(.859)^2$ = _____   (d) $\sqrt{.917}$ = _____

a) .67
b) .14
c) 4.81

---

111.   Convert each fraction to a decimal number. Round to <u>four</u> decimal places.

a) $\frac{1}{60}$ = _____   b) $\frac{1}{75}$ = _____   c) $\frac{17}{300}$ = _____

a) .928   c) .738
b) .077   d) .958

a) .0167   b) .0133   c) .0567

SELF-TEST 3 (pages 27-36)

1. Round to hundreds: 7,248

_____

3. Round to ten-thousands: 2,705,100

_____

2. Round to millions: 38,970,000

_____

4. Round to tens: 541.7

_____

5. Round 659.82 to the nearest whole number. _____

6. Round to tenths: 18.361

_____

8. Round to millionths: .00006529

_____

7. Round to thousandths: .27018

_____

9. Round to hundredths: 4.15835

_____

10. Round 8.7099514 to four decimal places. _____

Do each problem on a calculator and round as directed.

11. Round to one decimal place.
    $4.59 + 17.8 + 1.935 =$ _____

14. Round to hundred-thousands.
    $(7,330)^2 =$ _____

12. Round to hundredths.
    $2.13 - .0827 =$ _____

15. Round to ten-thousandths.
    $4.0615 \div 1.7582 =$ _____

13. Round to thousands.
    $90.8 \times 513 \times 7.64 =$ _____

16. Round to two decimal places.
    $\sqrt{1.43} =$ _____

17. Convert $\frac{41}{65}$ to a decimal number. Round to three decimal places. _____

ANSWERS:

1. 7,200	6. 18.4	11. 24.3	16. 1.20
2. 39,000,000	7. .270	12. 2.05	17. .631
3. 2,710,000	8. .000065	13. 356,000	
4. 540	9. 4.16	14. 53,700,000	
5. 660	10. 8.7100	15. 2.3100	

SUPPLEMENTARY PROBLEMS - CHAPTER 1

Assignment 1

Name the place in which the "8" appears in each number below.

    1. 4,815   2. 382,690   3. 18,745,000,000 4. 938.53

Name the place in which the "5" appears in each number below.

    5. .068537   6. 9.7513   7. 82.567   8. .0017354

In 147,318,560 :   9. What digit is in the <u>millions</u> place?
                10. What digit is in the <u>tens</u> place?

In .059834 : 11. What digit is in the <u>thousandths</u> place?
            12. What digit is in the <u>hundred-thousandths</u> place?

Convert each fraction to a decimal number.

    13. $\dfrac{23}{100}$ 14. $\dfrac{80}{10,000}$ 15. $\dfrac{9}{1,000}$ 16. $\dfrac{460}{1,000,000}$

Convert each mixed number to a decimal number.

    17. $48\dfrac{3}{100}$ 18. $4\dfrac{739}{100,000}$ 19. $256\dfrac{7}{10}$ 20. $18\dfrac{31}{1,000}$

Write the number corresponding to each word name.

    21. twenty million, seventy-one thousand
    22. four hundred three thousand, six hundred twenty
    23. eight billion, one hundred million, sixty thousand
    24. forty-two hundred eight and two tenths

Write the number corresponding to each word name.

    25. thirty-two ten-thousandths
    26. three and nine thousandths
    27. fifteen and seventy hundredths
    28. twelve hundred hundred-thousandths
    29. point zero zero one zero seven
    30. one thousand five point nine three

Identify the <u>larger</u> number in each pair.

    31. 49,802 and 49,810   32. 6.035 and 6.2   33. .001 and .0002

    34. 1,013 and 1,009   35. 309.1 and 310.05   36. .051 and .05027

Assignment 2

Do these on a calculator. Do not round the answers.

    1. 5.357 - 3.196   2. 793.8 + 49.6   3. 587 + 956 + 849

    4. .09021 - .00753   5. 25,400 x 85   6. 325.5 ÷ 18.6

    7. 6.48 ÷ 225   8. .0044 x 1,500   9. .56 x 7.2 x 180

    10. .072 ÷ 480   11. $(310)^2$ 12. $\sqrt{823.69}$

    13. $\sqrt{.0676}$ 14. $(.82)^2$ 15. $\dfrac{2.37}{3.16}$

Continued on following page.

SUPPLEMENTARY PROBLEMS - Continued

Do these on a calculator. Do not round the answers.

16. 60 x .0012 x .25

17. $\dfrac{1,290}{.0344}$

18. 27.35 ÷ 0

19. 9,200 x 0

20. 4.2 x 3.5 x 5.6

21. 257 + (-316)

22. (-248) - 512

23. (-248) - (-512)

24. (-38.8) + (-96.5)

25. (-8.5)(-.74)

26. (.00275)(-160)

27. $\dfrac{-15.5}{-24.8}$

28. $\dfrac{-273}{.0312}$

29. $(-213)^2$

30. $\sqrt{-16}$

Find the reciprocal of each number.

31. 400

32. 12.5

33. .0002

Convert each fraction to a decimal number.

34. $\dfrac{23}{25}$

35. $\dfrac{5}{32}$

36. $\dfrac{264}{165}$

Assignment 3

Round the following numbers as directed.

1. Round 379,510 to thousands.
2. Round 85.276 to the nearest whole number.
3. Round 4.02053 to thousandths.
4. Round .00017038 to millionths.
5. Round 30,448,900 to hundred-thousands.
6. Round 17,852 to hundreds.
7. Round 523.09648 to two decimal places.
8. Round .618245 to three decimal places.
9. Round 108,720,000 to millions.
10. Round 4,096,500,000 to billions.
11. Round 5.87465 to hundredths.
12. Round .023967 to ten-thousandths.

Do these on a calculator. Round as directed.

13. Round to hundreds.

   7,853 + 4,741 + 9,075

14. Round to hundredths.

   .8596 x 23.57

15. Round to millions.

   306,000 ÷ .00512

16. Round to ten-thousands.

   $(555)^2$

17. Round to tenths.

   $\sqrt{6,090.2}$

18. Round to tens.

   401.3 - 157.6

19. Find the reciprocal of 68, rounded to four decimal places.

20. Find the reciprocal of .000486, rounded to hundreds.

Continued on following page.

SUPPLEMENTARY PROBLEMS - Continued

21. Round to hundred-thousands.

$$(2,154)^2$$

22. Round to thousands.

$$7.19 \times 26,800$$

23. Round to millionths.

$$31.7 \div 91,400$$

24. Round to thousandths.

$$\sqrt{.572}$$

25. Convert $\frac{37}{79}$ to a decimal number, rounded to ten-thousandths.

26. Convert $\frac{1}{263}$ to a decimal number, rounded to hundred-thousandths.

27. Round to thousandths.

$$720 \times .000152 \times 3.09$$

28. Round to tenths.

$$51.442 - 37.975$$

29. Round to hundredths.

$$\sqrt{5}$$

30. Round to millionths.

$$(.0792)^2$$

31. Round to thousands.

$$347,915 + 526,438$$

32. Round to hundreds.

$$\frac{673.8}{.0491}$$

Chapter 2

PERCENT, RATIO, PROPORTION

In this chapter, we will discuss percents, ratios, and proportions. Applied problems involving those three concepts are discussed and solved. A calculator can be used for most of the problems.

2-1   CONVERTING PERCENTS TO FRACTIONS AND DECIMAL NUMBERS

In this section, we will define "percents" and show the methods for converting percents to fractions and decimal numbers. A table of common percent-fraction-decimal equivalents is given.

1.   The word "percent" means "hundredth". Therefore, any percent is equal to a "hundredths" decimal fraction with the percent-value as its numerator. For example:

$$83\% = \frac{83}{100} \qquad 12\frac{1}{2}\% = \frac{12\frac{1}{2}}{100} \qquad 5.75\% = \frac{5.75}{100}$$

$$100\% = \frac{100}{100} \qquad 240\% = \frac{240}{100} \qquad .16\% = \frac{.16}{100}$$

Convert each percent to a "hundredths" decimal fraction.

a) 1% = _____   b) 150% = _____   c) .5% = _____

2.   Some whole-number percents can be converted to simple fractions by reducing the "hundredths" decimal fraction to lowest terms. For example:

$$50\% = \frac{50}{100} = \frac{1}{2} \qquad 80\% = \frac{80}{100} = \frac{4}{5}$$

Convert each percent to a simple fraction.

a) 25% = _____   b) 10% = _____   c) 60% = _____

a) $\frac{1}{100}$

b) $\frac{150}{100}$

c) $\frac{.5}{100}$

a) $\frac{1}{4}$ b) $\frac{1}{10}$ c) $\frac{3}{5}$

3. Some mixed-number percents can be converted to simple fractions by converting the numerator to an improper fraction and then dividing that fraction by 100. For example:

$$33\frac{1}{3}\% = \frac{33\frac{1}{3}}{100} = \frac{\frac{100}{3}}{100} = \frac{\cancel{100}^{1}}{3} \times \frac{1}{\cancel{100}_{1}} = \frac{1}{3}$$

$$66\frac{2}{3}\% = \frac{66\frac{2}{3}}{100} = \frac{\frac{200}{3}}{100} = \frac{200}{3} \times \frac{1}{100} = \underline{\qquad}$$

---

4. Any percent that is a multiple of 100% equals a whole number. That is:

$$100\% = \frac{100}{100} = 1 \qquad 200\% = \frac{200}{100} = 2 \qquad 500\% = \frac{500}{100} = \underline{\qquad}$$

$\dfrac{2}{3}$

---

5. To convert each percent below to a decimal number, we converted it to a decimal fraction and then divided the numerator by 100. (<u>Note</u>: To divide by 100, we can simply shift the decimal point two places to the left.)

$$83\% = \frac{83}{100} = {}_{\wedge}83. = .83$$

$$172\% = \frac{172}{100} = 1{}_{\wedge}72. = 1.72$$

$$6.5\% = \frac{6.5}{100} = {}_{\wedge}06.5 = .065$$

$$.19\% = \frac{.19}{100} = {}_{\wedge}00.19 = .0019$$

5

The same conversions can be made directly <u>by</u> <u>simply</u> <u>shifting</u> <u>the</u> <u>decimal</u> <u>point</u> <u>two</u> <u>places</u> <u>to</u> <u>the</u> <u>left</u> <u>and</u> <u>dropping</u> <u>the</u> <u>percent</u> <u>sign</u>. That is:

$$83\% = {}_{\wedge}83. = .83 \qquad\qquad 6.5\% = {}_{\wedge}06.5 = .065$$

$$172\% = 1{}_{\wedge}72. = 1.72 \qquad\qquad .19\% = {}_{\wedge}00.19 = .0019$$

Convert each percent below to a decimal number.

a) 7% = _____   b) 41% = _____   c) 215% = _____

d) 16.5% = _____   e) .05% = _____   f) .8% = _____

---

a) .07   b) .41   c) 2.15

d) .165   e) .0005   f) .008

6. After converting some percents to decimal numbers, we can drop one or two final 0's after the decimal point. For example:

$$30\% = {}_{\curvearrowleft}30. = .30 \text{ or } .3$$

$$180\% = 1{}_{\curvearrowleft}80. = 1.80 \text{ or } 1.8$$

$$400\% = 4{}_{\curvearrowleft}00. = 4.00 \text{ or } 4$$

Convert each percent to a decimal number.

a) 90% = _____   b) 200% = _____   c) 130% = _____

---

7. To convert a mixed-number percent to a decimal number, we convert it to a decimal-number percent first. For example:

$$12\frac{1}{2}\% = 12.5\% = .125 \qquad 9\frac{3}{4}\% = 9.75\% = .0975$$

Convert each percent to a decimal number.

a) $11\frac{3}{4}\% =$ _____   b) $5\frac{1}{2}\% =$ _____

a) .9

b) 2

c) 1.3

---

8. Percents smaller than 1% are sometimes written as fractions. To convert them to decimal numbers, we convert them to decimal-number percents first. For example:

$$\frac{1}{2}\% = .5\% = .005 \qquad \frac{3}{4}\% = .75\% = .0075$$

Convert each percent to a decimal number.

a) $\frac{1}{4}\% =$ _____   b) $\frac{4}{5}\% =$ _____

a) .1175   b) .055

---

9. To avoid confusing percents smaller than 1% with percents larger than 1%, we name them in a special way. That is:

.1% is called "one-tenth of 1%"

$\frac{3}{4}\%$ is called "three-fourths of 1%"

Write the percent corresponding to each word name.

a) fifteen hundredths of 1% = _____

b) one half of 1% = _____

a) .0025   b) .008

---

a) .15%

b) $\frac{1}{2}\%$

10. Some common percent-fraction-decimal equivalents are given in the table below. Because of their usefulness, they should be memorized.

$25\% = \dfrac{1}{4} = .25$	$33\dfrac{1}{3}\% = \dfrac{1}{3} = .33$	$20\% = \dfrac{1}{5} = .2$	$10\% = \dfrac{1}{10} = .1$
$50\% = \dfrac{1}{2} = .5$	$66\dfrac{2}{3}\% = \dfrac{2}{3} = .67$	$40\% = \dfrac{2}{5} = .4$	$30\% = \dfrac{3}{10} = .3$
$75\% = \dfrac{3}{4} = .75$		$60\% = \dfrac{3}{5} = .6$	$70\% = \dfrac{7}{10} = .7$
		$80\% = \dfrac{4}{5} = .8$	$90\% = \dfrac{9}{10} = .9$

Note: The decimal equivalents for $33\dfrac{1}{3}\%$ and $66\dfrac{2}{3}\%$ are rounded to the nearest hundredth.

From memory, convert each percent to a fraction.

a) $25\% = $ _____   b) $80\% = $ _____   c) $66\dfrac{2}{3}\% = $ _____

d) $90\% = $ _____   e) $50\% = $ _____   f) $33\dfrac{1}{3}\% = $ _____

a) $\dfrac{1}{4}$ b) $\dfrac{4}{5}$ c) $\dfrac{2}{3}$ d) $\dfrac{9}{10}$ e) $\dfrac{1}{2}$ f) $\dfrac{1}{3}$

2-2   CONVERTING DECIMAL NUMBERS TO PERCENTS

In this section, we will discuss the decimal-point-shift method for converting decimal numbers to percents.

11. Because "percent" means "hundredths", any "hundredths" decimal fraction can be converted directly to a percent. For example:

$$\frac{9}{100} = 9\% \qquad \frac{60}{100} = 60\% \qquad \frac{200}{100} = \underline{\hspace{2cm}}$$

12. Any "hundredths" decimal number can be converted to a percent by converting it to a "hundredths" decimal fraction first. For example:

200%

$$.05 = \frac{5}{100} = 5\% \qquad 1.37 = \frac{137}{100} = 137\%$$

The same conversions can be made directly by simply shifting the decimal point two places to the right and adding the percent sign. That is:

$$.05 = .05_\curvearrowright = 5\% \qquad 1.37 = 1.37_\curvearrowright = 137\%$$

Continued on following page.

12.   Continued

Convert to a percent.

a) .83 = _____   b) .01 = _____   c) 2.96 = _____

---

13.   Any "tenths" decimal number can be converted to a percent by converting
it to a "hundredths" decimal fraction first. For example:

$$.4 = \frac{4}{10} = \frac{40}{100} = 40\% \qquad 1.8 = \frac{18}{10} = \frac{180}{100} = 180\%$$

The decimal-point-shift method can be used for the same conversions.
That is:

$$.4 = .40 = 40\% \qquad 1.8 = 1.80 = 180\%$$

Convert to a percent.

a) .1 = _____   b) .9 = _____   c) 2.3 = _____

---

14.   Any whole number can be converted to a percent by converting it to a
"hundredths" decimal fraction first. For example:

$$1 = \frac{100}{100} = 100\% \qquad 3 = \frac{300}{100} = 300\%$$

The decimal-point-shift method can be used for the same conversions.
That is:

$$1 = 1.00 = 100\% \qquad 3 = 3.00 = 300\%$$

Convert to a percent.

a) 2 = _____   b) 4 = _____   c) 7 = _____

---

15.   The decimal-point-shift method can be used to convert "thousandths" and
"ten-thousandths" decimal numbers to percents. For example:

$$.497 = .49\,7 = 49.7\% \qquad .0975 = .09\,75 = 9.75\%$$

$$.003 = .00\,3 = .3\% \qquad .0018 = .00\,18 = .18\%$$

Convert each decimal number to a percent.

a) .065 = _____   b) .1325 = _____   c) .0009 = _____

2-3   CONVERTING FRACTIONS TO PERCENTS

In this section, we will discuss the methods for converting fractions to percents. Though the decimal-fraction method is shown, the division method is emphasized because it is the general method. A table of common fraction-percent equivalents is given.

16.   Some fractions can be converted to percents by converting them to equivalent "hundredths" decimal fractions. For example:

$$\frac{1}{2} = \frac{50}{100} = 50\% \qquad \frac{3}{4} = \frac{75}{100} = 75\% \qquad \frac{2}{5} = \frac{40}{100} = 40\%$$

Using the decimal-fraction method, complete these conversions.

a) $\frac{1}{10} = \frac{\boxed{}}{100} = $ _____   b) $\frac{11}{20} = \frac{\boxed{}}{100} = $ _____

17.   Each fraction below has been converted to a percent by the decimal-fraction method.

$$\frac{1}{4} = \frac{25}{100} = 25\% \qquad \frac{3}{5} = \frac{60}{100} = 60\%$$

The same conversions can be made by converting the fractions to decimal numbers first. To do so, we divide. An example is shown. Complete the other conversion.

$$\frac{1}{4} = 4\overline{)1.00} = 25\%$$
$$\begin{array}{r} .25 \\ \underline{-8} \\ 20 \\ \underline{-20} \end{array}$$

$\frac{3}{5} = 5\overline{)3.} = $ _____

Answers (right column):

a) $\frac{10}{100} = 10\%$

b) $\frac{55}{100} = 55\%$

18.   We used the division method to convert $\frac{1}{3}$ to a mixed-number percent below. Notice that we terminated the division by writing the "hundredths" remainder as a fraction in lowest terms. Convert $\frac{2}{3}$ to a mixed-number percent.

$$\frac{1}{3} = 3\overline{)1.00}^{\,33\frac{1}{3}} = 33\frac{1}{3}\%$$
$$\begin{array}{r} \underline{-9} \\ 10 \\ \underline{-9} \\ 1 \end{array}$$

$\frac{2}{3} = 3\overline{)2.} = $ _____

Answers (right column):

$5\overline{)3.0}^{\,.6 \text{ or } .60} = 60\%$

$3\overline{)2.00}^{\,.66\frac{2}{3}} = 66\frac{2}{3}\%$

19.   Some common fraction-percent equivalents are shown in the table below. Because of their usefulness, <u>they should be memorized</u>.

$\frac{1}{4} = 25\%$	$\frac{1}{3} = 33\frac{1}{3}\%$	$\frac{1}{5} = 20\%$	$\frac{1}{10} = 10\%$
$\frac{1}{2} = 50\%$	$\frac{2}{3} = 66\frac{2}{3}\%$	$\frac{2}{5} = 40\%$	$\frac{3}{10} = 30\%$
$\frac{3}{4} = 75\%$		$\frac{3}{5} = 60\%$	$\frac{7}{10} = 70\%$
		$\frac{4}{5} = 80\%$	$\frac{9}{10} = 90\%$

From memory, convert each fraction to a percent.

a) $\frac{3}{4} = $ _____   b) $\frac{1}{3} = $ _____

c) $\frac{2}{5} = $ _____   d) $\frac{9}{10} = $ _____

---

20.   Some fractions that are not in lowest terms can be converted to percents by reducing them to fractions whose equivalents are known. For example:

   Since $\frac{30}{60}$ can be reduced to $\frac{1}{2}$, $\frac{30}{60} = 50\%$

   Since $\frac{50}{75}$ can be reduced to $\frac{2}{3}$, $\frac{50}{75} = $ _____

a) 75%   b) $33\frac{1}{3}\%$

c) 40%   d) 90%

---

21.   Since the "division" method is the general method for converting a fraction to a percent and a calculator can be used to perform the division, we do not ordinarily waste much time trying to reduce the fraction to lower terms.

We used a calculator to convert the fraction below directly to a percent. Use a calculator for the other conversion.

   $\frac{17}{68} = .25 = 25\%$ $\frac{34}{40} = $ _____ $ = $ _____

$66\frac{2}{3}\%$

---

22.   When a fraction equals a decimal number that contains more than two decimal places, the fraction converts to a decimal-number percent. An example is shown. Use a calculator for the other conversion.

   $\frac{3}{16} = .1875 = 18.75\%$ $\frac{10}{16} = $ _____ $ = $ _____

.85 = 85%

.625 = 62.5%

23. When the quotient is larger than "1", the percent is larger than 100%. An example is shown. Use a calculator for the other conversion.

$$\frac{31}{25} = 1.24 = 124\% \qquad \frac{47}{20} = \underline{\hspace{2cm}} = \underline{\hspace{2cm}}$$

---

24. When the first non-zero digit of the quotient is beyond the "hundredths" place, the percent is smaller than 1%. An example is shown. Use a calculator for the other conversion.

$$\frac{3}{400} = .0075 = .75\% \qquad \frac{1}{500} = \underline{\hspace{2cm}} = \underline{\hspace{2cm}}$$

2.35 = 235%

---

25. When one or both terms of the fraction is a decimal number, we convert it to a percent in the usual way. An example is shown. Use a calculator for the other conversion.

$$\frac{8.4}{17.5} = .48 = 48\% \qquad \frac{5.65}{4.52} = \underline{\hspace{2cm}} = \underline{\hspace{2cm}}$$

.002 = .2%

---

26. When the division is non-terminating, we round to the nearest whole-number percent or to the nearest tenth or hundredth of a percent.

To round to the nearest whole-number percent, we round to two decimal places before converting. For example:

$$\frac{211}{375} = .56266667 = .56 = 56\%$$

Convert each fraction to the nearest whole-number percent.

a) $\frac{1.8}{17} = \underline{\hspace{2cm}}$ 　　　　b) $\frac{2,500}{1,930} = \underline{\hspace{2cm}}$

1.25 = 125%

---

27. To round to the nearest tenth of a percent, we round to three decimal places before converting. For example:

$$\frac{39}{550} = .07090909 = .071 = 7.1\%$$

Convert each fraction to the nearest tenth of a percent.

a) $\frac{1.59}{1.75} = \underline{\hspace{2cm}}$ 　　　　b) $\frac{6,000}{475,000} = \underline{\hspace{2cm}}$

a) 11%   b) 130%

---

a) 90.9%   b) 1.3%

28. To round to the nearest hundredth of a percent, we round to four decimal places before converting. For example:

$$\frac{13}{150} = .08666667 = .0867 = 8.67\%$$

Convert each fraction to the nearest hundredth of a percent.

a) $\frac{2.3}{35}$ = _____

b) $\frac{120}{13,000}$ = _____

---

a) 6.57%   b) .92%

2-4   OPERATIONS INVOLVING PERCENTS

In this section, we will discuss adding and subtracting percents and multiplying or dividing an ordinary number by a percent.

---

29. Percents can be added or subtracted just like ordinary numbers. For example:

25% + 50% = 75%   100% - 40% = _____

---

30. To multiply an ordinary number by a percent, we must convert the percent to a fraction or decimal number. If it is converted to a decimal number, we can use a calculator for the multiplication. For example:

60%

$$20\% \times 35 = \frac{1}{5} \times 35 = 7$$
$$20\% \times 35 = .2 \times 35 = 7$$

Use either method to complete these:

a) 70% x 40 = _____

b) 50% x 84 = _____

---

31. Since most percents are not equal to simple fractions, we ordinarily convert the percent to a decimal number to multiply. An example is shown. Use a calculator for the other multiplications.

a) 28   b) 42

$$27.5\% \times 3,500 = .275 \times 3,500 = 962.5$$

a) 182% x 47.5 = _____ x _____ = _____

b) .6% x 910 = _____ x _____ = _____

---

a) 1.82 x 47.5 = 86.45

b) .006 x 910 = 5.46

32.   To divide an ordinary number by a percent, we must also convert the percent to a fraction or decimal number. If it is converted to a decimal number, we can use a calculator for the division. For example:

$$\frac{45}{30\%} = \frac{45}{\frac{3}{10}} = \overset{15}{\cancel{45}} \times \frac{10}{\cancel{3}} = 150$$

$$\frac{45}{30\%} = \frac{45}{.3} = 150$$

Use either method to complete these:

a) $\dfrac{27}{10\%}$ = _____

b) $\dfrac{75}{50\%}$ = _____

---

33.   Since most percents are not equal to simple fractions, we ordinarily convert to a decimal number to divide. An example is shown. Use a calculator for the other divisions.

$$\frac{700}{35\%} = \frac{700}{.35} = 2,000$$

a) $\dfrac{8.4}{6\%}$ = _____ = _____

b) $\dfrac{278}{12.5\%}$ = _____ = _____

a) 270   b) 150

a) $\dfrac{8.4}{.06}$ = 140   b) $\dfrac{278}{.125}$ = 2,224

2-5   THE THREE TYPES OF PERCENT PROBLEMS

In this section, we will discuss the three basic types of percent problems. We will show how they can be solved by converting to equations and then solving the equations.

---

34.   Two examples of the first type of percent problems are shown below. We converted each to an equation. Notice that "is" means "=" and "of" means "x" or "times".

What is 50% of 46 ?
↓   ↓   ↓   ↓   ↓
n   = 50% x 46

80% of 20 is what?
↓   ↓   ↓   ↓   ↓
80% x 20 =   n

Translate each problem below to an equation.

a) What is 25% of 64.8 ?

b) $33\frac{1}{3}\%$ of 960 is what?

_____   _____

a) n = 25% x 64.8   b) $33\frac{1}{3}\%$ x 960 = n

35.   We translated one problem below to an equation. Translate the other problem to an equation.

Find 6.5% of $700.

$$6.5\% \times \$700 = n$$

Find 135% of 875.

---

36.   To solve the problem below, we translated to an equation and then solved for "n" by converting $33\frac{1}{3}\%$ to a fraction to multiply. Use the same method to solve the other problem.

What is $33\frac{1}{3}\%$ of 66 ?

$$n = 33\frac{1}{3}\% \times 66$$

$$n = \frac{1}{3}(66) = 22$$

25% of 48 is what?

$135\% \times 875 = n$

---

37.   Ordinarily we have to convert to a decimal number when multiplying by a percent. An example is shown. Use a calculator for the other problem.

Find 2.4% of 120.

$$2.4\% \times 120 = n$$

$$.024(120) = n$$

$$n = 2.88$$

What is 18% of $7,500 ?

$25\% \times 48 = n$

$\frac{1}{4}(48) = n$

$n = 12$

---

38.   Two examples of the second type of percent problems are shown below. We translated each to an equation.

17 is what percent of 80 ?

$$17 = n\% \times 80$$

What percent of 200 is 159 ?

$$n\% \times 200 = 159$$

Translate each problem below to an equation.

a) 20 is what percent of 50 ?

b) What percent of 90 is 60 ?

$n = 18\% \times \$7,500$
$n = .18 \times \$7,500$
$n = \$1,350$

---

39.   To solve equations like those obtained in the last frame, we divide the other number by the coefficient of "n%" and then convert the fraction to a percent. An example is shown. Solve the other equation.

$$n\% \times 90 = 60$$

$$n\% = \frac{60}{90} = \frac{2}{3} = 66\frac{2}{3}\%$$

20 = n% × 50

a) $20 = n\% \times 50$

b) $n\% \times 90 = 60$

$n\% = 40\%$, from: $\frac{20}{50} = \frac{2}{5}$

40. Ordinarily we have to use the division method to solve the second type of percent problem. An example is shown. Use a calculator for the other problem. Round to the nearest whole-number percent.

What percent of 70 is 98 ?   14.9 is what percent of 84.7 ?

$$n\% \times 70 = 98$$

$$n\% = \frac{98}{70} = 140\%$$

---

41. Two examples of the third type of percent problems are shown below.

25 is 12% of what?   80% of what is 62 ?
↓ ↓   ↓   ↓   ↓   ↓   ↓   ↓   ↓   ↓
25 = 12% x   n   80% x   n = 62

Translate each problem below to an equation.

a) 40 is $66\frac{2}{3}\%$ of what?   b) 50% of what is 20 ?

_____   _____

---

42. To solve equations like those obtained in the last frame, we can sometimes convert the percent to a fraction. An example is shown. Solve the other equation.

$$40 = 66\frac{2}{3}\% \times n \qquad\qquad 50\% \times n = 20$$

$$40 = \frac{2}{3}n$$

$$\frac{3}{2}\overset{20}{(\cancel{40})} = \frac{3}{2}\left(\frac{2}{3}n\right)$$

$$60 = 1n$$

$$n = 60$$

---

43. Usually we have to convert the percent to a decimal number to solve the third type of percent problem. An example is shown. Use a calculator for the other problem. Round to the nearest whole number.

22 is 2.5% of what?   12% of what is 200 ?

$$22 = 2.5\% \times n$$

$$22 = .025 \times n$$

$$n = \frac{22}{.025} = 880$$

Right column answers:

$n\% = 18\%$, from $\dfrac{14.9}{84.7}$

a) $40 = 66\frac{2}{3}\% \times n$

b) $50\% \times n = 20$

n = 40, from:

$$\frac{1}{2}n = 20$$

$$2\left(\frac{1}{2}n\right) = 2(20)$$

$$n = 40$$

n = 1,667 , from: .12n = 200

$$n = \frac{200}{.12}$$

44.   Use a calculator to solve these.

   a) Find $7\frac{1}{2}\%$ of \$600 .   b) 39.6 is what percent of 28.3 ?
                                            (Round to the nearest whole number
                                                                  percent.)

---

45.   Use a calculator for these:

   a) What percent of 9,500 is 27 ?   b) 75 is 45% of what?
       (Round to the nearest hundredth   (Round to the nearest
       of a percent.)   whole number.)

a) \$45, from:

   $7\frac{1}{2}\% \times \$600 = n$

b) 140%, from:

   $39.6 = n\% \times 28.3$

---

46.   Use a calculator for these:

   a) What is .4% of 375 ?   b) 125% of what is 75 ?

a) .28%, from:

   $n\% \times 9,500 = 27$

b) 167, from:

   $75 = 45\% \times n$

---

a) 1.5, from: b) 60, from:

   $n = .4\% \times 375$ $125\% \times n = 75$

SELF-TEST 4 (pages 40-53)

Convert each percent to a fraction in lowest terms.

1.   40% = _____

2. $66\frac{2}{3}$% = _____

Convert each percent to a decimal number.

3.   15.3% = _____

4.   .07% = _____

Convert each decimal number to a percent.

5.   .002 = _____

6.   1.8 = _____

Convert each fraction to a percent.

7. $\frac{3}{4}$ = _____

8. $\frac{36}{32}$ = _____

9. Convert $\frac{2.63}{74.8}$ to the nearest hundredth of a percent. _____

10. Convert $\frac{8,190}{3,620}$ to the nearest whole number percent. _____

11. 80% x 450 = _____

12. $\frac{63}{25\%}$ = _____

13. .15% x 4,200 = _____

14. $\frac{360}{6.4\%}$ = _____

15. What is 6.25% of $2,500 ?

16. 240 is what percent of 400 ?

17. 20% of what is 70 ?

18. 520 is 8% of what?

19. Find .5% of 9,260.

20. What percent of 16,400 is 390 ?
     (Round to the nearest hundredth
     of a percent.)

ANSWERS:

1. $\frac{2}{5}$

2. $\frac{2}{3}$

3. .153

4. .0007

5. .2%

6. 180%

7. 75%

8. 112.5%

9. 3.52%

10. 226%

11. 360

12. 252

13. 6.3

14. 5,625

15. $156.25

16. 60%

17. 350

18. 6,500

19. 46.3

20. 2.38%

2-6   APPLIED PROBLEMS INVOLVING PERCENTS

In this section, we will discuss some applied problems involving percents. We will show how they can be solved by rewording, translating to an equation, and then solving the equation.

---

47.   To solve the problem below, we reworded it, translated the rewording to an equation, and then solved the equation. Use the same method for the other problem.

How many pounds of zinc are there in 50 pounds of an alloy if the alloy is 30% zinc?

Rewording: What is 30% of 50 ?

Equation: n = 30% x 50

Solution: n = .3(50) = 15

Therefore, there are 15 pounds of zinc in the alloy.

On a 40-item test, a student had 90% of the items correct. How many did she have correct?

Rewording: _____

Equation: _____

Solution: _____

---

48.   We used the rewording-equation method to solve the problem below. Use the same method for the other problem. Round to the nearest tenth of a percent.

Of 268 motors tested, 37 were defective. What percent were defective?

Rewording: 37 is what percent of 268 ?

Equation: 37 = n% x 268

Solution: $n\% = \dfrac{37}{268} = .138 = 13.8\%$

Therefore 13.8% of the motors were defective.

There are 14 grams of nitrogen in 63 grams of nitric acid. In terms of weight, what percent of nitric acid is nitrogen?

Rewording: _____

Equation: _____

Solution: _____

What is 90% of 40 ?

n = 90% x 40

n = 36 items correct

---

14 is what percent of 63 ?
14 = n% x 63
$n\% = \dfrac{14}{63} = 22.2\%$

49. We used the rewording-equation method to solve the problem below. Use the same method for the other problem.

How much money must be invested at 5% annual interest to earn $500 interest annually?

Rewording: $500 is 5% of what?

Equation: $500 = 5% x n

Solution: $n = \dfrac{\$500}{.05} = \$10,000$

Therefore, $10,000 must be invested to earn $500 annually.

A stainless steel alloy contains 8% nickel. How many pounds of the alloy can be made from 560 pounds of nickel?

Rewording: _____

Equation: _____

Solution: _____

---

50. Use the "rewording" hints to solve the problems in this frame and the next two frames.

a) If the sales tax rate is 4%, how much tax is paid for a $425.95 TV set. (Round to the nearest cent.)

(Hint: 4% of $425.95 is what?)

b) If a student got 26 items correct on a 32-item test, what percent grade did he receive? (Round to the nearest whole number percent.)

(Hint: 26 is what percent of 32?)

560 is 8% of what?

560 = 8% x n

$n = \dfrac{560}{.08} = 7,000$ pounds

---

51. a) An alloy weighing 32.7 kilograms contains 25.3 kilograms of aluminum. What percent of the alloy is aluminum? (Round to the nearest tenth of a percent.)

(Hint: 25.3 is what percent of 32.7?)

b) If iron rust is 70% iron, how much rust can be formed from 40 grams of iron? (Round to the nearest whole number.)

(Hint: 40 is 70% of what?)

a) $17.04

b) 81%

52. a) What is the annual interest on a \$2,000 loan if the interest rate is $10\frac{1}{2}\%$?

(Hint: $10\frac{1}{2}\%$ of \$2,000 is what?)

b) An ore contains .036% titanium. How much ore is needed to get 500 pounds of titanium? (Round to the nearest thousand.)

(Hint: .036% of what is 500?)

a)	77.4%
b)	57 grams

---

53. Make up your own "rewording" hints to solve the problems in this frame and the next two frames.

a) How many items would a student have to get correct to get 80% on a 40-item test?

b) If the annual interest on a \$500 loan is \$60, what is the interest rate?

a)	\$210
b)	1,389,000 pounds

---

54. a) If .92% of a lot of 17,500 precision ball bearings were defective, how many were defective?

b) An alloy contains 18.2% chromium. How much of the alloy can be made from 150 pounds of chromium? (Round to the nearest whole number.)

a)	32 items, from:
	80% of 40 is what?
b)	12%, from:
	\$60 is what percent of \$500?

---

55. a) A sample of polluted lake water contains "87 parts per million" of mercury. What percent is mercury?

a)	161 defective, from:
	.92% of 17,500 is what?
b)	824 pounds, from:
	18.2% of what is 150?

Continued on following page.

55.   Continued

b) A student received a grade of 71% on a test. If she had 25 items correct,
how many items did the test contain? (Round to the nearest whole
number.)

a) .0087%, from:	b) 35 items, from:
87 is what percent of 1,000,000?	71% of what is 25?

2-7   SOLVING FRACTIONAL EQUATIONS AND PROPORTIONS

In this section, we will discuss the methods for solving simple fractional equations and proportions and show
how a calculator can be used in the solutions.

56.   To clear the fraction and solve the equation below, we multiplied both
sides by "3", the denominator of the fraction. Solve the other equation by
multiplying both sides by "7".

$$\frac{x}{3} = 9 \qquad\qquad 5 = \frac{y}{7}$$

$$3\left(\frac{x}{3}\right) = 3(9)$$

$$x = 27$$

57.   To solve the equation below, we used a calculator to multiply 8.4 and 7.5 .   y = 35
Use a calculator to solve the other equation.

$$\frac{x}{8.4} = 7.5 \qquad\qquad 6.44 = \frac{y}{1.25}$$

$$8.4\left(\frac{x}{8.4}\right) = 8.4(7.5)$$

$$x = 63$$

58.   To clear the fraction below, we multiplied both sides by "t", the   y = 8.05
denominator of the fraction. Then we solved 12 = 3t by dividing
12 by 3. Use the same steps to solve the other equation.

$$\frac{12}{t} = 3 \qquad\qquad 6 = \frac{42}{x}$$

$$t\left(\frac{12}{t}\right) = t(3)$$

$$12 = 3t$$

$$t = \frac{12}{3} = 4$$

59.   To solve the equation below, we used a calculator to divide 4.9 by 1.4 .
       Use a calculator to solve the other equation.

$$\frac{4.9}{x} = 1.4 \qquad\qquad 9.04 = \frac{24.86}{y}$$

$$x\left(\frac{4.9}{x}\right) = x(1.4)$$

$$4.9 = 1.4x$$

$$x = \frac{4.9}{1.4} = 3.5$$

| $x = 7$ |

---

60.   Solve each equation. Round each root to hundredths.

       a) $\dfrac{x}{3.93} = 2.18$ \qquad\qquad b) $14.7 = \dfrac{86.1}{y}$

| $y = 2.75$ |

---

61.   A proportion is an equation with only one fraction on each side. To clear
       the fractions in a proportion, we can multiply both sides by both denomina-
       tors at the same time. For example:

$$\frac{x}{5} = \frac{3}{7} \qquad\qquad\qquad \frac{9}{2} = \frac{4}{y}$$

$$(5)(7)\left(\frac{x}{5}\right) = (5)(7)\left(\frac{3}{7}\right) \qquad (2)(y)\left(\frac{9}{2}\right) = (2)(y)\left(\frac{4}{y}\right)$$

$$\frac{(5)(7)(x)}{5} = \frac{(5)(7)(3)}{7} \qquad \frac{(2)(y)(9)}{2} = \frac{(2)(y)(4)}{y}$$

$$\left(\frac{5}{5}\right)(7)(x) = \left(\frac{7}{7}\right)(5)(3) \qquad \left(\frac{2}{2}\right)(y)(9) = \left(\frac{y}{y}\right)(2)(4)$$

$$(1)(7x) = (1)(15) \qquad\qquad (1)(9y) = (1)(8)$$

$$7x = 15 \qquad\qquad\qquad 9y = 8$$

However, there is a shortcut called "<u>cross-multiplication</u>" that can be used
to clear the fractions in a proportion. The "cross-multiplication" shortcut
for each proportion above is shown and discussed below.

$$\frac{x}{5} \diagup\!\!\!\!\diagdown \frac{3}{7} \qquad\qquad \frac{9}{2} \diagup\!\!\!\!\diagdown \frac{4}{y}$$

$$(x)(7) = (5)(3) \qquad\qquad (9)(y) = (2)(4)$$

$$7x = 15 \qquad\qquad\qquad 9y = 8$$

<u>Note</u>: 1) To get <u>the left side</u> of each non-fractional equation, we
        multiplied the <u>first numerator</u> and the <u>second denominator</u>.
        That is:

            To get "7x", we multiplied "x" and "7".
            To get "9y", we multiplied "9" and "y".

     2) To get <u>the right side</u> of each non-fractional equation, we
        multiplied the <u>first denominator</u> and the <u>second numerator</u>.
        That is:

            To get "15", we multiplied "5" and "3".
            To get " 8 ", we multiplied "2" and "4".

Continued on following page.

| a) $x = 8.57$ |
| b) $y = 5.86$ |

61. Continued

Use cross-multiplication to clear the fractions in each proportion below.

a) $\dfrac{8}{x} = \dfrac{3}{4}$ b) $\dfrac{1}{15} = \dfrac{y}{2}$ c) $\dfrac{12}{7} = \dfrac{5}{t}$

_____   _____   _____

---

62. To solve the proportion below, we cross-multiplied to clear the fractions and then solved "3x = 24". Use the same steps to solve the other proportion.

$$\dfrac{x}{6} = \dfrac{4}{3}$$ $$\dfrac{2}{5} = \dfrac{y}{15}$$

$$(x)(3) = (6)(4)$$

$$3x = 24$$

$$x = \dfrac{24}{3} = 8$$

a) $(8)(4) = (x)(3)$
       $32 = 3x$

b) $(1)(2) = (15)(y)$
       $2 = 15y$

c) $(12)(t) = (7)(5)$
       $12t = 35$

---

63. To solve the proportion below, we used a calculator to divide 210 by 50. Use a calculator to solve the other proportion.

$$\dfrac{7}{x} = \dfrac{50}{30}$$ $$\dfrac{4}{5} = \dfrac{10}{y}$$

$$(7)(30) = (x)(50)$$

$$210 = 50x$$

$$x = \dfrac{210}{50} = 4.2$$

$y = 6$, from:

   $30 = 5y$

---

64. Following the example, round the other root to hundredths.

$$\dfrac{3}{7} = \dfrac{x}{5}$$ $$\dfrac{9}{8} = \dfrac{7}{y}$$

$$(3)(5) = (7)(x)$$

$$15 = 7x$$

$$x = \dfrac{15}{7} = 2.14$$

$y = 12.5$, from:

   $4y = 50$

---

65. To solve the proportion below, we used a calculator twice: first to multiply 17 and 65 and then to divide 1,105 by 26. Use a calculator to solve the other proportion.

$$\dfrac{17}{26} = \dfrac{x}{65}$$ $$\dfrac{13}{34} = \dfrac{71.5}{x}$$

$$(17)(65) = (26)(x)$$

$$1,105 = 26x$$

$$x = \dfrac{1,105}{26} = 42.5$$

$y = 6.22$, from $\dfrac{56}{9}$

66.   We used a calculator twice to solve the proportion below.

$$\frac{46}{41} = \frac{57.5}{x}$$

$$(46)(x) = (41)(57.5)$$

$$46x = 2,357.5$$

$$x = 51.25$$

We can solve the same proportion by doing the multiplication and division in one combined operation on a calculator. To do so, we write the solution as we have done below.

$$\frac{46}{41} = \frac{57.5}{x}$$

$$(46)(x) = (41)(57.5)$$

$$x = \frac{(41)(57.5)}{46}$$

Then to perform $\frac{(41)(57.5)}{46}$ in one combined operation on a calculator, follow the steps below. Notice that the $\boxed{=}$ key is not pressed after the multiplication.

Enter	Press	Display
41	$\boxed{\text{x}}$	41.
57.5	$\boxed{\div}$	2,357.5
46	$\boxed{=}$	51.25

Is 51.25 the same solution we obtained above? _____

x = 187, from $\frac{2,431}{13}$

67.   To solve the proportion below in one combined operation on a calculator, we set up the solution as we have done.

$$\frac{x}{41.25} = \frac{7.4}{16.5}$$

$$(x)(16.5) = (41.25)(7.4)$$

$$x = \frac{(41.25)(7.4)}{16.5}$$

Then to perform $\frac{(41.25)(7.4)}{16.5}$ on a calculator, follow these steps:

Enter	Press	Display
41.25	$\boxed{\text{x}}$	41.25
7.4	$\boxed{\div}$	305.25
16.5	$\boxed{=}$	18.5

Therefore the solution of the proportion is: x = _____

Yes

68. To solve the proportion below in one combined operation on a calculator, we set up the solution as we have done.

$$\frac{0.875}{0.175} = \frac{x}{215}$$

$$(0.875)(215) = (0.175)(x)$$

$$x = \frac{(0.875)(215)}{0.175}$$

Then to perform $\frac{(0.875)(215)}{0.175}$ on a calculator, follow these steps:

Enter	Press	Display
.875	x	0.875
215	÷	188.125
.175	=	1075.

Therefore, the solution of the proportion is: x = _____

x = 18.5

---

69. Following the steps in the last frame, solve each proportion.

a) $\frac{1.75}{1.40} = \frac{w}{2,600}$ b) $\frac{14.8}{N} = \frac{0.216}{0.756}$

x = 1,075

---

70. Solve each proportion. Round to the indicated place.

a) Round to tenths.   b) Round to thousandths.

$$\frac{0.617}{A} = \frac{2.09}{93.4}$$ $$\frac{56,800}{730} = \frac{3.27}{d}$$

a) w = 3,250 , from:

$$w = \frac{(1.75)(2,600)}{1.40}$$

b) N = 51.8 , from:

$$N = \frac{(14.8)(0.756)}{0.216}$$

---

a) A = 27.6

b) d = .042

SELF-TEST 5 (pages 54-62)

1. A student got 61 items correct on a 72-item final exam. Find the student's percent grade. (Round to the nearest tenth of a percent.)

2. A bronze alloy contains 8.25% tin. How much tin is there in 2,750 pounds of the alloy? (Round to the nearest whole number.)

3. If sea water contains .13% magnesium, how many kilograms of sea water will contain 10 kilograms of magnesium? (Round to the nearest hundred.)

4. In a day's production of 163,800 metal cans, 7,420 defective cans were scrapped. What percent were scrapped? (Round to the nearest hundredth of a percent.)

Solve each equation. 5. $\dfrac{99}{x} = 4.5$

x = _____

6. $.75 = \dfrac{y}{14.4}$

y = _____

Solve each proportion. Round to the indicated place.

7. Round to thousands.

$$\frac{2,960}{.0518} = \frac{x}{3.44}$$

x = _____

8. Round to two decimal places.

$$\frac{.351}{p} = \frac{49.2}{803}$$

p = _____

9. Round to tenths.

$$\frac{786}{26,500} = \frac{1.14}{t}$$

t = _____

ANSWERS:

1. 84.7%

2. 227 pounds

3. 7,700 kilograms

4. 4.53%

5. x = 22

6. y = 10.8

7. x = 197,000

8. p = 5.73

9. t = 38.4

2-8   RATIOS

In this section, we will discuss ratios and the use of "ratio" language.

---

71. A "ratio" is a comparison of quantities by means of a division. The division is frequently written as a fraction. For example: If the length and width of a rectangle are 7 feet and 4 feet, the ratio of length to width is $\frac{7 \text{ feet}}{4 \text{ feet}}$ or simply $\frac{7}{4}$. If there are 3 pounds of copper and 8 pounds of tin in an alloy, the ratio of copper to tin is $\frac{3 \text{ pounds}}{8 \text{ pounds}}$ or simply _____ .	
72. Instead of using a fraction to express a ratio, we sometimes use the word "to" or a colon ":". For example: If a driving gear has 35 teeth and the driven gear has 12 teeth, the "teeth" ratio is $\frac{35}{12}$ or "35 to 12" or "35:12". If one process takes 11 seconds and a second process takes 20 seconds, the time ratio is $\frac{11}{20}$ or "_____ to _____" or "_____ : _____".	$\frac{3}{8}$
73. When ratio language is used, the ratio is ordinarily reduced to lowest terms if possible. For example: The ratio of 25 grams to 15 grams is $\frac{25}{15}$ or $\frac{5}{3}$ or "5 to 3". The ratio of 20 seconds to 40 seconds is $\frac{20}{40}$ or $\frac{1}{2}$ or "_____ to _____".	11 to 20 or 11:20
74. When a ratio reduces to a whole number, we sometimes write "1" as the denominator to make it clear that we are comparing two quantities. For example: The ratio of 25 meters to 5 meters is $\frac{25}{5}$ or $\frac{5}{1}$ or "5 to 1". The ratio of 30 ounces to 15 ounces is $\frac{30}{15}$ or $\frac{2}{1}$ or "_____ to _____".	1 to 2
75. Ratios are also used to compare decimal-number quantities. For example: The ratio of 38.9 grams to 17.3 grams is $\frac{38.9}{17.3}$. The ratio of 1.43 seconds to 7.66 seconds is _____ .	2 to 1

76. Since a ratio is a fraction (or division), any ratio can be converted to an ordinary number by performing the division. An example is shown. Use a calculator to convert the other ratios to ordinary numbers.

   The ratio of 24 feet to 2 feet is $\frac{24}{2}$ or 12.

   a) The ratio of 29 seconds to 116 seconds is $\frac{29}{116}$ or _____ .

   b) The ratio of 98 grams to 28 grams is $\frac{98}{28}$ or _____ .

1.43
7.66

77. When a ratio is converted to an ordinary number, it still stands for a comparison of two quantities. To show that fact, it is helpful to write the ordinary number over "1". For example:

   The ratio of 60 meters to 15 meters is $\frac{60}{15}$ or 4.
   A ratio of "4" means $\frac{4}{1}$ or "4 to 1".

   The ratio of 12 grams to 24 grams is $\frac{12}{24}$ or .5 .
   A ratio of ".5" means $\frac{.5}{1}$ or ".5 to 1".

   The ratio of 9.2 seconds to 1.6 seconds is $\frac{9.2}{1.6}$ or 5.75 .
   A ratio of "5.75" means $\frac{5.75}{1}$ or "_____ to _____".

a) .25

b) 3.5

78. For each problem below, set up the ratio and convert it to an ordinary number.

   a) The height and base of a triangle are 21 inches and 15 inches. What is the height-to-base ratio?

   b) A driving gear with 68 teeth is meshed with a driven gear with 24 teeth. What is the gear ratio? That is, what is the ratio of the number of teeth in the driving gear to the number of teeth in the driven gear? (Round to hundredths.)

5.75 to 1

79. For each problem below, set up the ratio and convert it to an ordinary number.

   a) An electrical signal of 0.835 volts is accompanied by a noise voltage of 0.0355 volts. What is the signal-to-noise ratio? (Round to tenths.)

a) $\frac{21 \text{ inches}}{15 \text{ inches}} = 1.4$

b) $\frac{68 \text{ teeth}}{24 \text{ teeth}} = 2.83$

Continued on following page.

79. Continued

   b) A drive shaft with a rotational speed of 825 rpm (revolutions per
       minute) drives a second shaft at a rotational speed of 543 rpm.
       What is the speed ratio? That is, what is the ratio of the rotational
       speed of the drive shaft to the rotational speed of the second
       shaft? (Round to hundredths.)

---

80. Remember that a ratio stated as an ordinary number still stands for a
     comparison of two quantities. For example:

     A gear ratio of 3.45 means "3.45 to 1".

     A signal-to-noise ratio of 27.6 means "_____ to _____".

a) $\dfrac{0.835 \text{ volts}}{0.0355 \text{ volts}} = 23.5$

b) $\dfrac{825 \text{ rpm}}{543 \text{ rpm}} = 1.52$

---

81. Two ratios are equivalent if they both reduce to the same lowest-terms
     ratio. For example, the two ratios below are equivalent because they
     both reduce to "3 to 4".

         15 grams to 20 grams   3 to 4 $\left(\text{from } \dfrac{15}{20} = \dfrac{3}{4}\right)$

         30 grams to 40 grams   3 to 4 $\left(\text{from } \dfrac{30}{40} = \dfrac{3}{4}\right)$

     By reducing to lowest terms, determine whether each pair of ratios is
     equivalent or not.

         a) 10 feet to 8 feet   b) 10 seconds to 25 seconds
             40 feet to 32 feet   15 seconds to 35 seconds

27.6 to 1

---

82. Two ratios are also equivalent if they both convert to the same ordinary
     number. For example, the two ratios below are equivalent because they
     both convert to 5.5 or "5.5 to 1".

         77 watts to 14 watts   5.5 to 1 $\left(\text{from } \dfrac{77}{14} = 5.5\right)$

         121 watts to 22 watts   5.5 to 1 $\left(\text{from } \dfrac{121}{22} = 5.5\right)$

     By converting to an ordinary number, determine whether each pair of
     ratios is equivalent or not.

         a) 735 rpm to 525 rpm   b) 0.572 volts to 0.0275 volts
             600 rpm to 375 rpm   0.676 volts to 0.0325 volts

a) Equivalent, since
    both equal "5 to 4"

b) Not equivalent

---

a) Not equivalent

b) Equivalent, since
    they both equal
    20.8 (or 20.8 to 1)

83.   Ratios are ordinarily used to compare quantities with the same units.
However, when a fraction (or division) is used to compare quantities
with different units, we can also think of the fraction as a ratio. For
example:

If a spring is stretched 1.3 inches by a force of 21 pounds, we can
use a fraction (or ratio) to compare the stretch to the force.
We get:

$$\frac{1.3 \text{ inches}}{21 \text{ pounds}} \quad \text{or a ratio of "1.3 to 21".}$$

If there are 55 grams of a gas in a volume of 22.4 liters, we can
use a fraction (or ratio) to compare the weight to the volume. We get:

$$\frac{55 \text{ grams}}{22.4 \text{ liters}} \quad \text{or a ratio of "} \underline{\hspace{2cm}} \text{ to } \underline{\hspace{2cm}} \text{".}$$

---

55 to 22.4

2-9   APPLIED PROBLEMS INVOLVING PROPORTIONS

In this section, we will discuss some applied problems that can be solved by setting up and solving a
proportion.

84.   A proportion is an equation containing two equivalent ratios. For example:

$$\frac{5 \text{ inches}}{6 \text{ pounds}} = \frac{10 \text{ inches}}{12 \text{ pounds}} \quad \text{is a proportion, since } \frac{5}{6} = \frac{10}{12}$$

Which of the following are proportions? $\underline{\hspace{2cm}}$

a) $\dfrac{20 \text{ grams}}{10 \text{ liters}} = \dfrac{60 \text{ grams}}{30 \text{ liters}}$ b) $\dfrac{100 \text{ miles}}{2 \text{ hours}} = \dfrac{200 \text{ miles}}{3 \text{ hours}}$

85.   To solve the problem below, we set up a proportion <u>with</u> units and then
solved the same proportion <u>without</u> units. State the solution in terms of
the original problem.

A car travels 100 miles in 2 hours. At that rate, how many hours
would it take to travel 350 miles?

$$\frac{100 \text{ miles}}{2 \text{ hours}} = \frac{350 \text{ miles}}{x \text{ hours}} \quad \text{or} \quad \frac{100}{2} = \frac{350}{x}$$

$$100x = 700$$

$$x = 7$$

Therefore, it would take $\underline{\hspace{1.5cm}}$ hours to travel 350 miles
at that rate.

Only (a)

86. We used the same method to solve the problem below. Notice that we used a calculator to solve the proportion and we rounded to the nearest tenth of a pound.

7 hours

If a spring is stretched 5.8 inches by a 21-pound force, what force is needed to stretch it 1.3 inches?

$$\frac{5.8 \text{ inches}}{21 \text{ pounds}} = \frac{1.3 \text{ inches}}{x \text{ pounds}} \quad \text{or} \quad \frac{5.8}{21} = \frac{1.3}{x}$$

$$(5.8)(x) = (21)(1.3)$$

$$x = \frac{(21)(1.3)}{5.8}$$

$$x = 4.7$$

Therefore, a force of _____ pounds is needed to stretch the spring 1.3 inches.

87. When using a proportion to solve a problem, the following steps should be followed to set up the proportion correctly.

4.7 pounds

1) Set up the known ratio on the left side.

2) Then make sure that the units in the ratio on the right side correspond to those of the known ratio.

For example, we set up the proportion correctly below.

If you need 3 grams of sodium chloride to make 1 liter of a solution, how many grams would you need to make 5 liters of the same solution?

$$\frac{3 \text{ grams}}{1 \text{ liter}} = \frac{x \text{ grams}}{5 \text{ liters}}$$

Set up the proportion for each problem below. Do not solve.

a) If a 3.25 inch length on a map represents 50 miles, how many miles would a 10 inch length represent?

b) If a car used 12.7 gallons of gas to travel 318 miles, how many gallons would it use to travel 500 miles?

a) $\dfrac{3.25 \text{ inches}}{50 \text{ miles}} = \dfrac{10 \text{ inches}}{x \text{ miles}}$

b) $\dfrac{12.7 \text{ gallons}}{318 \text{ miles}} = \dfrac{x \text{ gallons}}{500 \text{ miles}}$

88.   Use a proportion to solve each problem.

a) If an architectural drawing is scaled so that 0.75 inch represents 4 feet, what length represents 39 feet? (Round to two decimal places.)

b) If there are 147 pounds of zinc in 435 pounds of an alloy, how many pounds of zinc are there in 1,000 pounds of the alloy. (Round to the nearest whole number.)

---

89.   Use a proportion to solve each problem.

a) A copper wire 175 feet long has a resistance of 0.418 ohm. What is the resistance (in ohms) of 34 feet of the same wire? (Round to four decimal places.)

b) If 36.3 grams of oxygen are obtained by decomposing 92.7 grams of potassium chlorate, how many grams of potassium chlorate must be decomposed to get 100 grams of oxygen? (Round to the nearest whole number.)

a) 7.31 inches, from:

$$\frac{0.75}{4} = \frac{x}{39}$$

b) 338 pounds, from:

$$\frac{147}{435} = \frac{x}{1,000}$$

---

90.   Use a proportion to solve each problem.

a) A gear with a 12-inch diameter drives a gear with an 18-inch diameter. If the 18-inch gear has 48 teeth, how many teeth does the 12-inch gear have?

a) 0.0812 ohm, from:

$$\frac{175}{0.418} = \frac{34}{x}$$

b) 255 grams, from:

$$\frac{36.3}{92.7} = \frac{100}{x}$$

Continued on following page.

90.   Continued

   b) If a 20-gallon gas tank holds 75.7 liters, how many gallons would a
      50-liter gas tank hold? (Round to tenths.)

---

91.   Proportions can be used for conversions of units. Two examples are shown.

   There are 12 inches in 1 foot. How many inches are there in 15 feet?

   $$\frac{12 \text{ inches}}{1 \text{ foot}} = \frac{x \text{ inches}}{15 \text{ feet}} \quad \text{or} \quad \frac{12}{1} = \frac{x}{15}$$

   $$x = (12)(15) = 180$$

   Therefore, there are 180 inches in 15 feet.

   There are 60 seconds in 1 minute. Therefore, 1,500 seconds equal
   how many minutes?

   $$\frac{60 \text{ seconds}}{1 \text{ minute}} = \frac{1,500 \text{ seconds}}{x \text{ minutes}} \quad \text{or} \quad \frac{60}{1} = \frac{1,500}{x}$$

   $$60x = 1,500$$

   $$x = \frac{1,500}{60} = 25$$

   Therefore, 1,500 seconds equal 25 minutes.

Use a proportion to complete each conversion.

   a) If 1 ounce equals 28.35 grams, how many grams are there in
      2.75 ounces? (Round to two decimal places.)

   b) If 1 mile equals 1.6093 kilometers, how many miles are there in
      100 kilometers? (Round to hundredths.)

a) 32 teeth, from:

$$\frac{18}{48} = \frac{12}{x}$$

b) 13.2 gallons, from:

$$\frac{20}{75.7} = \frac{x}{50}$$

a) 77.96 grams, from: $\dfrac{1}{28.35} = \dfrac{2.75}{x}$

b) 62.14 miles, from: $\dfrac{1}{1.6093} = \dfrac{x}{100}$

2-10   APPLIED PROBLEMS INVOLVING RATIOS

In this section, we will discuss some applied problems involving ratios. Some of the problems can be solved by solving a proportion.

---

92.   Proportions can be used to solve problems involving ratios that are stated as fractions. An example is shown below.

If the ratio of length to width for a rectangle is 5 to 2, what is the length of the rectangle if its width is 17 feet?

$$\frac{5 \text{ feet}}{2 \text{ feet}} = \frac{x \text{ feet}}{17 \text{ feet}} \quad \text{or} \quad \frac{5}{2} = \frac{x}{17}$$

$$2x = 85$$

$$x = 42.5$$

Therefore, its length is 42.5 feet.

Using the same method, solve each problem below.

a) In an alloy, the ratio of zinc to copper is 3:10. How much copper is there if there are 65 pounds of zinc? (Round to the nearest whole number.)

b) If the ratio of the height of a triangle to its base is 7 to 4, find the height when the base is 22.5 centimeters. (Round to tenths.)

---

93.   Simple fractional equations can be used to solve problems involving ratios that are stated as ordinary numbers. An example is shown below.

If the driver-to-driven gear ratio is 1.5, how many teeth are there in the driving gear if there are 24 teeth in the driven gear?

$$1.5 = \frac{x \text{ teeth}}{24 \text{ teeth}} \quad \text{or} \quad 1.5 = \frac{x}{24}$$

$$x = (1.5)(24)$$

$$x = 36$$

Therefore, there are 36 teeth in the driving gear.

Continued on following page.

a) 217 pounds, from:

$$\frac{3}{10} = \frac{65}{x}$$

b) 39.4 centimeters, from:

$$\frac{7}{4} = \frac{x}{22.5}$$

93.   Continued

Using the same method, solve each problem below.

a) If the driver-to-driven speed ratio is 1.68, what is the rotational
speed in rpm of the drive shaft if the rotational speed of the
driven shaft is 525 rpm?

b) If the signal-to-noise ratio is 24.6, what is the noise voltage if the
signal voltage is 0.925 volts? (Round to four decimal places.)

---

94.   The formula for "efficiency" is shown below. As you can see, efficiency
is the ratio of power output to power input.

$$\text{Efficiency} = \frac{\text{Power Output}}{\text{Power Input}}$$

Efficiency is a ratio that is ordinarily stated as a percent. An example
is given below.

An electric motor has a power input of 1,500 watts and a power
output of 1,420 watts. What is its percent efficiency? (Round to
the nearest tenth of a percent.)

$$\text{Efficiency} = \frac{1,420 \text{ watts}}{1,500 \text{ watts}} = .947 = 94.7\%$$

Following the example, solve each problem. Round each percent to the
nearest tenth.

a) What is the efficiency of a gasoline engine if its power output is
240 horsepower and its power input is 955 horsepower?

b) What is the efficiency of an electrical transformer if it has an
input of 64,800 watts and an output of 62,600 watts?

a) 882 rpm, from:

$$1.68 = \frac{x \text{ rpm}}{525 \text{ rpm}}$$

b) 0.0376 volts, from:

$$24.6 = \frac{0.925 \text{ volts}}{x \text{ volts}}$$

95.   We used the "efficiency" formula to solve the problem below.

A gasoline engine has an efficiency of 24%. If its power output is 210 horsepower, find its power input.

$$24\% = \frac{210 \text{ horsepower}}{x \text{ horsepower}} \quad \text{or} \quad .24 = \frac{210}{x}$$

$$.24x = 210$$

$$x = \frac{210}{.24} = 875$$

Therefore its power input is 875 horsepower.

Following the example, solve each problem.

a) An electrical transformer has an efficiency of 98.2% and a power input of 67,500 watts. Find its power output in watts.

b) An electric motor has an efficiency of 92.5% and a power output of 1,280 watts. Find its power input in watts. (Round to the nearest whole number.)

a) 25.1%, from $\dfrac{240}{955}$

b) 96.6%, from $\dfrac{62,600}{64,800}$

a) 66,285 watts, from:

$$.982 = \frac{x}{67,500}$$

b) 1,384 watts, from:

$$.925 = \frac{1,280}{x}$$

SELF-TEST 6 (pages 63-73)

Write each of the following as a fraction in lowest terms.

1. The ratio of 60 grams to 24 grams. _____

2. The ratio of 48 seconds to 16 seconds. _____

Write each of the following as a decimal number.

3. The ratio of 10.8 volts to 2.4 volts. _____

4. The ratio of 2.35 centimeters to 3.76 centimeters. _____

5. If a 6 inch length on a road map represents 80 miles, what length represents 140 miles? _____

6. If a spring is stretched 0.27 inch by a 12-pound force, what stretch will be produced by a 20-pound force? _____

7. If 22.4 liters of carbon dioxide gas weigh 44 grams, find the weight of 8 liters. (Round to the nearest tenth.) _____

8. If a 250-foot length of copper wire weighs 4.95 pounds, what is the weight of a 75-foot length? (Round to two decimal places.) _____

9. The efficiency of an electric motor is 93.8%. If its power output is 746 watts, find its power input. (Round to the nearest watt.) _____

10. The width-to-height ratio of any TV screen is 4 to 3. If the width of a screen is 19.2 inches, find the height. _____

ANSWERS:

1. $\frac{5}{2}$ 5. 10.5 inches   9. 795 watts

2. $\frac{3}{1}$ 6. 0.45 inch   10. 14.4 inches

3. 4.5   7. 15.7 grams

4. .625   8. 1.49 pounds

SUPPLEMENTARY PROBLEMS - CHAPTER 2

Assignment 4

Convert each percent to a fraction in lowest terms.

1. 75%   2. 20%   3. $33\frac{1}{3}$%   4. 90%   5. 25%

Convert each percent to a decimal number or whole number.

6. 19%   7. 3.9%   8. .65%   9. 147%   10. 300%

Convert each decimal number or whole number to a percent.

11. .814   12. .0425   13. .006   14. 1.7   15. 5

Convert each fraction to a percent.

16. $\frac{1}{2}$ 17. $\frac{4}{5}$ 18. $\frac{3}{10}$ 19. $\frac{2}{3}$ 20. $\frac{59}{100}$

Convert to the nearest whole-number percent. | Convert to the nearest hundredth of a percent.

21. $\frac{14}{40}$ 22. $\frac{3.7}{21.9}$ 23. $\frac{243}{185}$ | 24. $\frac{76}{1,830}$ 25. $\frac{0.24}{506}$ 26. $\frac{1.79}{4.26}$

Solve the following problems.

27. 20 is what percent of 50?   28. What is 75% of 400?   29. 30% of what is 15?

30. What percent of 180 is 60?   31. Find 8.5% of $250.   32. 56 is 8% of what?

33. .25% of 800 is what?   34. 6% of what is 24?   35. 12 is what percent of 300?

Assignment 5

1. A cast iron alloy has 3.5% carbon. Find the amount of carbon in 500 pounds of the alloy.

2. If a student got 19 problems correct on a 23-problem test, what was her percent grade? (Round to the nearest whole number percent.)

3. An ore contains 0.8% uranium. How much uranium is there in 2,000 kilograms of the ore?

4. If 58.45 grams of sodium chloride contain 35.46 grams of chlorine, what percent is chlorine? (Round to the nearest hundredth of a percent.)

5. How much sales tax is paid for a $5,470 car if the sales tax rate is 5%?

6. If silver solder contains 75% silver, how much solder can be made from 500 grams of silver? (Round to the nearest gram.)

7. In making 260 kilograms of a bronze alloy, 223 kilograms of copper are used. What percent of the alloy is copper? (Round to the nearest tenth of a percent.)

8. How much money must be invested at 6% annual interest to earn $1,500 interest annually?

9. Sea water contains "380 parts per million" of potassium. What percent is potassium?

10. An ore contains 1.37% copper. How much ore is needed to get 100 pounds of copper? (Round to the nearest hundred.)

Solve. Round as directed.

11. $\dfrac{n}{12.5} = 60.8$ 12. $.625 = \dfrac{20}{x}$ 13. $\dfrac{y}{124} = \dfrac{27}{60}$ 14. $\dfrac{6.3}{3.5} = \dfrac{t}{.95}$

15. Round to tenths.   16. Round to hundredths.   17. Round to hundreds.

$\dfrac{2.61}{30.3} = \dfrac{w}{589}$ $\dfrac{9.35}{R} = \dfrac{5.09}{2.47}$ $\dfrac{728}{47,900} = \dfrac{51.6}{x}$

Assignment 6

1. The circumference of a circle is 62.8 centimeters and its diameter is 20 centimeters. Find the circumference-to-diameter ratio.

2. The width of a TV screen is 20 inches and its height is 15 inches. What is the width to height ratio (as a fraction in lowest terms)?

3. If a car used 7.3 gallons of gas to travel 235 miles, how far can it travel on 20 gallons of gas? (Round to the nearest mile.)

4. If 1 pound equals 453.6 grams, how many pounds are there in 1,000 grams? (Round to one decimal place.)

5. If 14.4 grams of carbon are burned, 52.8 grams of carbon dioxide are produced. How much carbon must be burned to produce 800 grams of carbon dioxide? (Round to the nearest gram.)

6. A 2.5-centimeter length on a drawing represents 3 meters. What length represents 10 meters? (Round to the nearest tenth.)

7. What force is needed to stretch a spring 2.4 inches if a 6-pound force stretches it 1.8 inches?

8. If 15 metal washers weigh 68 grams, how many washers are there in a lot weighing 971 grams?

9. If the driver-to-driven gear ratio is 2.40 and if the driving gear has 96 teeth, how many teeth are there in the driven gear?

10. If the length-to-width ratio of a rectangle is 7 to 4, find its length if its width is 12.4 centimeters.

11. An electrical transformer's efficiency is 97.8%. If its power input is 60 kilowatts, find its power output. (Round to the nearest tenth.)

Chapter 3

POWERS OF TEN AND SCIENTIFIC NOTATION

In this chapter, we will discuss powers of ten, scientific notation, and other power-of-ten forms of numbers. The laws of exponents for powers of ten are introduced. Operations with numbers in scientific notation and other power-of-ten forms are performed on a calculator. The decimal number system is related to powers of ten.

3-1   POWERS OF TEN

In this section, we will define what is meant by powers of ten with positive, zero, and negative exponents.

---

1.   Expressions like 10^2, 10^3, 10^6, 10^{-1}, 10^{-2}, and 10^{-5} are called "powers of ten". In each, 10 is called the base; the small number is called the "exponent". That is:

The following pattern is used to name powers of ten with positive exponents.

10^1 is called "10 to the first power" or "10 to the first".

10^2 is called "10 to the second power" or "10 to the second".

10^3 is called "10 to the third power" or "10 to the third".

Write the power of ten corresponding to each word name.

a) 10 to the fourth power = _____   b) 10 to the seventh = _____

---

2.   The words "squared" and "cubed" are frequently used for the second and third power of ten. That is:

We say "10 squared" instead of "10 to the second power".

We say "10 cubed" instead of "10 to the third power".

Therefore: 10 squared = 10^2 10 cubed = _____

---

a) 10^4 b) 10^7

10^3

3. Any power of ten with a positive exponent is a short way of writing a
    multiplication of 10's. The <u>exponent</u> tells us how many 10's to use.
    For example:

    $10^2 = 10 \cdot 10$ (The exponent tells us to multiply <u>two</u> 10's.)

    $10^5 = 10 \cdot 10 \cdot 10 \cdot 10 \cdot 10$ (The exponent tells us to multiply <u>five</u> 10's.)

    Write each power of ten as a multiplication of 10's.

    a) $10^3 = $ _____   b) $10^6 = $ _____

---

4. Powers of ten with positive exponents equal ordinary numbers like 10 ,
    100 , 1,000 , and so on. For example:

    $10^1 = 10$

    $10^2 = 10 \cdot 10 = 100$

    $10^3 = 10 \cdot 10 \cdot 10 = $ _____

a) $10 \cdot 10 \cdot 10$

b) $10 \cdot 10 \cdot 10 \cdot 10 \cdot 10 \cdot 10$

---

5. By examining the table below, you can see this fact: <u>The number of 0's
    in the ordinary number equals the exponent of the power of ten.</u>

10^6	=	1,000,000
10^5	=	100,000
10^4	=	10,000
10^3	=	1,000
10^2	=	100
10^1	=	10

    Using the above fact, convert these to an ordinary number.

    a) $10^7 = $ _____   b) $10^8 = $ _____

1,000

---

6. Using the number of 0's to determine the exponent, convert these to
    powers of ten.

    a) $100 = $ _____   b) $10,000,000 = $ _____

a) 10,000,000

b) 100,000,000

---

7. By examining the pattern at the right,
    you can see that the following definition
    makes sense for 10^0.

10^3	=	1,000
10^2	=	100
10^1	=	10
10^0	=	1

    $\boxed{10^0 = 1}$

a) 10^2 b) 10^7

Continued on following page.

7.   Continued

Convert each power of ten to an ordinary number and each ordinary number to a power of ten.

a) 10^0 = _____

d) 10^5 = _____

b) $10,000$ = _____

e) $10,000,000,000$ = _____

c) 10^1 = _____

f) 10^9 = _____

---

8.   When naming powers of ten with negative exponents, we can use the word "negative" or the word "minus". For example:

10^{-1} is called "10 to the negative-one power".
or "10 to the minus-one power".

10^{-4} is called "10 to the negative-four power".
or "10 to the _____ power".

a) 1
b) 10^4
c) 10

d) $100,000$
e) 10^{10}
f) $1,000,000,000$

---

9.   Sometimes the word "power" is omitted in the word names. For example:

10^{-2} is called "10 to the negative-two" or "10 to the minus-two".

10^{-5} is called "10 to the negative-five" or "10 to the _____".

minus-four

---

10.   The following definition is used for powers of ten with negative exponents.

$$10^{-a} = \frac{1}{10^a}$$

Therefore: $10^{-1} = \frac{1}{10^1}$ $10^{-3} = \frac{1}{10^3}$ 10^{-5} = _____

minus-five

---

11.   Because of the definition, powers of ten with negative exponents equal ordinary numbers like .1 , .01 , .001 , and so on. For example:

$$10^{-1} = \frac{1}{10^1} = \frac{1}{10} = .1$$

$$10^{-2} = \frac{1}{10^2} = \frac{1}{100} = .01$$

$$10^{-4} = \frac{1}{10^4} = \frac{1}{10,000} = \underline{\hspace{2cm}}$$

$\frac{1}{10^5}$

.0001

12. By examining the table below, you can see this fact: The <u>number</u> of <u>decimal</u> <u>places</u> <u>in</u> <u>each</u> <u>ordinary</u> <u>number</u> <u>equals</u> <u>the</u> <u>absolute</u> <u>value</u> <u>of</u> <u>the</u> <u>exponent</u>.

> 10^{-1} = .1
> 10^{-2} = .01
> 10^{-3} = .001
> 10^{-4} = .0001
> 10^{-5} = .00001
> 10^{-6} = .000001

Using the fact above, convert these to ordinary numbers.

a) 10^{-7} = _____   b) 10^{-8} = _____

---

13. Using the number of decimal places to determine the absolute value of the exponent, convert these to powers of ten.

a) .001 = _____   b) .000000001 = _____

a) .0000001
b) .00000001

---

14. Convert each power of ten to an ordinary number and each ordinary number to a power of ten.

a) 10^{-1} = _____   d) .00001 = _____

b) .01 = _____   e) 10^{-10} = _____

c) 10^{-4} = _____   f) .00000001 = _____

a) 10^{-3} b) 10^{-9}

---

15. Though we would not ordinarily do so, we can use the $\boxed{y^x}$ key on a calculator to convert a power of ten to an ordinary number. For example:

To show that $10^4 = 10,000$, follow these steps:

Enter	Press	Display
10	$\boxed{y^x}$	10.
4	$\boxed{=}$	10000.

To show that $10^{-6} = .000001$, follow these steps:

Enter	Press	Display
10	$\boxed{y^x}$	10.
6	$\boxed{+/-}$ $\boxed{=}$	0.000001

a) .1
b) 10^{-2}
c) .0001
d) 10^{-5}
e) .0000000001
f) 10^{-8}

16.   Though we would also not ordinarily do so, we can use the $\boxed{\log}$ key on a calculator to find the exponent of the power-of-ten form of a number. (Note: the word "logarithm" means "exponent".) For example:

To show that the exponent of the power-of-ten form of 1,000 is "3", follow these steps:

Enter	Press	Display
1,000	$\boxed{\log}$	3.

Therefore: $1,000 = 10^3$

To show that the exponent of the power-of-ten form of .00001 is -5, follow these steps:

Enter	Press	Display
.00001	$\boxed{\log}$	-5.

Therefore: $.00001 = 10^{-5}$

3-2   THE DECIMAL NUMBER SYSTEM AND POWERS OF TEN

In this section, we will relate the place-names in the decimal number system to powers of ten.

17.   In the decimal number system, any number can be written in an expanded form in which each term is a multiplication. For example:

$$5,617 = 5(1,000) + 6(100) + 1(10) + 7(1)$$
$$.843 = 8(.1) + 4(.01) + 3(.001)$$

Powers of ten can be used in the expanded forms above. That is:

$$5,617 = 5(10^3) + 6(10^2) + 1(10^1) + 7(10^0)$$
$$.843 = 8(10^{-1}) + 4(10^{-2}) + 3(10^{-3})$$

Write the ordinary number corresponding to each expanded form.

a) $3(10^3) + 9(10^2) + 4(10^1) + 8(10^0)$ = _____

b) $7(10^1) + 5(10^0) + 0(10^{-1}) + 6(10^{-2})$ = _____

c) $0(10^{-1}) + 0(10^{-2}) + 2(10^{-3}) + 1(10^{-4})$ = _____

a) 3,948

b) 75.06

c) .0021

18. In the diagram below, the names of the places in numbers are related to powers of ten.

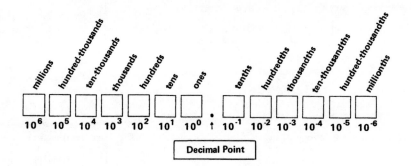

On the left side of the decimal point:

The name of the 10^0 place is <u>ones</u>.

The name of the 10^2 place is <u>hundreds</u>.

a) The name of the 10^3 place is _____.

b) The name of the 10^6 place is _____.

---

19. On the right side of the decimal point:

The name of the 10^{-1} place is <u>tenths</u>.

a) The name of the 10^{-3} place is _____.

b) The name of the 10^{-6} place is _____.

a) thousands

b) millions

---

20. Write the name for each of the following places.

a) 10^1 _____   c) 10^{-2} _____

b) 10^5 _____   d) 10^{-4} _____

a) thousandths

b) millionths

---

21. Write the power of ten corresponding to each place.

a) hundreds _____   c) thousandths _____

b) millions _____   d) millionths _____

a) tens
b) hundred-thousands
c) hundredths
d) ten-thousandths

a) 10^2 c) 10^{-3}
b) 10^6 d) 10^{-6}

22.   Other powers of ten and place names that are sometimes used are given in the tables below.

$$10^9 = \text{billions}$$
$$10^{12} = \text{trillions}$$

$$10^{-9} = \text{billionths}$$
$$10^{-12} = \text{trillionths}$$

Write the name for each place.

a) 10^{12} _____

b) 10^{-9} _____

---

23.   Write the power of ten corresponding to each place.

a) billions _____

b) trillionths _____

a) trillions

b) billionths

---

a) 10^9 b) 10^{-12}

3-3   LAWS OF EXPONENTS FOR POWERS OF TEN

In this section, we will discuss the laws of exponents for multiplying, dividing, squaring, and finding the square root of powers of ten.

24.   To multiply the powers of ten below, we converted each factor to an ordinary number and then converted each product back to a power of ten.

$$10^4 \times 10^2 = 10^6$$
$$\downarrow \quad \downarrow \quad \uparrow$$
$$10,000 \times 100 = 1,000,000$$

$$10^{-3} \times 10^1 = 10^{-2}$$
$$\downarrow \quad \downarrow \quad \uparrow$$
$$.001 \times 10 = .01$$

From the examples, you can see that the exponent of each product can be obtained by adding the exponents of the factors. That is:

$$10^4 \times 10^2 = 10^{4+2} = 10^6 \qquad 10^{-3} \times 10^1 = 10^{(-3)+1} = \underline{\quad\quad}$$

---

25.   To multiply powers of ten, we simply add their exponents. That is:

$$10^a \times 10^b = 10^{a+b}$$

10^{-2}

The above law of exponents holds for positive, negative, and "0" exponents. Therefore:

a) $10^3 \times 10^5 = 10^{3+5} = $ _____

c) $10^7 \times 10^{-2} = 10^{7+(-2)} = $ _____

b) $10^{-4} \times 10^0 = 10^{(-4)+0} = $ _____

d) $10^{-1} \times 10^{-6} = 10^{(-1)+(-6)} = $ _____

---

a) 10^8 b) 10^{-4} c) 10^5 d) 10^{-7}

26. To multiply more than two powers of ten, we also add their exponents. For example:

$$10^4 \times 10^{-3} \times 10^2 = 10^{4 + (-3) + 2} = 10^3$$

Use the law of exponents to complete these.

    a) $10^{-2} \times 10^7 \times 10^{-4} = $ _____   b) $10^3 \times 10^{-9} \times 10^4 = $ _____

---

27. To divide the powers of ten below, we converted each term to an ordinary number and then converted each quotient back to a power of ten.

a) 10^1 or 10

b) 10^{-2}

$$\frac{10^5}{10^2} = 10^3 \qquad\qquad \frac{10^{-3}}{10^1} = 10^{-4}$$

$$\frac{100,000}{100} = 1,000 \qquad\qquad \frac{.001}{10} = .0001$$

From the examples, you can see that the exponent of each quotient can be obtained by subtracting the exponent of the denominator from the exponent of the numerator. That is:

$$\frac{10^5}{10^2} = 10^{5-2} = 10^3 \qquad\qquad \frac{10^{-3}}{10^1} = 10^{(-3)-1} = \underline{\hspace{1cm}}$$

---

28. To divide powers of ten, we simply subtract the exponent of the denominator from the exponent of the numerator. That is:

10^{-4}

$$\boxed{\frac{10^a}{10^b} = 10^{a-b}}$$

The above law of exponents holds for positive, negative, and "0" exponents. Therefore:

    a) $\dfrac{10^8}{10^3} = 10^{8-3} = $ _____   c) $\dfrac{10^6}{10^{-1}} = 10^{6-(-1)} = $ _____

    b) $\dfrac{10^0}{10^4} = 10^{0-4} = $ _____   d) $\dfrac{10^{-7}}{10^{-5}} = 10^{(-7)-(-5)} = $ _____

---

29. Remember: To multiply powers of ten, we add the exponents.
               To divide powers of ten, we subtract the exponents.

a) 10^5 c) 10^7

b) 10^{-4} d) 10^{-2}

Complete: a) $10^{-5} \times 10^5 = $ _____   c) $10^{-4} \times 10^{-4} = $ _____

               b) $\dfrac{10^{-3}}{10^{-4}} = $ _____   d) $\dfrac{10^{-2}}{10^2} = $ _____

---

a) 10^0 or 1   b) 10^1 or 10   c) 10^{-8} d) 10^{-4}

30.   To perform each operation below, we began by substituting 10^1 for 10.

$$10 \times 10^4 = 10^1 \times 10^4 = 10^5 \qquad \frac{10^3}{10} = \frac{10^3}{10^1} = 10^2$$

Using the same steps, complete these:

a) $10^{-2} \times 10 =$ _____   b) $\dfrac{10}{10^{-6}} =$ _____

---

31.   To perform the division below, we began by substituting 10^0 for "1". Complete the other division.

$$\frac{1}{10^3} = \frac{10^0}{10^3} = 10^{-3} \qquad\qquad \frac{1}{10^{-4}} =$$ _____

a) 10^{-1} b) 10^7

---

32.   To square a power of ten, we simply multiply it by itself. For example:

$$(10^3)^2 = 10^3 \times 10^3 = 10^6 \qquad\qquad (10^{-4})^2 = 10^{-4} \times 10^{-4} = 10^{-8}$$

From the examples, you can see that we can square a power of ten by simply doubling its exponent. That is:

$$(10^3)^2 = 10^{2(3)} = 10^6 \qquad\qquad (10^{-4})^2 = 10^{2(-4)} =$$ _____

10^4

---

33.   To square a power of ten, we simply double its exponent. That is:

$$\boxed{\left(10^a\right)^2 = 10^{2a}}$$

Using the above law of exponents, complete these:

a) $(10^6)^2 = 10^{2(6)} =$ _____   c) $(10^{-1})^2 = 10^{2(-1)} =$ _____

b) $(10^1)^2 = 10^{2(1)} =$ _____   d) $(10^{-5})^2 = 10^{2(-5)} =$ _____

10^{-8}

---

34.   The square root of a power of ten is the power of ten whose square equals the original power of ten. For example:

Since $(10^4)^2 = 10^8$, $\sqrt{10^8} = 10^4$

Since $(10^{-3})^2 = 10^{-6}$, $\sqrt{10^{-6}} = 10^{-3}$

From the examples above, you can see that we can find the exponent of the square root by dividing the original exponent by 2. That is:

$$\sqrt{10^8} = 10^{\frac{8}{2}} = 10^4 \qquad\qquad \sqrt{10^{-6}} = 10^{\frac{-6}{2}} =$$ _____

a) 10^{12} c) 10^{-2}
b) 10^2 d) 10^{-10}

10^{-3}

35. To find the square root of a power of ten, <u>we</u> <u>simply</u> <u>divide</u> <u>its</u> <u>exponent</u> <u>by</u> 2.
That is:

$$\sqrt{10^a} = 10^{\frac{a}{2}}$$

Using the above law of exponents, complete these:

a) $\sqrt{10^{14}} = 10^{\frac{14}{2}} =$ _____   c) $\sqrt{10^{-2}} = 10^{\frac{-2}{2}} =$ _____

b) $\sqrt{10^2} = 10^{\frac{2}{2}} =$ _____   d) $\sqrt{10^{-8}} = 10^{\frac{-8}{2}} =$ _____

---

36. Remember: To <u>square</u> a power of ten, we <u>double</u> its exponent.

To <u>find</u> <u>the</u> <u>square</u> <u>root</u> of a power of ten, we <u>divide</u>
<u>its</u> <u>exponent</u> <u>by</u> 2.

Complete: a) $(10^4)^2 =$ _____   c) $(10^{-10})^2 =$ _____

b) $\sqrt{10^4} =$ _____   d) $\sqrt{10^{-10}} =$ _____

a) 10^7 c) 10^{-1}

b) 10^1 d) 10^{-4}

---

37. When the exponent of the power of ten is an <u>odd</u> number, the exponent of
its square root is a <u>fraction</u>. For example:

$$\sqrt{10^3} = 10^{\frac{3}{2}} \qquad\qquad \sqrt{10^{-5}} = 10^{-\frac{5}{2}}$$

We will not discuss powers of ten with fractional exponents in this chapter.
We will do so, however, in a later chapter.

a) 10^8 c) 10^{-20}

b) 10^2 d) 10^{-5}

---

38. Four laws of exponents for powers of ten are summarized below.

$$10^a \times 10^b = 10^{a+b}$$

$$(10^a)^2 = 10^{2a}$$

$$\frac{10^a}{10^b} = 10^{a-b}$$

$$\sqrt{10^a} = 10^{\frac{a}{2}}$$

Using the above laws, complete these:

a) $10^{-3} \times 10^9 =$ _____   e) $(10^9)^2 =$ _____

b) $\dfrac{10^1}{10^6} =$ _____   f) $\dfrac{10^{-5}}{10^{-9}} =$ _____

c) $(10^{-7})^2 =$ _____   g) $\sqrt{10^{-22}} =$ _____

d) $\sqrt{10^{16}} =$ _____   h) $10^{-1} \times 10^{-2} \times 10^{-8} =$ _____

39.   To simplify some fractional expressions, we use the laws of exponents for both multiplication and division. For example:

To simplify below, we multiplied in the numerator before dividing.

$$\frac{10^{-4} \times 10^{9}}{10^{2}} = \frac{10^{5}}{10^{2}} = 10^{5-2} = 10^{3}$$

To simplify below, we multiplied in the denominator before dividing.

$$\frac{10^{3}}{10^{8} \times 10^{-3}} = \frac{10^{3}}{10^{5}} = 10^{3-5} = 10^{-2}$$

To simplify below, we multiplied in both the numerator and denominator before dividing.

$$\frac{10^{-5} \times 10^{-2}}{10^{-3} \times 10^{6}} = \frac{10^{-7}}{10^{3}} = 10^{(-7)-3} = 10^{-10}$$

Using the same steps, simplify these:

a) $\dfrac{10^{-2} \times 10^{7}}{10^{-4}}$ = _____

c) $\dfrac{10^{6} \times 10^{-9}}{10^{-5} \times 10^{10}}$ = _____

b) $\dfrac{10^{-4}}{10^{-5} \times 10^{2}}$ = _____

d) $\dfrac{10^{7} + 10^{-1} \times 10^{4}}{10^{-9} \times 10^{7}}$ = _____

a) 10^{6} e) 10^{18}

b) 10^{-5} f) 10^{4}

c) 10^{-14} g) 10^{-11}

d) 10^{8} h) 10^{-11}

a) 10^{9} b) 10^{-1} c) 10^{-8} d) 10^{12}

3-4   RECIPROCALS AND DIVISIONS OF POWERS OF TEN

In this section, we will discuss the reciprocals of powers of ten and then show how reciprocals can be used to convert a division of powers of ten to a multiplication.

40.   Two powers of ten are a pair of reciprocals if their product is "+1".
Therefore:

10^{3} and 10^{-3} are reciprocals, since $10^{3} \times 10^{-3} = 10^{0} = +1$

10^{5} and 10^{-5} are reciprocals, since $10^{5} \times 10^{-5} = 10^{0} = +1$

From the examples, you can see that $\underline{10^{a}}$ and $\underline{10^{-a}}$ are reciprocals. That is:

10^{-1} is the reciprocal of 10^{1}.

10^{7} is the reciprocal of 10^{-7}.

Write the reciprocal of each power of ten.

a) 10^{-9} _____   b) 10^{13} _____   c) 10^{8} _____   d) 10^{-12} _____

a) 10^{9} b) 10^{-13} c) 10^{-8} d) 10^{12}

41. Any division of powers of ten can be converted to a multiplication <u>by</u> <u>multiplying</u> <u>the</u> <u>numerator</u> <u>by</u> <u>the</u> <u>reciprocal</u> <u>of</u> <u>the</u> <u>denominator</u>. That is:

$$\frac{10^5}{10^2} = 10^5 \text{ x (the reciprocal of } 10^2) = 10^5 \text{ x } 10^{-2}$$

   a) $\frac{10^{-3}}{10^{-4}} = 10^{-3}$ x (the reciprocal of 10^{-4}) $= 10^{-3}$ x _____

   b) $\frac{10^{-1}}{10^7} = 10^{-1}$ x (the reciprocal of 10^7) $= 10^{-1}$ x _____

---

42. The following conversion pattern should be obvious.

$$\boxed{\frac{10^a}{10^b} = 10^a \text{ x (the reciprocal of } 10^b) = 10^a \text{ x } 10^{-b}}$$

Using the above pattern, complete these conversions.

   a) $\frac{10^7}{10^3} = 10^7$ x _____   c) $\frac{10^{-1}}{10^{-5}} = 10^{-1}$ x _____

   b) $\frac{10^4}{10^{-2}} = 10^4$ x _____   d) $\frac{10^{-10}}{10^{12}} = 10^{-10}$ x _____

a) 10^{-3} x $\underline{10^4}$

b) 10^{-1} x $\underline{10^{-7}}$

---

43. Following the example, complete each division <u>after</u> converting to multiplication.

$$\frac{10^8}{10^6} = 10^8 \text{ x } 10^{-6} = 10^2$$

   a) $\frac{10^2}{10^7} =$ _____ x _____ = _____   c) $\frac{10^{-2}}{10^4} =$ _____ x _____ = _____

   b) $\frac{10^3}{10^{-1}} =$ _____ x _____ = _____   d) $\frac{10^{-9}}{10^{-8}} =$ _____ x _____ = _____

a) x 10^{-3} c) x 10^5

b) x 10^2 d) x 10^{-12}

---

44. There are two ways to perform a division like $\frac{10^5}{10^3}$. They are:

$$\frac{10^5}{10^3} = 10^{5-3} = 10^2 \qquad\qquad \frac{10^5}{10^3} = 10^5 \text{ x } 10^{-3} = 10^2$$

Using either method, complete these:

   a) $\frac{10^2}{10^9} =$ _____   b) $\frac{10^{-7}}{10^3} =$ _____   c) $\frac{10^{-5}}{10^{-5}} =$ _____

a) 10^2 x 10^{-7} = 10^{-5}

b) 10^3 x 10^1 = 10^4

c) 10^{-2} x 10^{-4} = 10^{-6}

d) 10^{-9} x 10^8 = 10^{-1}

---

a) 10^{-7} b) 10^{-10} c) 10^0 or 1

45. More complicated divisions can also be converted to multiplications <u>by multiplying</u> <u>the</u> <u>numerator</u> <u>by</u> <u>the</u> <u>reciprocal</u> <u>of</u> <u>each</u> <u>factor</u> <u>in</u> <u>the</u> denominator. That is:

$$\frac{10^8 \times 10^{-3}}{10^5} = 10^8 \times 10^{-3} \times 10^{-5}$$

$$\frac{10^{-2}}{10^7 \times 10^{-4}} = 10^{-2} \times 10^{-7} \times 10^4$$

$$\frac{10^6 \times 10^{-1}}{10^{-5} \times 10^9} = 10^6 \times 10^{-1} \times \underline{\hspace{1cm}} \times \underline{\hspace{1cm}}$$

---

46. Following the example, complete each division after converting to multiplication.

	$\times 10^5 \times 10^{-9}$

$$\frac{10}{10^3 \times 10^{-1}} = 10^1 \times 10^{-3} \times 10^1 = 10^{-1}$$

a) $\dfrac{10^{10}}{10^9 \times 10^{-7}} = \underline{\hspace{1cm}} \times \underline{\hspace{1cm}} \times \underline{\hspace{1cm}} = \underline{\hspace{1cm}}$

b) $\dfrac{10^4 \times 10^{-8}}{10^{-10} \times 10^3} = \underline{\hspace{1cm}} \times \underline{\hspace{1cm}} \times \underline{\hspace{1cm}} \times \underline{\hspace{1cm}} = \underline{\hspace{1cm}}$

a) $10^{10} \times 10^{-9} \times 10^7 = 10^8$ b) $10^4 \times 10^{-8} \times 10^{10} \times 10^{-3} = 10^3$

SELF-TEST 7 (pages 76-89)

Convert each power of ten to an ordinary number.

1. $10^3 = \underline{\hspace{2cm}}$ 2. $10^{-5} = \underline{\hspace{2cm}}$ 3. $10^0 = \underline{\hspace{1.5cm}}$

Convert each number to a power of ten.

4. $1,000,000 = \underline{\hspace{1.5cm}}$ 5. $.001 = \underline{\hspace{1.5cm}}$ 6. $10 = \underline{\hspace{1.5cm}}$

Write the power of ten corresponding to each place name.

7. ten-thousands $\underline{\hspace{1.5cm}}$ 8. hundredths $\underline{\hspace{1.5cm}}$

Write the place name corresponding to each power of ten.

9. 10^{-3} $\underline{\hspace{3cm}}$ 10. 10^6 $\underline{\hspace{3cm}}$

Continued on following page.

SELF-TEST 7 - Continued

Write each answer as a power of ten.

11. $10^3 \times 10^{-5} \times 10^{-1} = $ _____

12. $\dfrac{10^2}{10^{-4}} = $ _____

13. $\dfrac{10^9 \times 10^{-6}}{10^4} = $ _____

14. $\dfrac{10^{-3} \times 10^3}{10^4 \times 10^{-2}} = $ _____

15. $(10^{-3})^2 = $ _____

16. $\sqrt{10^4} = $ _____

17. The reciprocal of 10^{-4} is _____ (a power of ten).

Do each division by converting to a multiplication first.

18. $\dfrac{10^5}{10^{-2}} = 10^5 \times$ _____ $ = $ _____

19. $\dfrac{10^{-6}}{10^2 \times 10^{-3}} = 10^{-6} \times$ _____ $\times$ _____ $ = $ _____

ANSWERS:

1. 1,000	6. 10^1	11. 10^{-3}	16. 10^2
2. .00001	7. 10^4	12. 10^6	17. 10^4
3. 1	8. 10^{-2}	13. 10^{-1}	18. $10^5 \times \underline{10^2} = \underline{10^7}$
4. 10^6	9. thousandths	14. 10^{-2}	19. $10^{-6} \times \underline{10^{-2}} \times \underline{10^3} = \underline{10^{-5}}$
5. 10^{-3}	10. millions	15. 10^{-6}	

3-5   MULTIPLYING AND DIVIDING ORDINARY NUMBERS BY POWERS OF TEN

In this section, we will discuss the decimal-point-shift method for multiplying and dividing ordinary numbers by powers of ten.

47.   Since $10^1 = 10$, $10^2 = 100$, and $10^3 = 1,000$, we can use the decimal-point-shift method to multiply by those powers of ten. The exponent tells us how many places to shift the decimal point to the right. For example:

$10^1 \times .56 = .5\,6 = 5.6$ (Shifted one place.)
                                        (Exponent is "1".)

$10^2 \times .39 = .39 = 39$ (Shifted two places.)
                                        (Exponent is "2".)

$10^3 \times 9.8 = 9.800 = 9,800$ (Shifted three places.)
                                        (Exponent is "3".)

Continued on following page.

47.   Continued

Use the decimal-point-shift method for these:

a) 10^1 x .35 = _____   c) 10^1 x 470 = _____

b) 10^2 x .00508 = _____   d) 10^3 x 6.5 = _____

---

48.   The decimal-point-shift method can also be used to multiply by powers of ten with larger exponents. The number of places shifted depends on the exponent. For example:

10^4 x 2.56 = 2.5600 = 25,600   (Shifted four places.)

10^7 x .0029 = .0029000 = 29,000   (Shifted seven places.)

Use the decimal-point-shift method for these:

a) 10^6 x .00098 = _____   b) 10^8 x .000125 = _____

| a) 3.5 | c) 4,700 |
| b) .508 | d) 6,500 |

---

49.   The decimal-point-shift method can also be used when the power of ten is the second factor. For example:

775 x 10^3 = 775.000 = 775,000

.094 x 10^5 = .09400 = 9,400

Use the decimal-point-shift method for these:

a) 1.5 x 10^1 = _____   b) .00007 x 10^6 = _____

| a) 980 | b) 12,500 |

---

50.   Since $10^0 = 1$, we do not shift the decimal point when multiplying by 10^0. Therefore, the product is identical to the other factor. That is:

10^0 x 55 = 55   1.39 x 10^0 = _____

| a) 15 | b) 70 |

---

51.   Since $10^{-1} = .1$, $10^{-2} = .01$, and $10^{-3} = .001$, we can use the decimal-point-shift method to multiply by those powers of ten. The absolute value of the exponent tells us how many places to shift the decimal point to the left. For example:

10^{-1} x 3.8 = 3.8 = .38 $\left(\begin{array}{l}\text{Shifted one place.}\\ \text{Exponent is "-1".}\end{array}\right)$

10^{-2} x 7,500 = 7,500. = 75 $\left(\begin{array}{l}\text{Shifted two places.}\\ \text{Exponent is "-2".}\end{array}\right)$

10^{-3} x .91 = 000.91 = .00091 $\left(\begin{array}{l}\text{Shifted three places.}\\ \text{Exponent is "-3".}\end{array}\right)$

Continued on following page.

1.39

51.   Continued

Use the decimal-point-shift method for these:

a) 10^{-1} x 87 = _____   c) 10^{-1} x 2,400 = _____

b) 10^{-2} x 2.94 = _____   d) 10^{-3} x .58 = _____

---

52.   The decimal-point-shift method can also be used to multiply by powers of ten with larger negative exponents. The number of places shifted depends on the absolute value of the exponent. For example:

a) 8.7   c) 240

b) .0294   d) .00058

10^{-4} x 36 = .0036. = .0036   (Shifted four places.)

10^{-7} x 90,000,000 = 9.0,000,000. = 9   (Shifted seven places.)

Use the decimal-point-shift method for these:

a) 10^{-6} x 391 = _____   b) 10^{-8} x 45,000,000 = _____

---

53.   The same method can be used when the power of ten is the second factor. For example:

a) .000391   b) .45

$$510 \times 10^{-3} = .510. = .51$$

$$1,200 \times 10^{-5} = .01,200. = .012$$

Use the decimal-point-shift method for these:

a) 58 x 10^{-1} = _____   b) 2.4 x 10^{-6} = _____

---

54.   Don't confuse the direction of the decimal-point-shift when multiplying by powers of ten.

a) 5.8   b) .0000024

If the exponent is positive, we shift to the right since we are multiplying by 10 , 100 , 1,000 , and so on. That is:

$$39 \times 10^{1} = 39.0. = 390 \qquad 5.3 \times 10^{3} = 5.300. = 5,300$$

If the exponent is negative, we shift to the left since we are multiplying by .1 , .01 , .001 , and so on. That is:

$$39 \times 10^{-1} = 3.9. = 3.9 \qquad 5.3 \times 10^{-3} = .005.3 = .0053$$

Use the decimal-point-shift method for these:

a) 72 x 10^{2} = _____   c) .48 x 10^{4} = _____

b) 72 x 10^{-2} = _____   d) .48 x 10^{-4} = _____

---

a) 7,200   b) .72   c) 4,800   d) .000048

55.   Any division of an ordinary number by a power of ten can be performed by converting to a multiplication. For example:

$$\frac{289}{10^2} = 289 \times \text{(the reciprocal of } 10^2) = 289 \times 10^{-2} = 2.89$$

$$\frac{75}{10^{-3}} = 75 \times \text{(the reciprocal of } 10^{-3}) = 75 \times 10^3 = \underline{\hspace{2cm}}$$

---

56.   Convert each division to a multiplication.

    a) $\frac{77.6}{10^{-5}} = 77.6 \times \underline{\hspace{2cm}}$ b) $\frac{.068}{10^4} = .068 \times \underline{\hspace{2cm}}$

75,000

---

57.   Following the example, complete each division by converting to a multiplication.

$$\frac{456}{10^4} = \qquad 456 \times 10^{-4} \qquad = \qquad .0456$$

    a) $\frac{.078}{10^{-2}} = \underline{\hspace{1.5cm}} \times \underline{\hspace{1.5cm}} = \underline{\hspace{1.5cm}}$

    b) $\frac{.6}{10^3} = \underline{\hspace{1.5cm}} \times \underline{\hspace{1.5cm}} = \underline{\hspace{1.5cm}}$

    c) $\frac{8.9}{10^{-4}} = \underline{\hspace{1.5cm}} \times \underline{\hspace{1.5cm}} = \underline{\hspace{1.5cm}}$

a) $\times 10^5$ b) $\times 10^{-4}$

---

58.   When one factor in a multiplication is a number like 10 , 100 , 1,000 , or .1 , .01 , .001 , that factor can be converted to a power of ten. For example:

$$66 \times 1,000 = 66 \times 10^3 = 66,000$$
$$.01 \times .75 = 10^{-2} \times .75 = .0075$$

Following the examples, complete these:

    a) $9.86 \times 100 = 9.86 \times \underline{\hspace{2cm}} = \underline{\hspace{2cm}}$

    b) $.001 \times 4,500 = \underline{\hspace{2cm}} \times 4,500 = \underline{\hspace{2cm}}$

a) $.078 \times 10^2 = 7.8$

b) $.6 \times 10^{-3} = .0006$

c) $8.9 \times 10^4 = 89,000$

---

59.   To perform each division below, we converted the denominator to a power of ten.

$$\frac{7,289}{100} = \frac{7,289}{10^2} = 7,289 \times 10^{-2} = 72.89$$

$$\frac{.55}{.001} = \frac{.55}{10^{-3}} = .55 \times 10^3 = 550$$

Continued on following page.

a) $9.86 \times 10^2 = 986$

b) $10^{-3} \times 4,500 = 4.5$

59.   Continued

Following the examples, complete these:

a) $\dfrac{157,000}{10,000} =$ _____

b) $\dfrac{2.5}{.000001} =$ _____

---

60.   Though we would not ordinarily do so, we can use the $\boxed{y^x}$ key on a calculator to multiply an ordinary number by a power of ten. For example:

To do $2.78 \times 10^3 = 2,780$, follow these steps:

Enter	Press	Display
2.78	$\boxed{x}$	2.78
10	$\boxed{y^x}$	10.
3	$\boxed{=}$	2780.

To do $14.7 \times 10^{-2} = .147$, follow these steps:

Enter	Press	Display
14.7	$\boxed{x}$	14.7
10	$\boxed{y^x}$	10.
2	$\boxed{+/-}$ $\boxed{=}$	0.147

a) 15.7

b) 2,500,000

---

61.   Though we would also not ordinarily do so, we can use the $\boxed{y^x}$ key to divide an ordinary number by a power of ten. For example:

To do $\dfrac{649}{10^2} = 6.49$, follow these steps:

Enter	Press	Display
649	$\boxed{\div}$	649.
10	$\boxed{y^x}$	10.
2	$\boxed{=}$	6.49

To do $\dfrac{.75}{10^{-4}} = 7,500$, follow these steps:

Enter	Press	Display
.75	$\boxed{\div}$	0.75
10	$\boxed{y^x}$	10.
4	$\boxed{+/-}$ $\boxed{=}$	7500.

3-6   SCIENTIFIC NOTATION

Any number can be written in "scientific notation". That is, it can be written as a multiplication of a number between 1 and 10 and a power of ten. We will discuss "scientific notation" in this section.

62.   In the tables below, some whole numbers and decimal numbers are written in "scientific notation". Notice these points about "scientific notation":

1) The first factor is a number between 1 and 10.
2) The second factor is a power of ten.

Whole Number	Scientific Notation
520,000	5.2×10^5
37,000	3.7×10^4
1,600	1.6×10^3
480	4.8×10^2
95	9.5×10^1

Decimal Number	Scientific Notation
.47	4.7×10^{-1}
.018	1.8×10^{-2}
.0092	9.2×10^{-3}
.00053	5.3×10^{-4}
.000066	6.6×10^{-5}

↑ A number between 1 and 10   ↑ A power of ten

↑ A number between 1 and 10   ↑ A power of 10

Which of the following are written in scientific notation? That is, which are multiplications of a number between 1 and 10 and a power of ten?

a) 25×10^8 b) 3.4×10^5 c) 9.1×10^{-3} d) $.67 \times 10^{-7}$

Only (b) and (c)

63.   To convert a number written in scientific notation to its ordinary form, we simply perform the multiplication by the decimal-point-shift method. For example:

$$2.7 \times 10^3 = 2.700 = 2,700$$

$$9.6 \times 10^{-5} = .00009.6 = .000096$$

Convert each of these to ordinary numbers.

a) 3×10^4 = _____   c) 2.54×10^2 = _____

b) 3×10^{-4} = _____   d) 2.54×10^{-2} = _____

a) 30,000   c) 254

b) .0003   d) .0254

64. A number written in scientific notation can stand for a very large or very small number. For example:

$$5.1 \times 10^7 = 5.\underline{1000000} = 51,000,000$$

$$3.9 \times 10^{-9} = \underline{000000003}.9 = .0000000039$$

Convert each of these to an ordinary number.

a) $7.13 \times 10^9 =$ _____ b) $5 \times 10^{-7} =$ _____

---

65. To convert a number larger than "1" to scientific notation, we must find both the first factor and the exponent for the power of ten. Since the first factor <u>must be a number between 1 and 10</u>, we can find it by writing a caret ($_\wedge$) <u>after the first digit on the left</u>. That is:

For $7_\wedge 3,000$, the first factor is 7.3.

For $9_\wedge 00,000$, the first factor is 9.

For $1_\wedge 8.4$, the first factor is 1.84.

Having found the first factor, we can find the exponent of the power of ten <u>by counting the number of places from the caret to the decimal point</u>. That is:

$7_\wedge 3,000. = 7.3 \times 10^4$ (<u>Four</u> places to the decimal point)

$8_\wedge 00,000. = 8 \times 10^5$ (<u>Five</u> places to the decimal point)

$9 1_\wedge .7 = 9.17 \times 10^1$ (<u>One</u> place to the decimal point)

Write each number in scientific notation.

a) $47,000 =$ _____ x _____   c) $88.55 =$ _____ x _____

b) $200 =$ _____ x _____   d) $9,160,000 =$ _____ x _____

Answer column:

a) $7,130,000,000$

b) $.0000005$

---

66. To convert a number smaller than "1" to scientific notation, we must again find both the first factor and the exponent for the power of ten. Since the first factor must be <u>a number between 1 and 10</u>, we can find it by writing a caret after the <u>first non-zero digit in the number</u>. That is:

For $.06_\wedge 8$, the first factor is 6.8.

For $.1_\wedge 95$, the first factor is 1.95.

For $.00007_\wedge$, the first factor is 7.

Answer column:

a) 4.7×10^4

b) 2×10^2

c) 8.855×10^1

d) 9.16×10^6

Continued on following page.

66.   Continued

Having found the first factor, we can find the exponent of the power of ten
by counting the number of places from the caret to the decimal point.
Since we count to the left, the exponent is negative.   That is:

$.06\!\!\stackrel{\frown}{}\!8 = 6.8 \times 10^{-2}$ (Two places to the decimal point)

$.1\!\!\stackrel{\frown}{}\!95 = 1.95 \times 10^{-1}$ (One place to the decimal point)

$.00007\!\!\stackrel{\frown}{} = 7 \times 10^{-5}$ (Five places to the decimal point)

Write each number in scientific notation.

a) .00025 = _____ x _____   c) .06718 = _____ x _____

b) .4 = _____ x _____   d) .0000019 = _____ x _____

---

67.   In scientific notation, the exponent can be either positive or negative.

If the number is larger than "1", we count to the right from the
caret to the decimal point. Therefore, the exponent is positive.
For example:

$$9\!\!\stackrel{\wedge}{}1,300 = 9.13 \times 10^4$$

If the number is smaller than "1", we count to the left from the
caret to the decimal point. Therefore, the exponent is negative.
For example:

$$.009\!\!\stackrel{\wedge}{}13 = 9.13 \times 10^{-3}$$

Write each number in scientific notation.

a) 5,600 = _____ x _____   c) 300,000,000 = _____ x _____

b) .056 = _____ x _____   d) .000000003 = _____ x _____

a) 2.5×10^{-4}

b) 4×10^{-1}

c) 6.718×10^{-2}

d) 1.9×10^{-6}

---

68.   When the original number is quite large or quite small, the absolute value
of the exponent is quite large. For example:

$$7\!\!\stackrel{\wedge}{}1,800,000,000 = 7.18 \times 10^{10}$$
$$.000000009\!\!\stackrel{\wedge}{}46 = 9.46 \times 10^{-9}$$

Write each number in scientific notation.

a) 259,000,000 = _____

b) .0000000184 = _____

a) 5.6×10^3

b) 5.6×10^{-2}

c) 3×10^8

d) 3×10^{-9}

a) 2.59×10^8

b) 1.84×10^{-8}

69. The name "scientific" notation is used because it is a convenient way of writing the very large and very small measurements that occur in science. For example:

    The speed of light is 2.998×10^8 meters per second.

    The diameter of a large molecule is 1.7×10^{-7} centimeter.

    Using the decimal-point-shift method, convert each measurement above to an ordinary number.

    a) 2.998×10^8 meters per second = _____ meters per second

    b) 1.7×10^{-7} centimeter = _____ centimeter

---

70. Two more measurements from science are given below.

    A light-year (the distance light travels in one year) is $5,870,000,000,000$ miles.

    One cycle of a television broadcast signal takes $.00000000481$ second.

    Ordinarily measurements of that size would be expressed in scientific notation. Convert them to scientific notation below.

    a) $5,870,000,000,000$ miles = _____ miles

    b) $.00000000481$ second = _____ second

    a) $299,800,000$

    b) $.00000017$

---

71. A number in scientific notation can be named by simply substituting a place-name for the power of ten. Some examples are shown.

    $$3.58 \times 10^3 = 3.58 \text{ thousand}$$
    $$8.71 \times 10^{-2} = 8.71 \text{ hundredths}$$
    $$5.49 \times 10^{-6} = 5.49 \text{ millionths}$$

    Write the correct name in each blank.

    a) $4.5 \times 10^6 = 4.5$ _____   c) $9.4 \times 10^{-3} = 9.4$ _____

    b) $1.7 \times 10^{-1} = 1.7$ _____   d) $3.8 \times 10^{-5} = 3.8$ _____

    a) 5.87×10^{12} miles

    b) 4.81×10^{-9} second

---

72. Write the correct name in each blank.

    a) $7.8 \times 10^9 = 7.8$ _____   c) $4.1 \times 10^{12} = 4.1$ _____

    b) $6.5 \times 10^{-9} = 6.5$ _____   d) $1.9 \times 10^{-12} = 1.9$ _____

    a) 4.5 million

    b) 1.7 tenths

    c) 9.4 thousandths

    d) 3.8 hundred-
        thousandths

---

a) 7.8 <u>billion</u>   b) 6.5 <u>billionths</u>   c) 4.1 <u>trillion</u>   d) 1.9 <u>trillionths</u>

3-7   CALCULATOR ANSWERS IN SCIENTIFIC NOTATION

When the number of digits in a calculator answer is more than the capacity of the display, the answer is displayed in scientific notation. We will discuss answers of that type in this section.

---

73.   When a calculator displays a number in scientific notation, <u>only the exponent</u> of the power of ten is shown. The exponent appears on the right side. For example:

The display $\boxed{2.54 \quad 12}$ means 2.54×10^{12}.

The display $\boxed{6.81 \quad -11}$ means 6.81×10^{-11}.

Ordinarily a "0" is printed in front of a one-digit exponent. For example:

The display $\boxed{4.08 \quad 08}$ means 4.08×10^{8}.

The display $\boxed{7.66 \quad -07}$ means _____.

---

74.   When the number of digits in a calculator answer is more than the capacity of the display, the answer is displayed in scientific notation. Two examples are given.

7.66×10^{-7}

Try $8,300,000 \times 250,000$ on a calculator.

Enter	Press	Display
8,300,000	$\boxed{\times}$	8300000.
250,000	$\boxed{=}$	2.075 12

Try $\dfrac{.00115}{18,400,000}$ on a calculator.

Enter	Press	Display
.00115	$\boxed{\div}$	0.00115
18,400,000	$\boxed{=}$	6.25 -11

To get ordinary numbers as answers for the operations above, you have to convert from scientific notation. That is:

$$8,300,000 \times 250,000 = 2.075 \times 10^{12} = 2,075,000,000,000$$

$$\frac{.00115}{18,400,000} = 6.25 \times 10^{-11} = \underline{\hspace{5cm}}$$

---

75.   Complete. Report each answer in scientific notation.

a) $2,600,000 \times 14,500,000 =$ _____

b) $\dfrac{.00098}{14,000,000} =$ _____

c) $(900,000)^2 =$ _____

d) $(.0000012)^2 =$ _____

.0000000000625

---

a) 3.77×10^{13} b) 7×10^{-11} c) 8.1×10^{11} d) 1.44×10^{-12}

3-8   CONVERTING OTHER MULTIPLICATIONS TO SCIENTIFIC NOTATION

In this section, we will discuss the method for converting multiplications like 415×10^7 or $.028 \times 10^{-5}$ to scientific notation.

76. To convert 415×10^7 and $.028 \times 10^{-5}$ to scientific notation, we convert the first factor to scientific notation and then multiply the powers of ten. That is: $$415 \times 10^7 = (4.15 \times 10^2) \times 10^7 = 4.15 \times 10^9$$ $$.028 \times 10^{-5} = (2.8 \times 10^{-2}) \times 10^{-5} = \underline{\quad} \times \underline{\quad}$$	
77. Following the example, convert the other multiplication to scientific notation. $$25.6 \times 10^{-5} = (2.56 \times 10^1) \times 10^{-5} = 2.56 \times 10^{-4}$$ $$81.9 \times 10^{-8} = (8.19 \times 10^1) \times 10^{-8} = \underline{\quad} \times \underline{\quad}$$	2.8×10^{-7}
78. Following the example, convert the other multiplication to scientific notation. $$.329 \times 10^{12} = (3.29 \times 10^{-1}) \times 10^{12} = 3.29 \times 10^{11}$$ $$.704 \times 10^8 = (7.04 \times 10^{-1}) \times 10^8 = \underline{\quad} \times \underline{\quad}$$	8.19×10^{-7}
79. Convert each multiplication to scientific notation. a) $270 \times 10^{10} = $ _____ b) $.055 \times 10^{-6} = $ _____ c) $49.7 \times 10^{-4} = $ _____ d) $.611 \times 10^5 = $ _____	7.04×10^7
80. We can also name other power-of-ten forms of numbers by substituting a place-name for the power of ten. For example: $$275 \times 10^6 = 275 \text{ million}$$ $$39.9 \times 10^{-3} = 39.9 \text{ thousandths}$$ Write the correct name in each blank. a) $76 \times 10^2 = 76$ _____ c) $83 \times 10^{-1} = 83$ _____ b) $51 \times 10^3 = 51$ _____ d) $67 \times 10^{-6} = 67$ _____	a) 2.7×10^{12} b) 5.5×10^{-8} c) 4.97×10^{-3} d) 6.11×10^4

a) 76 hundred   b) 51 thousand   c) 83 tenths   d) 67 millionths

81.   Write the correct name in each blank.

a) $16.5 \times 10^9 = 16.5$ _____   b) $268 \times 10^{-12} = 268$ _____

a) 16.5 billion   b) 268 trillionths

SELF-TEST 8 (pages 89-100)

Perform each multiplication and division.

1. $.762 \times 10^4 = $ _____   2. $\dfrac{37.5}{10^3} = $ _____   3. $\dfrac{.084}{10^{-1}} = $ _____

Convert from scientific notation to an ordinary number.

4. $1.39 \times 10^5 = $ _____

5. $8.2 \times 10^{-3} = $ _____

Convert each number to scientific notation.

6. $.0007 = $ _____

7. $5{,}140{,}000 = $ _____

8. One second equals 2.778×10^{-4} hour. Write this time as an ordinary number. _____

9. A communications satellite can process $4{,}200{,}000{,}000{,}000$ bits of information daily. Write this quantity in scientific notation. _____

Use a calculator for these. Report each answer in scientific notation.

10. $\dfrac{.0000172}{21{,}500{,}000}$ 11. $(2{,}700{,}000)^2$ 12. $35{,}000{,}000 \times 8{,}000$

= _____   = _____   = _____

Convert each multiplication to scientific notation.

13. $9{,}000 \times 10^4$ 14. $.296 \times 10^{-5}$ 15. 41.4×10^{-3} 16. $.0083 \times 10^6$

= _____   = _____   = _____   = _____

ANSWERS:

1. 7,620	5. .0082	9. 4.2×10^{12} bits	13. 9×10^7
2. .0375	6. 7×10^{-4}	10. 8×10^{-13}	14. 2.96×10^{-6}
3. .84	7. 5.14×10^6	11. 7.29×10^{12}	15. 4.14×10^{-2}
4. 139,000	8. .0002778 hour	12. 2.8×10^{11}	16. 8.3×10^3

3-9   MULTIPLYING AND DIVIDING NUMBERS IN SCIENTIFIC NOTATION

In this section, we will discuss the procedure for multiplying and dividing numbers in scientific notation on a calculator.

---

82. To multiply a number in scientific notation by an ordinary number, we multiply the two ordinary-number factors. For example: $$4 \times (2 \times 10^5) = (4 \times 2) \times 10^5 = 8 \times 10^5$$ $$3 \times (3 \times 10^{-8}) = (3 \times 3) \times 10^{-8} = \underline{\hspace{2cm}}$$	
83. When the first factor of the product is not a number between 1 and 10, we convert the product to scientific notation. For example: $$20 \times (7.1 \times 10^6) = 142 \times 10^6 = (1.42 \times 10^2) \times 10^6 = 1.42 \times 10^8$$ $$.04 \times (5.3 \times 10^{-5}) = .212 \times 10^{-5} = (2.12 \times 10^{-1}) \times 10^{-5} = \underline{\hspace{2cm}}$$	9×10^{-8}
84. To multiply two numbers in scientific notation, we multiply the two ordinary-number factors and the two powers of ten. For example: $$(3 \times 10^4) \times (2 \times 10^5) = (3 \times 2) \times (10^4 \times 10^5) = 6 \times 10^9$$ $$(2 \times 10^{-3}) \times (4 \times 10^{-7}) = (2 \times 4) \times (10^{-3} \times 10^{-7}) = \underline{\hspace{2cm}}$$	2.12×10^{-6}
85. When the first factor of the product is not a number between 1 and 10, we convert the product to scientific notation. For example: $$(5 \times 10^{-2}) \times (3.1 \times 10^5) = 15.5 \times 10^3 = (1.55 \times 10^1) \times 10^3 = 1.55 \times 10^4$$ $$(9.2 \times 10^{-7}) \times (8 \times 10^4) = 73.6 \times 10^{-3} = (7.36 \times 10^1) \times 10^{-3} = \underline{\hspace{2cm}}$$	8×10^{-10}
86. A calculator can be used to multiply numbers in scientific notation. To enter a number in scientific notation on a calculator, we use the $\boxed{\text{EE}}$ key. ($\boxed{\text{EE}}$ means "Enter Exponent".) Following the steps below, enter 6.5×10^8 and 1.9×10^{-5} .	7.36×10^{-2}

Enter	Press	Display
6.5	$\boxed{\text{EE}}$	6.5 00
8		6.5 08

Note: Press $\boxed{\text{C}}$ to clear the display.

Enter	Press	Display
1.9	$\boxed{\text{EE}}$	1.9 00
5	$\boxed{+/-}$	1.9 -05

Note: Press $\boxed{\text{C}}$ to clear the display.

87.   To perform $3 \times (2.4 \times 10^7) = 7.2 \times 10^7$ on a calculator, follow the steps below.

Enter	Press	Display
3	$\boxed{\text{x}}$	3.
2.4	$\boxed{\text{EE}}$	2.4 00
7	$\boxed{=}$	7.2 07

88.   To perform $1.2 \times (4.3 \times 10^{-7}) = 5.16 \times 10^{-7}$ on a calculator, follow the steps below.

Enter	Press	Display
1.2	$\boxed{\text{x}}$	1.2
4.3	$\boxed{\text{EE}}$	4.3 00
7	$\boxed{+/-}$ $\boxed{=}$	5.16 -07

89.   To perform $(3.2 \times 10^8) \times (1.4 \times 10^{-4}) = 4.48 \times 10^4$ on a calculator, follow these steps.

Enter	Press	Display
3.2	$\boxed{\text{EE}}$	3.2 00
8	$\boxed{\text{x}}$	3.2 08
1.4	$\boxed{\text{EE}}$	1.4 00
4	$\boxed{+/-}$ $\boxed{=}$	4.48 04

90.   If $(8 \times 10^{-9}) \times (7 \times 10^6)$ is performed without a calculator, the product has to be converted to scientific notation. That is:

$$(8 \times 10^{-9}) \times (7 \times 10^6) = 56 \times 10^{-3} = (5.6 \times 10^1) \times 10^{-3} = 5.6 \times 10^{-2}$$

However, when the same multiplication is performed on a calculator, the product is automatically reported in scientific notation. Do the same multiplication below.

Enter	Press	Display
8	$\boxed{\text{EE}}$	8. 00
9	$\boxed{+/-}$ $\boxed{\text{x}}$	8. -09
7	$\boxed{\text{EE}}$	7. 00
6	$\boxed{=}$	5.6 -02

91.   Use a calculator for these.

   a) $4.8 \times (7.93 \times 10^{12})$ = _____

   b) $(6.7 \times 10^{-4}) \times (8.38 \times 10^{-7})$ = _____

a) 3.8064×10^{13} b) 5.6146×10^{-10}

92. In the last frame, the calculator reported the first factor of the product with four decimal places. We frequently round the first factor <u>to two</u> <u>decimal places</u>. For example:

      $\downarrow$
    3.8064×10^{13} is rounded to 3.81×10^{13}

      $\downarrow$
    5.6146×10^{-10} is rounded to 5.61×10^{-10}

Use a calculator for these. Round to two decimal places.

  a) $27.6 \times (1.44 \times 10^{-12})$ = _____

  b) $(7.14 \times 10^{-10}) \times (6.76 \times 10^{13})$ = _____

---

93. To divide a number in scientific notation by an ordinary number, we divide the ordinary numbers. For example:

$$\frac{6 \times 10^7}{2} = \frac{6}{2} \times 10^7 = 3 \times 10^7$$

$$\frac{9.3 \times 10^{-9}}{3} = \frac{9.3}{3} \times 10^{-9} = \text{_____}$$

a) 3.97×10^{-11}

b) 4.83×10^4

---

94. To divide two numbers in scientific notation, we divide the two ordinary numbers and the two powers of ten. For example:

$$\frac{8 \times 10^9}{2 \times 10^4} = \frac{8}{2} \times \frac{10^9}{10^4} = 4 \times 10^5$$

$$\frac{9 \times 10^{-8}}{3 \times 10^{-6}} = \frac{9}{3} \times \frac{10^{-8}}{10^{-6}} = \text{_____}$$

3.1×10^{-9}

---

95. Following the example, convert the other quotient below to scientific notation.

$$\frac{3.01 \times 10^7}{7} = .43 \times 10^7 = (4.3 \times 10^{-1}) \times 10^7 = 4.3 \times 10^6$$

$$\frac{6.66 \times 10^{10}}{8.88 \times 10^{20}} = .75 \times 10^{-10} = \text{_____}$$

3×10^{-2}

7.5×10^{-11}

96.   A calculator can also be used for divisions involving numbers in scientific notation. When a calculator is used, <u>the</u> <u>quotient</u> <u>is</u> <u>automatically</u> <u>reported</u> <u>in</u> <u>scientific</u> <u>notation</u>.

To perform $\dfrac{6.48 \times 10^{10}}{20} = 3.24 \times 10^9$, follow these steps:

Enter	Press	Display
6.48	EE	6.48 00
10	÷	6.48 10
20	=	3.24 09

To perform $\dfrac{1.24 \times 10^{-16}}{4.96 \times 10^{-12}} = 2.5 \times 10^{-5}$, follow these steps:

Enter	Press	Display
1.24	EE	1.24 00
16	+/− ÷	1.24 −16
4.96	EE	4.96 00
12	+/− =	2.5 −05

To perform $\dfrac{935}{2.2 \times 10^4} = 4.25 \times 10^{-2}$, follow these steps:

Enter	Press	Display
935	÷	935.
2.2	EE	2.2 00
4	=	4.25 −02

97.   Use a calculator for these. Round to two decimal places.

a) $\dfrac{9.03 \times 10^{-7}}{269} =$ _____

b) $\dfrac{5.18 \times 10^8}{9.06 \times 10^{-6}} =$ _____

c) $\dfrac{256}{4.18 \times 10^{-3}} =$ _____

98.   A first electrical impulse has a duration of 3.72×10^{-9} second. A second electrical impulse has a duration of 2.27×10^{-6} second. State the ratio of the first impulse to the second impulse as an ordinary number rounded to hundred-thousandths.

a) 3.36×10^{-9}

b) 5.72×10^{13}

c) 6.12×10^4

.00164

3-10   SQUARES AND SQUARE ROOTS OF NUMBERS IN SCIENTIFIC NOTATION

In this section, we will discuss the procedure for squaring and finding the square root of a number in scientific notation on a calculator.

---

99.   To square a number in scientific notation, we simply square the ordinary-number part and the power of ten. For example:

$$(2 \times 10^8)^2 = (2)^2 \times (10^8)^2 = 4 \times 10^{16}$$

$$(3 \times 10^{-5})^2 = (3)^2 \times (10^{-5})^2 = \underline{\qquad\qquad}$$

---

100.   When the square of the first factor is not a number between 1 and 10, we convert the square to scientific notation. For example:

$$(4.5 \times 10^3)^2 = (4.5)^2 \times (10^3)^2 = 20.25 \times 10^6 = 2.025 \times 10^7$$

$$(8.1 \times 10^{-7})^2 = (8.1)^2 \times (10^{-7})^2 = 65.61 \times 10^{-14} = \underline{\qquad\qquad}$$

9×10^{-10}

---

101.   When a calculator is used to square a number in scientific notation, the square is automatically reported in scientific notation.

Perform $(6.8 \times 10^6)^2 = 4.624 \times 10^{13}$ below.

Enter	Press	Display
6.8	EE	6.8 00
6	x^2	4.624 13

Perform $(5.1 \times 10^{-11})^2 = 2.601 \times 10^{-21}$ below.

Enter	Press	Display
5.1	EE	5.1 00
11	+/− x^2	2.601 −21

6.561×10^{-13}

---

102.   To find the square root of a number in scientific notation, we find the square root of the ordinary number and the square root of the power of ten. For example:

$$\sqrt{4 \times 10^{10}} = \sqrt{4} \times \sqrt{10^{10}} = 2 \times 10^5$$

$$\sqrt{9 \times 10^{-6}} = \sqrt{9} \times \sqrt{10^{-6}} = \underline{\qquad\qquad}$$

3×10^{-3}

103.   The square root of a number in scientific notation can also be found on a calculator.

Perform $\sqrt{1.44 \times 10^{12}} = 1.2 \times 10^6$ below.

Enter	Press	Display
1.44	EE	1.44 00
12	$\sqrt{x}$	1.2 06

Perform $\sqrt{5.76 \times 10^{-16}} = 2.4 \times 10^{-8}$ below.

Enter	Press	Display
5.76	EE	5.76 00
16	+/- $\sqrt{x}$	2.4 -08

104.   To find the square root of a number in scientific notation without a calculator, the power-of-ten factor must have an <u>even</u> exponent. If it has an <u>odd</u> exponent, we must convert it to a form with an <u>even</u> exponent. Two examples are shown.

$$\sqrt{1.6 \times 10^5} = \sqrt{1.6 \times 10^1 \times 10^4} = \sqrt{16 \times 10^4} = \sqrt{16} \times \sqrt{10^4} = 4 \times 10^2$$

$$\sqrt{4.9 \times 10^{-7}} = \sqrt{4.9 \times 10^1 \times 10^{-8}} = \sqrt{49 \times 10^{-8}} = \sqrt{49} \times \sqrt{10^{-8}} = 7 \times 10^{-4}$$

However, when a calculator is used, all of the conversions are done by the calculator. Use a calculator to find the same two square roots below.

Enter	Press	Display
1.6	EE	1.6 00
5	$\sqrt{x}$	4. 02

Enter	Press	Display
4.9	EE	4.9 00
7	+/- $\sqrt{x}$	7. -04

105.   Use a calculator for these. Round to two decimal places.

a) $(9.55 \times 10^9)^2 = $ _____

b) $(1.99 \times 10^{-12})^2 = $ _____

c) $\sqrt{7.38 \times 10^{13}} = $ _____

d) $\sqrt{2.75 \times 10^{-11}} = $ _____

a) 9.12×10^{19} c) 8.59×10^6

b) 3.96×10^{-24} d) 5.24×10^{-6}

106. Numbers like $275,000,000,000$ and $.000000000648$ contain more digits than the capacity of the display. Therefore, to enter those numbers on a calculator, they must be converted to scientific notation. That is:

$$275,000,000,000 \text{ must be entered as } 2.75 \times 10^{11}$$

$$.000000000648 \text{ must be entered as } 6.48 \times 10^{-10}$$

Perform each operation. When necessary, convert a number to scientific notation to enter it on a calculator. Round all answers to two decimal places.

a) $71,800,000,000 \times .000499 = $ _____

b) $\dfrac{.00000000364}{27,000} = $ _____

c) $(3,750,000,000,000)^2 = $ _____

d) $\sqrt{.0000000000155} = $ _____

---

a) 3.58×10^7, from:

$$(7.18 \times 10^{10}) \times .000499$$

b) 1.35×10^{-13}, from:

$$\dfrac{3.64 \times 10^{-9}}{27,000}$$

c) 1.41×10^{25}, from:

$$(3.75 \times 10^{12})^2$$

d) 3.94×10^{-6}, from:

$$\sqrt{1.55 \times 10^{-11}}$$

3-11   OPERATIONS INVOLVING OTHER POWER-OF-TEN FORMS OF NUMBERS

Sometimes numbers are written in a power-of-ten form in which the first factor is not a number between 1 and 10. We will discuss operations with those types of numbers in this section.

---

107. Numbers like 375×10^3 or $.045 \times 10^{-6}$ are not in scientific notation because the first factor is not a number between 1 and 10. The $\boxed{\text{EE}}$ key can be used to enter such numbers on a calculator. Do so below.

Enter	Press	Display
375	$\boxed{\text{EE}}$	375. 00
3		375. 03

Note: Press $\boxed{\text{C}}$ to clear the display.

Enter	Press	Display
.045	$\boxed{\text{EE}}$	0.045 00
6	$\boxed{+/-}$	0.045 -06

Note: Press $\boxed{\text{C}}$ to clear the display.

108.   To perform $(250 \times 10^3) \times (74 \times 10^{-8}) = 1.85 \times 10^{-1}$ on a calculator, follow the steps below.

Enter	Press	Display
250	EE	250. 00
3	x	2.5 05
74	EE	74. 00
8	+/- =	1.85 -01

109.   Do these on a calculator. Round to two decimal places.

a) $47.8 \times (645 \times 10^7) =$ _____

b) $(66.9 \times 10^{-5}) \times (.025 \times 10^{-3}) =$ _____

110.   To perform $\dfrac{343 \times 10^5}{.098 \times 10^{-2}} = 3.5 \times 10^{10}$ on a calculator, follow these steps.

a) 3.08×10^{11}

b) 1.67×10^{-8}

Enter	Press	Display
343	EE	343. 00
5	÷	3.43 07
.098	EE	0.098 00
2	+/- =	3.5 10

111.   Do these on a calculator. Round to two decimal places.

a) $\dfrac{749 \times 10^8}{50.7} =$ _____

b) $\dfrac{.625}{.033 \times 10^{-10}} =$ _____

112.   To perform $(840 \times 10^5)^2 = 7.056 \times 10^{15}$ on a calculator, follow these steps.

a) 1.48×10^9

b) 1.89×10^{11}

Enter	Press	Display
840	EE	840. 00
5	x^2	7.056 15

113.   To perform $\sqrt{256 \times 10^{-8}} = 1.6 \times 10^{-3}$ on a calculator, follow these steps.

Enter	Press	Display
256	EE	256. 00
8	+/- $\sqrt{x}$	1.6 -03

114.   Do these on a calculator. Round to two decimal places.

  a) $(.0147 \times 10^{-6})^2 = $ _____

  b) $\sqrt{.099 \times 10^{10}} = $ _____

---

a) 2.16×10^{-16} b) 3.15×10^4

3-12   PROPORTIONS INVOLVING NUMBERS IN SCIENTIFIC NOTATION

In this section, we will discuss the method for solving proportions involving numbers in scientific notation. Some applied problems are included.

115.   To isolate the letter in the proportion below, we cross-multiplied and then divided the right side by 6.05×10^6 .

$$\frac{6.05 \times 10^6}{16.7} = \frac{5.66 \times 10^9}{N}$$

$$(6.05 \times 10^6)(N) = (16.7)(5.66 \times 10^9)$$

$$N = \frac{(16.7)(5.66 \times 10^9)}{6.05 \times 10^6}$$

To complete the solution on a calculator in one combined operation, follow the steps below.

Enter	Press	Display	
16.7	x	16.7	
5.66	EE	5.66	00
9	÷	9.4522	10
6.05	EE	6.05	00
6	=	1.5623	04

Rounding the first factor in scientific notation to hundredths, the solution of the proportion is:

N = _____ x _____   or   _____

---

116.   To isolate the letter in the proportion below, we cross-multiplied and then divided by 3.84×10^6 .

$$\frac{2.57}{3.84 \times 10^6} = \frac{N}{6.29 \times 10^4}$$

$$(2.57)(6.29 \times 10^4) = (3.84 \times 10^6)(N)$$

$$N = \frac{(2.57)(6.29 \times 10^4)}{3.84 \times 10^6}$$

Use a calculator to complete the solution. Round the first factor in scientific notation to two decimal places.

N = _____ x _____   or   _____

1.56×10^4 or 15,600

117.   Following the steps in the last two frames, solve the following proportion. Round the first factor in scientific notation to two decimal places.

$$\frac{y}{9.56 \times 10^{-12}} = \frac{0.332}{7.49 \times 10^{-9}}$$

4.21×10^{-2} or .0421

118.   To solve the problem below, we set up and solved a proportion involving numbers in scientific notation.

If it takes 1 second for light to travel 1.86×10^5 miles, how long does it take light to travel from the sun to earth if that distance is 9.30×10^7 miles?

$$\frac{1 \text{ second}}{1.86 \times 10^5 \text{ miles}} = \frac{N \text{ seconds}}{9.30 \times 10^7 \text{ miles}} \quad \text{or} \quad \frac{1}{1.86 \times 10^5} = \frac{N}{9.30 \times 10^7}$$

$$9.30 \times 10^7 = (1.86 \times 10^5)(N)$$

$$N = \frac{9.30 \times 10^7}{1.86 \times 10^5}$$

$$N = 5 \times 10^2 = 500$$

Therefore, it takes light 500 seconds to travel from the sun to the earth.

Following the example above, use a proportion to solve the following problem.

If a current of 7.2×10^{-4} ampere is obtained when 1 volt is applied to a circuit, how many volts must be applied to get a current of 1.5×10^{-3} ampere? (Round to two decimal places.)

$y = 4.24 \times 10^{-4}$
or .000424

119.   To solve the problem below, we set up and solved a proportion.

If there are 6.02×10^{23} molecules in 22.4 liters of gas, how many molecules are there in 1 liter of the same gas? (Report the answer in scientific notation with the first factor rounded to two decimal places.)

$$\frac{6.02 \times 10^{23} \text{ molecules}}{22.4 \text{ liters}} = \frac{N \text{ molecules}}{1 \text{ liter}} \quad \text{or} \quad \frac{6.02 \times 10^{23}}{22.4} = \frac{N}{1}$$

$$6.02 \times 10^{23} = 22.4N$$

$$N = \frac{6.02 \times 10^{23}}{22.4}$$

$$N = 2.69 \times 10^{22}$$

Therefore, there are 2.69×10^{22} molecules in 1 liter of the gas.

2.08×10^0 volts or
2.08 volts, from:

$$\frac{7.2 \times 10^{-4}}{1} = \frac{1.5 \times 10^{-3}}{N}$$

Continued on following page.

119.   Continued

Use a proportion to solve the following problem.

The ratio of the speed of light in a vacuum to the speed of light in water
is 4 to 3. If the speed of light in a vacuum is 3×10^8 meters per
second, what is the speed of light in water?

2.25×10^8 meters per second, from: $\dfrac{4}{3} = \dfrac{3 \times 10^8}{N}$

SELF-TEST 9 (pages 101-111)

Use a calculator for Problems 1-6. Report each answer in scientific notation.

1. $\dfrac{5.52 \times 10^{-6}}{1.84 \times 10^{-2}} =$ _____

2. $(4.7 \times 10^4)^2 =$ _____

3. $\sqrt{8.1 \times 10^{-5}} =$ _____

4. $320 \times (3.75 \times 10^3)$

= _____

5. $(1.8 \times 10^{-4}) \times (9.5 \times 10^2)$

= _____

6. $\dfrac{7.37 \times 10^5}{13.4} =$ _____

Use a calculator for Problems 7-12. Report each answer in scientific notation with the first factor rounded
to two decimal places.

7. $(927 \times 10^5) \times (.214 \times 10^{-3})$

= _____

8. $\dfrac{60.3 \times 10^6}{.0882 \times 10^3} =$ _____

9. $.0518 \times (137 \times 10^{-4})$

= _____

10. $\dfrac{5,190}{20.6 \times 10^8} =$ _____

11. $\sqrt{.0124 \times 10^9}$

= _____

12. $(26.5 \times 10^{-2})^2$

= _____

Use a calculator to solve these proportions. Report each answer in scientific notation with the first factor
rounded to two decimal places.

13. $\dfrac{7.46 \times 10^8}{528} = \dfrac{N}{1.73}$

N = _____

14. $\dfrac{3.26 \times 10^{-3}}{84.9} = \dfrac{7.18 \times 10^{-9}}{y}$

y = _____

ANSWERS:

1. 3×10^{-4}
2. 2.209×10^9
3. 9×10^{-3}
4. 1.2×10^6

5. 1.71×10^{-1}
6. 5.5×10^4
7. 1.98×10^4
8. 6.84×10^5

9. 7.10×10^{-4}
10. 2.52×10^{-6}
11. 3.52×10^3
12. 7.02×10^{-2}

13. $N = 2.44 \times 10^6$
14. $y = 1.87 \times 10^{-4}$

SUPPLEMENTARY PROBLEMS - CHAPTER 3

Assignment 7

Convert each power of ten to an ordinary number.

1. 10^6 2. 10^{-2} 3. 10^4 4. 10^{-6} 5. 10^1

Convert each number to a power of ten.

6. 1,000   7. .001   8. 100,000,000   9. .1

10. .0000001   11. 100,000   12. 10   13. 1

Write the power of ten corresponding to each place name.

14. thousands   15. millionths   16. millions   17. thousandths

Write the place name corresponding to each power of ten.

18. 10^2 19. 10^9 20. 10^{-1} 21. 10^{-4}

Using the laws of exponents, write each answer as a power of ten.

22. $10^{-5} \times 10^3$ 23. $10^4 \times 10^{-3} \times 10$ 24. $\dfrac{10^6}{10^2}$ 25. $\dfrac{10^{-2}}{10^4}$

26. $(10^2)^2$ 27. $(10^{-1})^2$ 28. $(10^0)^2$ 29. $\sqrt{10^6}$ 30. $\sqrt{10^{-2}}$

31. $\dfrac{10^2 \times 10^{-5}}{10^3}$ 32. $\dfrac{10^{-1} \times 10^8}{10}$ 33. $\dfrac{10^0}{10^4 \times 10^{-6}}$ 34. $\dfrac{10^3}{10^{-2} \times 10^5}$

35. $\dfrac{10 \times 10^5}{10^4 \times 10^0}$ 36. $\dfrac{10^{-1} \times 10^6}{10^5 \times 10^3}$ 37. $\dfrac{10^3 \times 10^{-4}}{10^{-3} \times 10^2}$ 38. $\dfrac{10^{-2} \times 10^{-5}}{10^8 \times 10^{-1}}$

Assignment 8

Convert each of the following to an ordinary number.

1. $10^{-2} \times 4,930$ 2. $10^3 \times .76$ 3. 59.2×10^{-6} 4. 200×10^4

5. $\dfrac{2,150}{10^2}$ 6. $\dfrac{36.2}{10^{-3}}$ 7. $\dfrac{.008}{10^{-4}}$ 8. $\dfrac{.94}{10^3}$

Convert from scientific notation to an ordinary number.

9. 8.2×10^3 10. 1×10^5 11. 5×10^{-3} 12. 4.6×10^{-6}

13. 6.108×10^{-1} 14. 2.92×10^0 15. 3.07×10^6 16. 5.814×10^2

Convert each number to scientific notation.

17. .02   18. .00076   19. 49,000   20. 10,000,000

21. 915   22. 183,200   23. .005604   24. .218

Convert each multiplication to scientific notation.

25. 500×10^3 26. 12.7×10^5 27. $.39 \times 10^{-2}$ 28. $.06 \times 10^{-6}$

29. $4,210 \times 10^{-3}$ 30. 98×10^{-5} 31. $.0088 \times 10^8$ 32. $.2075 \times 10^4$

Do these problems on a calculator. Report each answer in scientific notation with the first factor rounded to two decimal places.

33. $5,830,000 \times 927,000$ 34. $\dfrac{.00000319}{6,520,000}$ 35. $(7,853,000)^2$

Continued on following page.

36. In chemistry, the quantity 602,000,000,000,000,000,000,000 molecules is called "Avogadro's number". Write this quantity in scientific notation.

37. The diameter of the earth is 1.2756×10^4 kilometers. Write this quantity as an ordinary number.

38. The wavelength of an x-ray is 7.82×10^{-7} centimeter. Write this length as an ordinary number.

39. One second equals .000011574 day. Write this time in scientific notation.

40. The "half life" of radioactive <u>uranium 238</u> is 4.5×10^{10} years. Write this time as an ordinary number.

Assignment 9

Do each problem on a calculator. Report each answer in scientific notation with the first factor rounded to <u>two decimal places</u>.

1. $(6 \times 10^4) \times (7 \times 10^6)$
2. $(9.2 \times 10^{-3}) \times (5.8 \times 10^{-6})$
3. $(1.84 \times 10^8) \times (7.59 \times 10^{-5})$

4. $\dfrac{3 \times 10^{-2}}{7 \times 10^3}$
5. $\dfrac{8.06 \times 10^5}{1.55 \times 10^{-3}}$
6. $\dfrac{2.4 \times 10^{-5}}{6.7 \times 10^{-1}}$
7. $\dfrac{4.91 \times 10^6}{9.25 \times 10^3}$

8. $\dfrac{2,950}{73.2 \times 10^3}$
9. $\dfrac{.00504}{962 \times 10^{-6}}$
10. $\dfrac{214 \times 10^8}{32.3}$
11. $\dfrac{.0167 \times 10^{-5}}{.00622}$

12. $723 \times (12.5 \times 10^3)$
13. $.0308 \times (7,190 \times 10^{-8})$
14. $23.7 \times (.814 \times 10^5)$

15. $(41.2 \times 10^{-6}) \times (.564 \times 10^2)$
16. $(8,310 \times 10^3) \times (9.44 \times 10^{-5})$
17. $(609 \times 10^{-9}) \times (377 \times 10^{-3})$

18. $\dfrac{48.3 \times 10^4}{91.7 \times 10^5}$
19. $\dfrac{.0512 \times 10^{-5}}{363 \times 10^2}$
20. $\dfrac{7,480 \times 10^{-2}}{.116 \times 10^6}$
21. $\dfrac{.00552 \times 10^5}{9.15 \times 10^{-3}}$

22. $\sqrt{15.8 \times 10^{-5}}$
23. $\sqrt{.395 \times 10^9}$
24. $(7.07 \times 10^3)^2$
25. $(.00655 \times 10^{-4})^2$

Solve each proportion. Report each answer in scientific notation with the first factor rounded to <u>two decimal places</u>.

26. $\dfrac{1,850}{29.3} = \dfrac{N}{5.93 \times 10^6}$
27. $\dfrac{4.12 \times 10^{-3}}{740} = \dfrac{7.37 \times 10^{-6}}{x}$
28. $\dfrac{878}{y} = \dfrac{9.69 \times 10^3}{5.15 \times 10^{-3}}$

29. If 1 inch equals 2.54×10^{-2} meter, how many meters are there in 50 inches?

30. If a 12°C temperature rise causes a length increase of 9.74×10^{-3} centimeter in a metal rod, what temperature rise will cause a length increase of 1.50×10^{-2} centimeter?

Chapter 4

FORMULA EVALUATION

Formulas are frequently used in science, technology, statistics, geometry, and other branches of mathematics to state relationships and solve applied problems. In this chapter, we will discuss the meaning of formulas and show how a calculator can be used for the combined operations involved in formula evaluation. Major emphasis is given to the numerical part of formula evaluation. Some evaluations with "xy" equations are included. When necessary, different instructions are given for calculators with parentheses symbols and for calculators without parentheses symbols.

4-1   FORMULAS INVOLVING ADDITION, SUBTRACTION, AND MULTIPLICATION

In this section, we will show how a calculator can be used to evaluate formulas involving addition, subtraction, and multiplication.

1.   Formulas are a concise way of stating the relationships between physical quantities. For example:

   $\boxed{C = K - 273°}$ is a concise way of stating the relationship between degrees-Kelvin (K) and degrees-Celsius (C).

   $\boxed{s = vt}$ is a concise way of stating the relationship between distance traveled (s), average velocity (v), and time traveled (t) for a moving object.

By substituting numbers in place of the letters, we can use formulas to solve applied problems. This process is called "formula evaluation". For example:

Using $\boxed{C = K - 273°}$, we can find the number of degrees-Celsius that is equivalent to any number of degrees-Kelvin. That is:

   27°C is equivalent to 300°K, since:

   $C = K - 273 = 300° - 273° = 27°$

Using $\boxed{s = vt}$, we can find the distance traveled by an object if we know its average velocity and the time traveled. That is:

   If its average velocity is 100 kilometers per hour and the time traveled is 5 hours, the distance traveled is 500 kilometers, since:

   $s = vt = (100)(5) = \underline{\hspace{2cm}}$

500

114

2.   Some evaluations require only an addition, a subtraction, or a multiplication.
     For example:

     In $\boxed{M = V + 100}$, find M when V = 50 .

          M = V + 100 = 50 + 100 = 150

     In $\boxed{C = A - L}$, find C when A = 30 and L = 10 .

          C = A - L = 30 - 10 = 20

     In $\boxed{E = IR}$, find E when I = 10 and R = 7.5 .

          E = IR = (10)(7.5) = 75

A calculator can be used for evaluations like those above. Use a calculator
for these.

a) In $\boxed{a = b + c}$, find "a" when b = 1.78 and c = 4.66 .   a = _____

b) In $\boxed{R = V - T}$, find R when V = 47.1 and T = 26.9 .   R = _____

c) In $\boxed{P = 0.433h}$, find P when h = 50 .   P = _____

---

3.   The evaluation below involves a three-factor multiplication.

          In $\boxed{V = LWH}$, find V when L = 8 , W = 6 , and H = 10 .

               V = LWH = (8)(6)(10)

                   = (48)(10) = 480

To use a calculator for the same evaluation, follow the steps below.

Enter	Press	Display
8	$\boxed{\times}$	8.
6	$\boxed{\times}$	48.
10	$\boxed{=}$	480.

Use a calculator for this evaluation.

     In $\boxed{V = LWH}$, when L = 7.5, W = 4.6, and H = 6.9,   V = _____

---

4.   Use a calculator for these. Round to one decimal place.

a) In $\boxed{P = 62.4h}$, when h = 7.7 ,   P = _____

b) In $\boxed{V = LWH}$, when L = 12.4, W = 9.3, and H = 10.7,   V = _____

---

a) a = 6.44

b) R = 20.2

c) P = 21.65

V = 238.05

a) P = 480.5

b) V = 1,233.9

5.   The evaluation below involves a multiplication and an addition.

   In $\boxed{a = bc + d}$, find "a" when $b = 3$, $c = 4$, and $d = 20$.

$$a = bc + d = \underset{\downarrow}{(3)(4)} + 20$$
$$= 12 + 20 = 32$$

To perform the same evaluation in one process on a calculator, follow these steps.

Enter	Press	Display
3	$\boxed{\text{x}}$	3.
4	$\boxed{+}$	12.
20	$\boxed{=}$	32.

Use the same process on a calculator for this evaluation.

   In $\boxed{V = RS + H}$, when $R = 2.5$, $S = 8.4$, and $H = 9.9$, $V = \underline{\hspace{1cm}}$

$V = 30.9$

6.   The evaluation below also involves a multiplication and an addition.

   In $\boxed{H = E + PV}$, find H when $E = 25$, $P = 2$, and $V = 5$.

$$H = E + PV = 25 + \underset{\downarrow}{(2)(5)}$$
$$= 25 + 10 = 35$$

To perform the same evaluation in one process on a calculator, follow these steps.

Enter	Press	Display
25	$\boxed{+}$	25.
2	$\boxed{\text{x}}$	2.
5	$\boxed{=}$	35.

Use the same process on a calculator for this evaluation.

   In $\boxed{E = e + Ir}$, when $e = 27.6$, $I = 2.8$, and $r = 4.5$, $E = \underline{\hspace{1cm}}$

---

a) $E = 40.2$

7.   The evaluation below involves two multiplications and an addition.

   In $\boxed{P = 2L + 2W}$, find P when $L = 7$ and $W = 5$.

$$P = 2L + 2W = \underset{\downarrow}{2(7)} + \underset{\downarrow}{2(5)}$$
$$= 14 + 10 = 24$$

Continued on following page.

7.   Continued

To perform the same evaluation in one process on a calculator, follow these steps.

Enter	Press	Display
2	X	2.
7	+	14.
2	X	2.
5	=	24.

Use the same process on a calculator for this evaluation.

In $\boxed{H = 1.5V + 2.5Q}$, when V = 6.4 and Q = 5.8 , H = _____

---

8.   Use a calculator for these. Round to tenths.

$H = 24.1$

a) In $\boxed{h = 1.5v + s}$, when v = 6.45 and s = 9.03 , h = _____

b) In $\boxed{H = E + PV}$, when E = 7.58 , P = 1.44 , and V = 3.89 ,

H = _____

c) In $\boxed{Q = cd + ef}$, when c = 2.33 , d = 8.09 , e = 5.47 , and f = 7.91 ,

Q = _____

---

a) h = 18.7   b) H = 13.2   c) Q = 62.1

4-2   FORMULAS INVOLVING A SINGLE DIVISION

In this section, we will show how a calculator can be used to evaluate formulas involving a single division.

---

9.   The evaluation below involves only a simple division.

In $\boxed{I = \dfrac{E}{R}}$, find I when E = 40 and R = 8 .

$$I = \frac{E}{R} = \frac{40}{8} = 5$$

Use a calculator for these. Round to two decimal places.

a) In $\boxed{L = \dfrac{A}{W}}$, when A = 9.68 and W = 1.43 , L = _____

b) In $\boxed{h = \dfrac{P}{0.433}}$, when P = 2.77 , h = _____

---

a) L = 6.77   b) h = 6.40

10.   The evaluation below also involves only a simple division.

In $\boxed{t = \dfrac{1}{f}}$, find "t" when f = 4 .

$$t = \frac{1}{f} = \frac{1}{4} = .25$$

The evaluation above can be done on a calculator either by dividing 1 by 4 or by entering 4 and pressing the reciprocal key $\boxed{1/x}$. Use either method for these.

In $\boxed{t = \dfrac{1}{f}}$:   a) when f = 10 , t = _____

b) when f = 25 , t = _____

---

11.   The evaluation below involves a multiplication (in the numerator) and then a division.

In $\boxed{b = \dfrac{2A}{h}}$, find "b" when A = 20 and h = 5 .

$$b = \frac{2A}{h} = \frac{2(20)}{5} = \frac{40}{5} = 8$$

To perform the same evaluation in one process on a calculator, follow these steps.

Enter	Press	Display
2	$\boxed{\times}$	2.
20	$\boxed{\div}$	40.
5	$\boxed{=}$	8.

Use the same process on a calculator for this evaluation.

In $\boxed{v = \dfrac{Ftg}{m}}$, when F = 20 , t = 15 , g = 32 , and m = 25 , v = _____

---

12.   To perform evaluations with formulas like $A = \frac{1}{2}bh$ or $A = \frac{1}{3}Bh$, we can rewrite them to get a division on the right side. That is:

Instead of $A = \frac{1}{2}bh$, we can use $A = \dfrac{bh}{2}$.

Instead of $A = \frac{1}{3}Bh$, we can use $A = \dfrac{Bh}{3}$.

a) In $\boxed{A = \dfrac{1}{2}bh}$, when b = 6.4 and h = 8.5 , A = _____

b) In $\boxed{A = \dfrac{1}{3}Bh}$, when B = 30 and h = 6 , A = _____

---

a) t = 0.1

b) t = 0.04

v = 384

13. The evaluation below involves a division and an addition.

In $\boxed{L = \dfrac{M}{P} + X}$, find L when $M = 8$, $P = 2$, and $X = 10$.

$$L = \frac{M}{P} + X = \frac{8}{2} + 10 = 4 + 10 = 14$$

To perform the same evaluation in one process on a calculator, follow these steps.

Enter	Press	Display
8	$\div$	8.
2	$+$	4.
10	$=$	14.

Use the same process on a calculator for this evaluation.

In $\boxed{F = \dfrac{9C}{5} + 32°}$, when $C = 55°$, $F = \underline{\hspace{2cm}}$

a) A = 27.2

b) A = 60

14. The evaluation below involves a subtraction and a division.

In $\boxed{X = L - \dfrac{M}{P}}$, find X when $L = 20$, $M = 10$, and $P = 2$.

$$X = L - \frac{M}{P} = 20 - \frac{10}{2} = 20 - 5 = 15$$

To perform the same evaluation in one process on a calculator, follow these steps.

Enter	Press	Display
20	$-$	20.
10	$\div$	10.
2	$=$	15.

Use the same process on a calculator for this evaluation.

In $\boxed{r = R - \dfrac{2RF}{w}}$, when $R = 100$, $F = 25$, and $w = 80$, $r = \underline{\hspace{2cm}}$

F = 131°

15. Use a calculator for these. Round to tenths.

a) In $\boxed{V = \dfrac{4st}{a}}$, when $s = 50.5$, $t = 10$, and $a = 32.2$, $V = \underline{\hspace{2cm}}$

b) In $\boxed{A = \dfrac{1}{3}Bh}$, when $B = 18.7$ and $h = 5.5$, $A = \underline{\hspace{2cm}}$

c) In $\boxed{X = L - \dfrac{M}{P}}$, when $L = 75.9$, $M = 48.7$, and $P = 12.3$,

$$X = \underline{\hspace{2cm}}$$

r = 37.5

16.   The evaluation below involves a subtraction and a division.

In $\boxed{P = \dfrac{H - E}{V}}$, find P when $H = 30$, $E = 12$, and $V = 3$.

$$P = \frac{H - E}{V} = \frac{30 - 12}{3} = \frac{18}{3} = 6$$

If we simply perform the subtraction and then divide immediately on a calculator, we get 26 instead of 6 because the calculator performs the evaluation below. Try it.

$$P = H - \frac{E}{V} = 30 - \frac{12}{3} = 30 - 4 = 26$$

Therefore, to perform the top evaluation correctly on a calculator, we must press the "equals" key $\boxed{=}$ to complete the subtraction <u>before</u> performing the division. The correct steps are:

Enter	Press	Display
30	$\boxed{-}$	30.
12	$\boxed{=}$ $\boxed{\div}$	18.
3	$\boxed{=}$	6.

Using the same steps, do this evaluation on a calculator. Be sure to press $\boxed{=}$ to complete the subtraction before dividing.

In $\boxed{I = \dfrac{E - e}{r}}$, when $E = 16.7$, $e = 12.5$, and $r = 1.2$, $I = $ _____

a) $V = 62.7$

b) $A = 34.3$

c) $X = 71.9$

---

17.   The evaluation below involves an addition and a division.

In $\boxed{R = \dfrac{p + q}{2}}$, find R when $p = 8$ and $q = 10$.

$$R = \frac{p + q}{2} = \frac{8 + 10}{2} = \frac{18}{2} = 9$$

If we simply perform the addition and then divide immediately on a calculator, we get 13 instead of 9 because the calculator performs the evaluation below. Try it.

$$R = p + \frac{q}{2} = 8 + \frac{10}{2} = 8 + 5 = 13$$

Therefore, to perform the top evaluation correctly on a calculator, we must press $\boxed{=}$ to complete the addition before dividing. The steps are:

Enter	Press	Display
8	$\boxed{+}$	8.
10	$\boxed{=}$ $\boxed{\div}$	18.
2	$\boxed{=}$	9.

Continued on following page.

$I = 3.5$

17.   Continued

Using the same steps, do this evaluation on a calculator.

In $\boxed{a = \dfrac{b + c}{d}}$, when b = 154, c = 26, and d = 45,   a = _____

---

18.   Do these on a calculator. Be sure to press $\boxed{=}$ before dividing. Round to two decimal places.

a) In $\boxed{r = \dfrac{E - e}{I}}$,   when E = 8.94, e = 1.49, and I = 2.06,   r = _____

b) In $\boxed{V = \dfrac{S + T}{M}}$, when S = 3.66, T = 7.59, and M = 4.18,   V = _____

a = 4

---

19.   The evaluation below involves a multiplication, a subtraction, and a division.

In $\boxed{X = \dfrac{PL - M}{P}}$, find X when P = 2 , L = 10 , and M = 6 .

$$X = \frac{PL - M}{P} = \frac{2(10) - 6}{2} = \frac{20 - 6}{2} = \frac{14}{2} = 7$$

a) r = 3.62

b) V = 2.69

To perform the same evaluation in one process on a calculator, we use the steps below. Notice that we pressed $\boxed{=}$ to complete the subtraction before dividing.

Enter	Press	Display
2	$\boxed{x}$	2.
10	$\boxed{-}$	20.
6	$\boxed{=}$ $\boxed{\div}$	14.
2	$\boxed{=}$	7.

Using the same steps, do this evaluation on a calculator.

In $\boxed{C = \dfrac{5F - 160°}{9}}$, when F = 68° ,   C = _____

---

20.   Do these on a calculator. Press $\boxed{=}$ before dividing. Round to tenths.

a) In $\boxed{H = \dfrac{D - 4C}{F}}$, when D = 996, C = 112, and F = 13,   H = _____

b) In $\boxed{L = \dfrac{M + PX}{P}}$, when M = 88.7, P = 9.5, and X = 14.1,

L = _____

C = 20°

a) H = 42.2

b) L = 23.4

21.   To perform the evaluation below, we simply evaluate the fraction and then add the value of "h".

In $\boxed{T = \dfrac{a - b}{c} + h}$, find T when a = 10 , b = 4 , c = 2, and h = 5 .

$$T = \frac{10 - 4}{2} + 5 = \frac{6}{2} + 5 = 3 + 5 = 8$$

To perform the same evaluation on a calculator, we use the same general method. The steps are shown below. Notice that we do not have to press $\boxed{=}$ before adding 5.

Enter	Press	Display
10	$\boxed{-}$	10.
4	$\boxed{=}$ $\boxed{\div}$	6.
2	$\boxed{+}$	3.
5	$\boxed{=}$	8.

Use the same method for this evaluation. Round to tenths.

In $\boxed{M = \dfrac{R + S}{T} + V}$, when R = 47.5 , S = 16.9 , T = 4.9 , and

V = 20.3 , M = _____

M = 33.4

4-3   FORMULAS REQUIRING MORE THAN ONE DIVISION

When some evaluations are done on a calculator, more than one division is required. We will discuss evaluations of that type in this section.

22.   To perform the evaluation below, we can perform the multiplication in the denominator and then divide.

In $\boxed{a = \dfrac{b}{cd}}$, find "a" when b = 40 , c = 2 , and d = 5 .

$$a = \frac{b}{cd} = \frac{40}{(2)(5)} = \frac{40}{10} = 4$$

However, we can perform the same evaluation by dividing 40 by 2 and then dividing 20 by 5.

$$a = \frac{b}{cd} = \frac{\overset{20}{\cancel{40}}}{\underset{1}{(\cancel{2})(5)}} = \frac{20}{5} = 4$$

Continued on following page.

22.   Continued

The second method is used to perform the evaluation on a calculator. That is, we perform $40 \div 2 \div 5$ which requires two divisions. The steps are:

Enter	Press	Display
40	$\div$	40.
2	$\div$	20.
5	$=$	4.

Using the same method, do this one on a calculator.

In $\boxed{H = \dfrac{B}{VT}}$, when $B = 200$, $V = 4$, and $T = 10$, $H = $ _____

---

23.   Do these on a calculator. Round to two decimal places.

a) In $\boxed{v = \dfrac{s}{mp}}$, when $s = 17.8$, $m = 2.41$, and $p = 1.49$, $v = $ _____

b) In $\boxed{B = \dfrac{C}{DF}}$, when $C = 2.49$, $D = 1.33$, and $F = 0.87$, $B = $ _____

$H = 5$

---

24.   To perform the evaluation below, we can perform the multiplications in both terms and then divide.

In $\boxed{k = \dfrac{cd}{pq}}$, find "k" when $c = 8$, $d = 6$, $p = 4$, and $q = 3$.

$$k = \frac{cd}{pq} = \frac{(8)(6)}{(4)(3)} = \frac{48}{12} = 4$$

However, we can do the same evaluation by performing the multiplication in the numerator to get 48, dividing 48 by 4, and then dividing 12 by 3.

$$k = \frac{cd}{pq} = \frac{(8)(6)}{(4)(3)} = \frac{\overset{12}{\cancel{48}}}{\underset{1}{(\cancel{4})(3)}} = \frac{12}{3} = 4$$

The second method is used to perform the evaluation on a calculator. That is, we perform $(8)(6) \div 4 \div 3$ which requires two divisions. The steps are:

Enter	Press	Display
8	x	8.
6	$\div$	48.
4	$\div$	12.
3	$=$	4.

Continued on following page.

a) $v = 4.96$

b) $B = 2.15$

24.   Continued

Using the same steps, do this one on a calculator.

In $\boxed{e = \dfrac{FL}{AY}}$, when F = 50, L = 4, A = 2, and Y = 5, e = _____

---

25.   When a formula contains many letters (or variables), it is helpful to write the formula with the numbers substituted before doing the calculator operations. Do the evaluations below. Round to tenths.

a) In $\boxed{F = \dfrac{mv}{gt}}$, when m = 60.9, v = 87.5, g = 32.2, and t = 10.8,

$$F = \frac{(60.9)(87.5)}{(32.2)(10.8)} = \underline{\hspace{2cm}}$$

b) In $\boxed{b = \dfrac{1.5dh}{mq}}$, when d = 317, h = 549, m = 240, and q = 105,

$$b = \frac{(1.5)(317)(549)}{(240)(105)} = \underline{\hspace{2cm}}$$

---

26.   To perform the evaluation below, we can simplify the numerator and then divide it by 2, 3, and 5.

In $\boxed{V = \dfrac{PQ}{RST}}$, find V when P = 6, Q = 50, R = 2, S = 3, and T = 5 .

$$V = \frac{(6)(50)}{(2)(3)(5)} = \frac{\overset{150}{\cancel{300}}}{\cancel{(2)}(3)(5)} = \frac{\overset{50}{\cancel{150}}}{\cancel{(3)}(5)} = \frac{50}{5} = 10$$

Therefore, to perform it on a calculator, we do (6)(50) ÷ 2 ÷ 3 ÷ 5 which requires three divisions. The steps are:

Enter	Press	Display
6	$\boxed{\text{x}}$	6.
50	$\boxed{\div}$	300.
2	$\boxed{\div}$	150.
3	$\boxed{\div}$	50.
5	$\boxed{=}$	10.

Using the same method, do this one on a calculator.

In $\boxed{H = \dfrac{r}{xyz}}$, when r = 1,600, x = 40, y = 2, and z = 5, H = _____

---

e = 20

a) F = 15.3

b) b = 10.4

H = 4

27.   Do these on a calculator. Round to three decimal places.

   a) In $t = \dfrac{LM}{3.14SV}$,   when L = 5.07 ,   M = 2.99 ,   S = 4.75 ,   and

V = 3.24 ,   t = _____

   b) In $F = \dfrac{33,000H}{abT}$,   when H = 0.025 ,   a = 59.6 ,   b = 7.14 ,   and

T = 2.63 ,   F = _____

a) t = 0.314

b) F = 0.737

28.   To do the following evaluation on a calculator, two methods are described below.

   In $t = \dfrac{1}{av}$,   find "t" when a = 4 and v = 20 .

   1) We can enter "1" and then divide by both 4 and 20.
       The steps are:

Enter	Press	Display
1	÷	1.
4	÷	0.25
20	=	0.0125

   2) We can multiply 4 and 20 and then press the reciprocal key
       $\boxed{1/x}$ after pressing $\boxed{=}$. The steps are:

Enter	Press	Display
4	x	4.
20	= $\quad$ 1/x	0.0125

Using either method, do this evaluation.

   In $m = \dfrac{1}{2cd}$,   when c = 4 and d = 5 ,   m = _____

m = 0.025

29. To perform the evaluation below on a calculator, we simply evaluate the fraction and then add the value of x without pressing $\boxed{=}$. The steps are shown.

In $\boxed{L = \dfrac{2M}{WX} + X}$, find L when $M = 24$, $W = 4$, and $X = 3$.

Enter	Press	Display
2	$\boxed{x}$	2.
24	$\boxed{\div}$	48.
4	$\boxed{\div}$	12.
3	$\boxed{+}$	4.
3	$\boxed{=}$	7.

Using the same method, do this one on a calculator.

In $\boxed{t = \dfrac{ab}{cd} - c}$, when $a = 16$, $b = 10$, $c = 2$ and $d = 5$,

$t = \underline{\hspace{2cm}}$

30. Do these on a calculator. Round to tenths.

a) In $\boxed{V = \dfrac{S}{BT} - M}$, when $S = 187$, $B = 5.4$, $T = 2.3$, and $M = 4.9$, $V = \underline{\hspace{2cm}}$

b) In $\boxed{H = \dfrac{CD}{ABE} + B}$, when $C = 17.9$, $D = 48.3$, $A = 5.6$, $B = 10.4$, and $E = 2.7$, $H = \underline{\hspace{2cm}}$

31. To perform the evaluation below on a calculator, we enter the value for D, press $\boxed{-}$, and then evaluate the fraction in the usual way. The steps are shown.

In $\boxed{F = D - \dfrac{H}{ms}}$, find F when $D = 100$, $H = 60$, $m = 2$, and $s = 3$.

Enter	Press	Display
100	$\boxed{-}$	100.
60	$\boxed{\div}$	60.
2	$\boxed{\div}$	30.
3	$\boxed{=}$	90.

Continued on following page.

t = 14

a) V = 10.2

b) H = 15.9

31. Continued

Using the same method, do this one on a calculator.

In $\boxed{v = h + \dfrac{3I}{ab}}$, when $h = 75$, $I = 16$, $a = 4$, and $b = 2$,

$v = \underline{\hspace{2cm}}$

---

32. Do these on a calculator. Round to tenths.

a) In $\boxed{X = Q - \dfrac{cd}{1.5t}}$, when $Q = 79.8$, $c = 14.4$, $d = 22.9$,

and $t = 68.7$, $X = \underline{\hspace{2cm}}$

b) In $\boxed{k = p + \dfrac{q}{4st}}$, when $p = 41.3$, $q = 81.6$, $s = 2.6$, and

$t = 1.9$, $k = \underline{\hspace{2cm}}$

v = 81

---

33. To perform the evaluation below, we evaluate the numerator, press $\boxed{=}$ before dividing, and then divide by the values for C and D. The steps are shown.

In $\boxed{K = \dfrac{CD - FG}{CD}}$, find K when $C = 4$, $D = 25$, $F = 6$ and $G = 7$.

Enter	Press	Display
4	$\boxed{\text{x}}$	4.
25	$\boxed{-}$	100.
6	$\boxed{\text{x}}$	6.
7	$\boxed{=}$ $\boxed{\div}$	58.
4	$\boxed{\div}$	14.5
25	$\boxed{=}$	0.58

Using the same method, do this one.

In $\boxed{V = \dfrac{R + ab}{at}}$, when $R = 20$, $a = 2$, $b = 15$, and $t = 5$,

$V = \underline{\hspace{2cm}}$

a) X = 76.6

b) k = 45.4

V = 5

34.   Do these on a calculator. Round to hundredths.

a) In $\boxed{v = \dfrac{100 - st}{at}}$, when s = 2.96 , t = 7.24 , and a = 1.33 ,

$$v = \underline{\hspace{2cm}}$$

b) In $\boxed{H = \dfrac{CK + E}{CKD}}$, when C = 12.7 , K = 21.9 , E = 40.4 ,

and D = 6.4 , H = \underline{\hspace{2cm}}

a) v = 8.16   b) H = 0.18

SELF-TEST 10 (pages 114-128)

1. Find H. Round to tenths.

$\boxed{H = E + PV}$ E = 6.75

P = 9.48

V = 7.39

2. Find "b".

$\boxed{b = \dfrac{2A}{h}}$ A = 150

h = 12.5

3. Find F when C = 27 .

$\boxed{F = \dfrac{9C}{5} + 32}$

4. Find I. Round to hundredths.

$\boxed{I = \dfrac{E - e}{r}}$ E = 32.6

e = 17.9

r = 8.34

5. Find L. Round to one decimal place.

$\boxed{L = \dfrac{M + PX}{P}}$ M = 780

P = 42

X = 54

6. Find "w". Round to three decimal places.

$\boxed{w = \dfrac{1}{4ad}}$ a = 2.18

d = 1.37

7. Find G. Round to hundredths.

$\boxed{G = \dfrac{d + w}{bw}}$ d = 236

w = 505

b = 1.12

8. Find F. Round to two decimal places.

$\boxed{F = D - \dfrac{H}{ms}}$ D = 21.48

H = 159

m = 4.45

s = 2.63

ANSWERS: 1. H = 76.8   3. F = 80.6   5. L = 72.6   7. G = 1.31

2. b = 24   4. I = 1.76   6. w = .084   8. F = 7.89

4-4   FORMULAS INVOLVING SQUARES

In this section, we will show how a calculator can be used to evaluate formulas containing squares.

35.   To perform the evaluation below, we simply square the value of "s".

$$\boxed{A = s^2} \ , \ \text{when} \ s = 7 \ , \ A = (7)^2 = 49$$

To perform the above evaluation on a calculator, we simply enter 7 and press $\boxed{x^2}$.

Enter	Press	Display
7	$\boxed{x^2}$	49.

In $\boxed{A = s^2}$: a) When s = 1.5 , A = _____

b) When s = 10.4 , A = _____

36.   To perform the evaluation below, we square the value of I before multiplying.

In $\boxed{P = I^2R}$, find P when I = 3 and R = 10 .

$$P = I^2R = (3)^2(10) = (9)(10) = 90$$

To perform the same evaluation on a calculator, we simply press $\boxed{x^2}$ immediately after entering the value for I. The steps are:

Enter	Press	Display
3	$\boxed{x^2}$ $\boxed{x}$	9.
10	$\boxed{=}$	90.

Do this one on a calculator. (Note: Press $\boxed{x^2}$ immediately after entering the value for "w". Then press $\boxed{=}$ to get the final answer.)

In $\boxed{d = 4w^2}$, find "d" when w = 5 .   d = _____

37.   To perform the evaluation below, we square the values for x and y before multiplying.

In $\boxed{m = x^2y^2}$, find "m" when x = 5 and y = 3 .

$$m = x^2y^2 = (5^2)(3^2) = (25)(9) = 225$$

To perform the same evaluation on a calculator, we press $\boxed{x^2}$ immediately after entering the values for x and y. The steps are:

Enter	Press	Display
5	$\boxed{x^2}$ $\boxed{x}$	25.
3	$\boxed{x^2}$ $\boxed{=}$	225.

Continued on following page.

---

a) A = 2.25

b) A = 108.16

d = 100

37.   Continued

Using the same method, do this one on a calculator.

In $\boxed{a = b^2 c^2}$, when $b = 9$ and $c = 11$, $a =$ _____

---

38.   Do these on a calculator. Round to two decimal places.

a) In $\boxed{A = 0.7854 d^2}$, when $d = 2.4$, $A =$ _____

b) In $\boxed{v = h^2 q^2}$, when $h = 1.26$ and $q = 2.09$, $v =$ _____

$a = 9{,}801$

---

39.   When evaluating a formula containing a squared letter on a calculator, the same steps are used except that $\boxed{x^2}$ is pressed immediately after entering each value that is to be squared. Another example is given below.

In $\boxed{G = A - B^2 H^2}$, find G when $A = 100$, $B = 2$, and $H = 3$.

Enter	Press	Display
100	$\boxed{-}$	100.
2	$\boxed{x^2}$ $\boxed{\text{x}}$	4.
3	$\boxed{x^2}$ $\boxed{=}$	64.

Do these on a calculator. Round to tenths.

a) In $\boxed{t = h - b^2 c}$, when $h = 97.6$, $b = 4.8$, and $c = 1.6$,

t = _____

b) In $\boxed{b = 4c^2 F^2 - r}$, when $c = 0.15$, $F = 17.1$, and $r = 11.2$,

b = _____

a) $A = 4.52$

b) $v = 6.93$

---

40.   The calculator steps for the evaluation below are shown. Notice that we pressed $\boxed{x^2}$ immediately after entering the value for "t".

In $\boxed{s = \dfrac{at^2}{2}}$, find "s" when $a = 40$ and $t = 3$.

Enter	Press	Display
40	$\boxed{\text{x}}$	40.
3	$\boxed{x^2}$ $\boxed{\div}$	360.
2	$\boxed{=}$	180.

Do these on a calculator. Round to tenths.

a) In $\boxed{H = \dfrac{D^2 N}{2.5}}$, when $D = 3.58$ and $N = 5$, $H =$ _____

b) In $\boxed{P = \dfrac{E^2}{R}}$, when $E = 61.8$ and $R = 40$, $P =$ _____

a) $t = 60.7$

b) $b = 15.1$

41. The calculator steps for the evaluation below are shown. Notice again that
     we pressed $\boxed{x^2}$ immediately after entering the value for "d".

     In $\boxed{P = \dfrac{pL}{d^2}}$, find P when p = 10 , L = 16 , and d = 2 .

Enter	Press	Display
10	$\boxed{x}$	10.
16	$\boxed{\div}$	160.
2	$\boxed{x^2}$ $\boxed{=}$	40.

     Do these on a calculator. Round to hundredths.

     a) In $\boxed{R = \dfrac{P}{I^2}}$, when P = 52.7 and I = 6.35 , R = _____

     b) In $\boxed{t = \dfrac{a^2 T}{v^2}}$, when a = 21.9 , T = 4.5 , and v = 36.8 ,

     t = _____

     a) H = 25.6

     b) P = 95.5

42. The calculator steps for the evaluation below are shown. Notice that we
     simply subtracted the value for "d" after evaluating the fraction.

     In $\boxed{m = \dfrac{ab}{c^2} - d}$, find "m" when a = 6, b = 10, c = 2, and d = 8 .

Enter	Press	Display
6	$\boxed{x}$	6.
10	$\boxed{\div}$	60.
2	$\boxed{x^2}$ $\boxed{-}$	15.
8	$\boxed{=}$	7.

     Do this one on a calculator. Round to tenths.

     In $\boxed{P = \dfrac{mv^2}{r} - mg}$, when m = 6.24 , v = 30.9 , r = 12.5 , and

     g = 32.2 , P = _____

     a) R = 1.31

     b) t = 1.59

43. The calculator steps for the evaluations below are shown. Notice that we
     divided by both F^2 and "t".

     In $\boxed{h = \dfrac{W^2}{F^2 t}}$, find "h" when W = 12 , F = 2 , and t = 9 .

Enter	Press	Display
12	$\boxed{x^2}$ $\boxed{\div}$	144.
2	$\boxed{x^2}$ $\boxed{\div}$	36.
9	$\boxed{=}$	4.

     Continued on following page.

     P = 275.7

43.   Continued

Do these on a calculator. Round to tenths.

a) In $\boxed{s = \dfrac{v^2}{2g}}$, when $v = 49.7$ and $g = 32$, $s =$ _____

b) In $\boxed{P = \dfrac{rw}{2t^2}}$, when $r = 7.75$, $w = 89.8$, and $t = 4$, $P =$ _____

---

44.   To do the evaluation below on a calculator, two methods are described.

In $\boxed{m = \dfrac{1}{b^2 t^2}}$, find "m" when $b = 2$ and $t = 10$.

1) We can enter "1" and then divide by both 2^2 and 10^2.

2) We can multiply 2^2 and 10^2 and then press the reciprocal key $\boxed{1/x}$.

Using either method, complete the evaluation. $m =$ _____

---

a) $s = 38.6$

b) $P = 21.7$

---

45.   The calculator steps for the evaluation below are shown. Notice that we pressed $\boxed{=}$ to complete the subtraction in the numerator <u>before</u> dividing.

In $\boxed{r = \dfrac{s^2 d^2 - h}{2}}$, find "r" when $s = 3$, $d = 8$, and $h = 54$.

Enter	Press		Display
3	$\boxed{x^2}$	$\boxed{x}$	9.
8	$\boxed{x^2}$	$\boxed{-}$	576.
54	$\boxed{=}$	$\boxed{\div}$	522.
2	$\boxed{=}$		261.

Do these on a calculator. Round to hundredths.

a) In $\boxed{d = \dfrac{ab^2 - 4s^2}{b^2}}$, when $a = 7$, $b = 14$, and $s = 9$, $d =$ _____

b) In $\boxed{m = \dfrac{p^2 + q^2}{2v}}$, when $p = 5.98$, $q = 9.03$, and $v = 6$,

$m =$ _____

$m = 0.0025$

---

a) $d = 5.35$

b) $m = 9.78$

46. To perform the evaluation below, we simplify the expression in the parentheses <u>before</u> squaring.

In $\boxed{a = \left(\dfrac{b}{c}\right)^2}$, find "a" when $b = 20$ and $c = 4$.

$$a = \left(\dfrac{b}{c}\right)^2 = \left(\dfrac{20}{4}\right)^2 = (5)^2 = 25$$

To do the same evaluation on a calculator, we must press $\boxed{=}$ to complete the division <u>before</u> squaring. The steps are:

Enter	Press	Display
20	$\boxed{\div}$	20.
4	$\boxed{=}\ \boxed{x^2}$	25.

Do these on a calculator. Be sure to press $\boxed{=}$ to complete the operation within parentheses <u>before</u> squaring.

a) In $\boxed{b = (a - c)^2}$, when $a = 10$ and $c = 3$, b = _____

b) In $\boxed{x = \left(\dfrac{b - t}{c}\right)^2}$, when $b = 270$, $t = 150$, and $c = 10$, x = _____

---

a) b = 49   b) x = 144

4-5   FORMULAS INVOLVING SQUARE ROOTS

In this section, we will show how a calculator can be used to evaluate formulas containing square-root radicals.

---

47. To perform the evaluation below, we simply find the square root of the value of "A".

In $\boxed{s = \sqrt{A}}$, when $A = 25$, $s = 5$.

To perform the same evaluation on a calculator, we simply enter 25 and press $\boxed{\sqrt{x}}$.

Enter	Press	Display
25	$\boxed{\sqrt{x}}$	5.

In $\boxed{s = \sqrt{A}}$: a) when $A = 1.96$, s = _____

b) when $A = 462.25$, s = _____

---

a) s = 1.4

b) s = 21.5

48.   To perform the evaluation below, we multiply the values of P and R before finding the square root.

In $\boxed{E = \sqrt{PR}}$, find E when P = 4 and R = 9 .

$$E = \sqrt{PR} = \sqrt{(4)(9)} = \sqrt{36} = 6$$

To perform the same evaluation on a calculator, we press $\boxed{=}$ to complete the multiplication before pressing $\boxed{\sqrt{x}}$. The steps are:

Enter	Press	Display
4	$\boxed{x}$	4.
9	$\boxed{=}$ $\boxed{\sqrt{x}}$	6.

Do these on a calculator. Be sure to press $\boxed{=}$ before pressing $\boxed{\sqrt{x}}$.

a) In $\boxed{V = \sqrt{64h}}$, when h = 9 , V = _____

b) In $\boxed{v = \sqrt{2as}}$, when a = 50 and s = 121 , v = _____

49.   Whenever an evaluation involves a square-root radical, the operation within the radical is performed before finding the square root. For example:

In $\boxed{c = \sqrt{a^2 + b^2}}$, find "c" when a = 3 and b = 4 .

$$c = \sqrt{a^2 + b^2} = \sqrt{3^2 + 4^2} = \sqrt{9 + 16} = \sqrt{25} = 5$$

When performing the same evaluation on a calculator, we press $\boxed{=}$ to complete the operation within the radical <u>before</u> pressing $\boxed{\sqrt{x}}$. The steps are:

Enter	Press	Display
3	$\boxed{x^2}$ $\boxed{+}$	9.
4	$\boxed{x^2}$ $\boxed{=}$ $\boxed{\sqrt{x}}$	5.

Do these on a calculator. Round to two decimal places.

a) In $\boxed{v = \sqrt{w - e}}$, when w = 79.6 and e = 17.5 , v = _____

b) In $\boxed{d = \sqrt{b^2 - 4ac}}$, when b = 11, a = 3, and c = 4, d = _____

c) In $\boxed{P = \sqrt{Q^2 + R^2}}$, when Q = 5.66 and R = 7.59 , P = _____

a) V = 24

b) v = 110

a) v = 7.88

b) d = 8.54

c) P = 9.47

50. We pressed $\boxed{=}$ before pressing $\boxed{\sqrt{x}}$ to complete the evaluation below.

In $\boxed{t = \sqrt{\dfrac{2s}{a}}}$, find "t" when $s = 72$ and $a = 4$.

Enter	Press	Display
2	$\boxed{\times}$	2.
72	$\boxed{\div}$	144.
4	$\boxed{=}\ \boxed{\sqrt{x}}$	6.

Do these on a calculator. Round to hundredths.

a) In $\boxed{d = \sqrt{\dfrac{A}{0.7854}}}$, when $A = 60$, $d = $ _____

b) In $\boxed{d = \sqrt{\dfrac{pL}{R}}}$, when $p = 14.5$, $L = 27.2$, and $R = 18.8$,

$d = $ _____

c) In $\boxed{t = \sqrt{\dfrac{rw}{2P}}}$, when $r = 313$, $w = 46.8$, and $P = 898$,

$t = $ _____

51. Notice again that we pressed $\boxed{=}$ before pressing $\boxed{\sqrt{x}}$ to complete the evaluation below.

In $\boxed{C = \sqrt{1 - \dfrac{H}{h}}}$, find C when $H = 32$ and $h = 50$.

Enter	Press	Display
1	$\boxed{-}$	1.
32	$\boxed{\div}$	32.
50	$\boxed{=}\ \boxed{\sqrt{x}}$	0.6

Do these on a calculator. Round to two decimal places.

a) In $\boxed{d = \sqrt{\dfrac{m}{t} + 1}}$, when $m = 82.7$ and $t = 12.5$, $d = $ _____

b) In $\boxed{V = \sqrt{100 - \dfrac{b^2}{F}}}$, when $b = 37.5$ and $F = 68.5$, $V = $ _____

c) In $\boxed{t = \sqrt{\dfrac{a^2}{r^2} - 1}}$, when $a = 6.48$ and $r = 1.59$, $t = $ _____

a) d = 8.74

b) d = 4.58

c) t = 2.86

a) d = 2.76

b) V = 8.91

c) t = 3.95

52.   To do the evaluation below on a calculator, we pressed $\boxed{=}$ to complete the subtraction in the numerator before dividing and then we pressed $\boxed{=}$ before pressing $\boxed{\sqrt{x}}$.

In $\boxed{a = \sqrt{\dfrac{b-c}{c}}}$, find "a" when $b = 100$ and $c = 10$.

Enter	Press	Display
100	$\boxed{-}$	100.
10	$\boxed{=}$ $\boxed{\div}$	90.
10	$\boxed{=}$ $\boxed{\sqrt{x}}$	3.

Do these on a calculator. Round to three decimal places.

a) In $\boxed{v = \sqrt{\dfrac{F-D}{F}}}$, when $F = 87.1$ and $D = 14.8$, $v =$ _____

b) In $\boxed{t = \sqrt{\dfrac{1.5s^2 + V}{1.5}}}$, when $s = 0.65$ and $V = 0.49$, $t =$ _____

---

53.   To perform the evaluation below, we find $\sqrt{100}$ before multiplying.

In $\boxed{r = a\sqrt{w}}$, find "r" when $a = 7$ and $w = 100$.

$$r = a\sqrt{w} = 7\sqrt{100} = 7(10) = 70$$

To perform the same evaluation on a calculator, we simply press $\boxed{\sqrt{x}}$ after entering 100 . The steps are:

Enter	Press	Display
7	$\boxed{x}$	7.
100	$\boxed{\sqrt{x}}$ $\boxed{=}$	70.

Do these on a calculator by simply pressing $\boxed{\sqrt{x}}$ after entering the value for the radicand.

a) In $\boxed{R = P + \sqrt{Q}}$, when $P = 17$ and $Q = 81$, $R =$ _____

b) In $\boxed{b = \dfrac{\sqrt{d}}{c}}$, when $d = 64$ and $c = 2$, $b =$ _____

a) $v = 0.911$

b) $t = 0.866$

---

54.   Whenever a radical containing a single letter is part of an operation, we perform it in the usual way. However, we press $\boxed{\sqrt{x}}$ after entering the value of the radicand. For example:

In $\boxed{b = \dfrac{a\sqrt{x}}{c}}$, find "b" when $a = 8$, $x = 144$, and $c = 3$.

Enter	Press	Display
8	$\boxed{x}$	8.
144	$\boxed{\sqrt{x}}$ $\boxed{\div}$	96.
3	$\boxed{=}$	32.

a) $R = 26$

b) $b = 4$

Continued on following page.

54.   Continued

Do these on a calculator. Round to tenths.

a) In $\boxed{p = m + r\sqrt{t}}$, when m = 40.8, r = 11.6, and t = 19.9,

$$p = \underline{\hspace{1.5in}}$$

b) In $\boxed{F = \dfrac{D - S}{\sqrt{G}}}$, when D = 477, S = 199, and G = 30.3,

$$F = \underline{\hspace{1.5in}}$$

---

55.   To do the evaluation below on a calculator, two methods are described.

In $\boxed{t = \dfrac{1}{b\sqrt{m}}}$, find "t" when b = 5 and m = 100 .

1) We can enter "1" and then divide by both 5 and $\sqrt{100}$.

2) We can multiply 5 and $\sqrt{100}$, press $\boxed{=}$, and then press the reciprocal key $\boxed{1/x}$.

Using either method, complete the evaluation. $t = \underline{\hspace{1.2in}}$

a) p = 92.5

b) F = 50.5

---

56.   To do the evaluation below on a calculator, we simply evaluate the radical in the numerator and then divide by 4. The steps are shown:

In $\boxed{s = \dfrac{\sqrt{h + 2r}}{d}}$, find "s" when h = 50 , r = 7 , and d = 4 .

Enter	Press	Display
50	$\boxed{+}$	50.
2	$\boxed{\times}$	2.
7	$\boxed{=}$ $\boxed{\sqrt{x}}$ $\boxed{\div}$	8.
4	$\boxed{=}$	2.

Use a calculator for these. Round to hundredths.

a) In $\boxed{a = \dfrac{\sqrt{bc}}{d}}$, when b = 9.17 , c = 6.88 , and d = 1.29 ,

$$a = \underline{\hspace{1.2in}}$$

b) In $\boxed{M = \dfrac{\sqrt{P^2 - Q^2}}{3R}}$, when P = 97.8 , Q = 31.4 , and R = 7.66 ,

$$M = \underline{\hspace{1.2in}}$$

t = 0.02

a) a = 6.16

b) M = 4.03

57. When a radical containing an operation is a factor in a multiplication, we must evaluate the radical first on a calculator. That is:

Instead of $\boxed{a = b\sqrt{cd}}$, we must evaluate $\boxed{a = (\sqrt{cd})(b)}$

Instead of $\boxed{T = \dfrac{m}{b}\sqrt{\dfrac{d}{c}}}$, we must evaluate $\boxed{T = \left(\sqrt{\dfrac{d}{c}}\right)\left(\dfrac{m}{b}\right)}$

The calculator steps for the evaluation below are shown. Notice that we evaluated the radical first and then multiplied by 10.

In $\boxed{v = s\sqrt{\dfrac{p}{q}}}$, find "v" when $s = 10$, $p = 48$, and $q = 3$.

Enter	Press	Display
48	$\div$	48.
3	$=$ $\sqrt{x}$ x	4.
10	$=$	40.

Do these on a calculator by evaluating the radical first. Round to tenths.

a) In $\boxed{G = H\sqrt{ST}}$, when $H = 2.78$, $S = 5.19$, and $T = 8.77$,

$$G = \underline{\hspace{2cm}}$$

b) In $\boxed{w = 2g\sqrt{\dfrac{a}{r}}}$, when $g = 9.01$, $a = 68.9$, and $r = 12.6$,

$$w = \underline{\hspace{2cm}}$$

58. The calculator steps for the evaluation below are shown. Notice again that we evaluated the radical before multiplying by 24 and dividing by 3.

In $\boxed{T = \dfrac{m}{b}\sqrt{\dfrac{d}{c}}}$, find T when $m = 24$, $b = 3$, $d = 18$, and $c = 2$.

Enter	Press	Display
18	$\div$	18.
2	$=$ $\sqrt{x}$ x	3.
24	$\div$	72.
3	$=$	24.

Do this one on a calculator. Round to three decimal places.

In $\boxed{V = \dfrac{1.5}{H}\sqrt{\dfrac{P}{Q}}}$, when $H = 6.45$, $P = 7.03$, and $Q = 1.98$,

$$V = \underline{\hspace{2cm}}$$

a) $G = 18.8$

b) $w = 42.1$

V = 0.438

59. When a radical involving an operation is the denominator of a fraction, we evaluate the radical first and put that value in storage by pressing $\boxed{\text{STO}}$, enter the numerator, and then divide using $\boxed{\text{RCL}}$ to recall the value of the denominator. For example:

In $\boxed{K = \dfrac{L}{\sqrt{MT}}}$, find K when $L = 100$, $M = 2$, and $T = 8$.

Enter	Press	Display
2	$\boxed{\text{x}}$	2.
8	$\boxed{=}$ $\boxed{\sqrt{x}}$ $\boxed{\text{STO}}$	4.
100	$\boxed{\div}$ $\boxed{\text{RCL}}$ $\boxed{=}$	25.

Using the same general method, do these on a calculator. Round to hundredths.

a) In $\boxed{F = \dfrac{W}{\sqrt{ht}}}$, when $W = 87.6$, $h = 21.7$, and $t = 3.98$,

$$F = \underline{\hspace{2cm}}$$

b) In $\boxed{a = \dfrac{b}{\sqrt{\dfrac{c}{d}}}}$, when $b = 9.47$, $c = 48.6$, and $d = 17.9$,

$$a = \underline{\hspace{2cm}}$$

---

60. To do the evaluation below on a calculator, two methods are described.

In $\boxed{a = \dfrac{1}{\sqrt{p^2 + w^2}}}$, find "a" when $p = 9$ and $w = 12$.

1) We can evaluate the denominator, press $\boxed{\text{STO}}$, enter "1", and then press $\boxed{\div}$ $\boxed{\text{RCL}}$ to divide.

2) We can evaluate the denominator and then press the reciprocal key $\boxed{1/x}$.

Using either method, complete the evaluation. Round to thousandths.

$$a = \underline{\hspace{2cm}}$$

a) $F = 9.43$

b) $a = 5.75$

$a = 0.067$

4-6   FORMULAS INVOLVING CUBES AND CUBE ROOTS

In this section, we will define the cube and cube root of a number, show the calculator procedure for cubing a number or finding its cube root, and then do evaluations with formulas involving cubes and cube roots.

---

61.   To cube a number, we use the number as a factor <u>three</u> times. For example:

$$(2)^3 = (2)(2)(2) = 8 \qquad (50)^3 = (50)(50)(50) = 125,000$$

To cube a number on a calculator, we use the "power" key $\boxed{y^X}$. The steps for cubing each number above are shown below.

Enter	Press	Display
2	$\boxed{y^X}$	2.
3	$\boxed{=}$	8.

Enter	Press	Display
50	$\boxed{y^X}$	50.
3	$\boxed{=}$	125,000.

Using the $\boxed{y^X}$ key, complete each cubing below.

a) $(17)^3 = $ _____   b) $(115)^3 = $ _____

---

62.   Using the same method, complete these.

a) $(4.5)^3 = $ _____   b) $(0.12)^3 = $ _____

| a) 4,913 |
| b) 1,520,875 |

---

63.   Complete these. Round to five decimal places.

(a) $(0.149)^3 = $ _____   b) $(0.075)^3 = $ _____

| a) 91.125 |
| b) 0.001728 |

---

64.   When a very large or very small number is cubed, the answer is reported in scientific notation. Do these. Round the first factor to two decimal places.

a) $(8,590)^3 = $ _____   b) $(0.000625)^3 = $ _____

| a) 0.00331 |
| b) 0.00042 |

---

65.   To perform evaluations with the formula below, we simply cube the value of "s".

In $\boxed{V = s^3}$: a) find V when s = 8.5 .   V = _____

b) find V when s = 47 .   V = _____

| a) 6.34×10^{11} |
| b) 2.44×10^{-10} |

66. The cube root of a number is <u>one</u> <u>of</u> <u>three</u> <u>equal</u> <u>factors</u> whose product is the number. For example:

The cube root of 64 is 4, because $(4)(4)(4) = 64$

The cube root of $1,728$ is 12, because $(12)(12)(12) = 1,728$

The operation "find the cube root of" is expressed by the symbol $\sqrt[3]{}$
That is:

The cube root of $64 = \sqrt[3]{64} = 4$

The cube root of $1,728 = \sqrt[3]{1,728} = 12$

The procedure for finding the cube roots above on a calculator depends on the particular calculator used.

1) If the calculator has a "root" key $\boxed{\sqrt[x]{y}}$, we use that key.
The steps are shown below.

Enter	Press	Display
64	$\boxed{\sqrt[x]{y}}$	64.
3	$\boxed{=}$	4.

Enter	Press	Display
1,728	$\boxed{\sqrt[x]{y}}$	1728.
3	$\boxed{=}$	12.

2) If the calculator does not have a "root" key, we use the "inverse" key $\boxed{INV}$ together with the "power" key $\boxed{y^x}$. The steps are shown below.

Enter	Press	Display
64	$\boxed{INV}$ $\boxed{y^x}$	64.
3	$\boxed{=}$	4.

Enter	Press	Display
1,728	$\boxed{INV}$ $\boxed{y^x}$	1728.
3	$\boxed{=}$	12.

a) V = 614.125

b) V = 103,823

67. Using either $\boxed{\sqrt[x]{y}}$ or $\boxed{INV}$ $\boxed{y^x}$, complete these.

a) $\sqrt[3]{512}$ = _____   b) $\sqrt[3]{19,683}$ = _____

68. Do these. Round to tenths.

a) $\sqrt[3]{182,400}$ = _____   b) $\sqrt[3]{19,400}$ = _____

a) 8   b) 27

a) 56.7   b) 26.9

69. Do these. Round to hundredths.

    a) $\sqrt[3]{6.31}$ = _____

    b) $\sqrt[3]{47.5}$ = _____

---

70. Do these. Round to three decimal places.

    a) $\sqrt[3]{0.646}$ = _____

    b) $\sqrt[3]{0.0509}$ = _____

a) 1.85   b) 3.62

---

71. To perform evaluations with the formula below, we simply find the cube root of V.

    In $\boxed{s = \sqrt[3]{V}}$: a) find "s" when V = 2,744 .   s = _____

    b) find "s" when V = 27,000 .   s = _____

a) 0.864   b) 0.371

a) s = 14   b) s = 30

SELF-TEST 11 (pages 129-142)

1. Find A when d = 23.8 . Round to the nearest whole number.

   $\boxed{A = 0.7854d^2}$

2. Find R. Round to hundredths.

   $\boxed{R = \dfrac{pL}{a^2}}$ $\begin{aligned} p &= 0.87 \\ L &= 295 \\ a &= 7.34 \end{aligned}$

3. Find "m". Round to one decimal place.

   $\boxed{m = \dfrac{t^2 + w^2}{2v}}$ $\begin{aligned} t &= 15.4 \\ w &= 22.8 \\ v &= 6.13 \end{aligned}$

4. Find Z. Round to the nearest whole number.

   $\boxed{Z = \sqrt{R^2 + X^2}}$ $\begin{aligned} R &= 570 \\ X &= 810 \end{aligned}$

5. Find "v". Round to thousandths.

   $\boxed{v = \sqrt{\dfrac{F - D}{F}}}$ $\begin{aligned} F &= 3.96 \\ D &= 1.09 \end{aligned}$

6. Find "t". Round to two decimal places.

   $\boxed{t = \dfrac{h}{w}\sqrt{\dfrac{m}{p}}}$ $\begin{aligned} h &= 46.1 \\ w &= 17.7 \\ m &= 230 \\ p &= 350 \end{aligned}$

7. Find V when s = 86.8 . Round to thousands.

   $\boxed{V = s^3}$

8. Find "s" when V = 2,840 . Round to tenths.

   $\boxed{s = \sqrt[3]{V}}$

ANSWERS: 1. A = 445   3. m = 61.7   5. v = 0.851   7. V = 654,000

2. R = 4.76   4. Z = 990   6. t = 2.11   8. s = 14.2

4-7   SUBSCRIPTS IN FORMULAS

Some formulas contain letters or numbers as subscripts. We will discuss evaluations with formulas of that type in this section.

72.   When it makes sense to use the same letter more than once in a formula, subscripts are used. Either letters or numbers can be used as the subscripts. Two examples are discussed.

The formula below shows the relationship between average velocity (or speed), original velocity, and final velocity when acceleration is constant.

$$v_{av} = \frac{v_o + v_f}{2}$$

where: v_{av} = average velocity

v_o = original velocity

v_f = final velocity

The formula below shows the relationship between the total resistance in an electric circuit and the three separate resistances in the circuit.

$$R_t = R_1 + R_2 + R_3$$

where: R_t = total resistance

R_1 = first resistance

R_2 = second resistance

R_3 = third resistance

Evaluations with formulas containing subscripts are performed in the usual way. For example:

In $v_{av} = \frac{v_o + v_f}{2}$, find "v_{av}" when $v_o = 20$ and $v_f = 40$.

$$v_{av} = \frac{v_o + v_f}{2} = \frac{20 + 40}{2} = \frac{60}{2} = 30$$

In $R_t = R_1 + R_2 + R_3$, find R_t when $R_1 = 10$, $R_2 = 15$, and $R_3 = 30$.

$$R_t = R_1 + R_2 + R_3 = 10 + 15 + 30 = \underline{\hspace{2cm}}$$

73.   The abbreviation "sub" is used when naming a letter with a subscript. For example:

V_2 is called "V sub 2".

V_t is called "V sub t".

Write a letter with a subscript for each of these.

a) F sub 1 _____   b) v sub f _____   c) R sub 3 _____

$R_t = 55$

74.   Use a calculator for these.

    a) In $\boxed{X = X_L - X_C}$, when $X_L = 27.8$ and $X_C = 14.9$,

$$X = \underline{\hspace{2cm}}$$

    b) In $\boxed{v_1 = v_2 - at}$, when $v_2 = 91.6$, $a = 10.4$, and $t = 3$,

$$v_1 = \underline{\hspace{2cm}}$$

    c) In $\boxed{E = IR_1 + IR_2}$, when $I = 10$, $R_1 = 12.5$, and $R_2 = 17.8$,

$$E = \underline{\hspace{2cm}}$$

a) F_1

b) v_f

c) R_3

75.   Use a calculator for these. Round to tenths.

    a) In $\boxed{T_1 = \dfrac{V_1 T_2}{V_2}}$, when $V_1 = 10.4$, $T_2 = 15$, and $V_2 = 12.7$,

$$T_1 = \underline{\hspace{2cm}}$$

    b) In $\boxed{t_1 = t_2 - \dfrac{H}{ms}}$, when $t_2 = 50$, $H = 9.68$, $m = 1.5$, and

$$s = 2.03 , \quad t_1 = \underline{\hspace{2cm}}$$

    c) In $\boxed{P_1 = \dfrac{P_2 V_2 T_1}{V_1 T_2}}$, when $P_2 = 40.7$, $V_2 = 20.5$, $T_1 = 30$,

$$V_1 = 34.9 , \quad \text{and} \quad T_2 = 55 , \quad P_1 = \underline{\hspace{2cm}}$$

a) $X = 12.9$

b) $v_1 = 60.4$

c) $E = 303$

76.   Use a calculator for these. Round to hundredths.

    a) In $\boxed{I_a = \dfrac{E_1 - E_0}{R_a}}$, when $E_1 = 27.5$, $E_0 = 15.8$, and

$$R_a = 10.5 , \quad I_a = \underline{\hspace{2cm}}$$

    b) In $\boxed{c_2 = \dfrac{K - c_1 a}{b}}$, when $K = 100$, $c_1 = 2.5$, $a = 24.8$, and

$$b = 7.5 , \quad c_2 = \underline{\hspace{2cm}}$$

a) $T_1 = 12.3$

b) $t_1 = 46.8$

c) $P_1 = 13.0$

a) $I_a = 1.11$

b) $c_2 = 5.07$

77.   Use a calculator for these. Round to two decimal places.

a) In $\boxed{t_1 = \dfrac{mst_2 - H}{ms}}$, find t_1 when $m = 50$, $s = 19.7$, $t_2 = 10$,
and $H = 65.5$.

$$t_1 = \underline{\hspace{1.5cm}}$$

b) In $\boxed{h = \dfrac{P - p_1}{w} + h_1}$, find "h" when $P = 47.6$, $p_1 = 35.9$,
$w = 2.46$, and $h_1 = 1.59$.

$$h = \underline{\hspace{1.5cm}}$$

---

78.   Don't confuse the subscript "2" with the "2" that is the symbol for squaring.

The subscript "2" is written <u>slightly</u> <u>below</u> the <u>letter</u>.
It usually means "the second". For example:

v_2 means "the second velocity".

The symbol for squaring is written <u>slightly</u> <u>above</u> the <u>letter</u>.
That is:

v^2 means "square the value of v".

a) Does t_2 or t^2 mean "square the value of t"? $\underline{\hspace{1.5cm}}$

b) Does V_2 or V^2 mean "the second volume"? $\underline{\hspace{1.5cm}}$

a) $t_1 = 9.93$

b) $h = 6.35$

---

79.   In $F = \dfrac{m_1 m_2}{rd^2}$:   a) Does m_2 mean "the second mass" or
"the mass squared"? $\underline{\hspace{3cm}}$

b) Does d^2 mean "the second distance" or
"the distance squared"? $\underline{\hspace{2.5cm}}$

a) t^2

b) V_2

---

80.   Use a calculator for these.

a) In $\boxed{I_c = I_a - d^2 m}$, when $I_a = 87.5$, $d = 1.25$, and
$m = 12$, $I_c = \underline{\hspace{1.5cm}}$

b) In $\boxed{F = \dfrac{m_1 m_2}{rd^2}}$, when $m_1 = 17.5$, $m_2 = 13.5$, $r = 10.8$,
and $d = 2.5$, $F = \underline{\hspace{1.5cm}}$

a) the second mass

b) the distance squared

---

a) $I_c = 68.75$

b) $F = 3.5$

81. In the formulas below, letters with subscripts are squared. Use a calculator for the evaluations.

a) In $\boxed{I_1 = \dfrac{I_2(d_2)^2}{(d_1)^2}}$, when $I_2 = 15$, $d_2 = 12$, and $d_1 = 20$,

$$I_1 = \underline{\hspace{2cm}}$$

b) In $\boxed{s = \dfrac{(v_f)^2 - (v_o)^2}{2g}}$, when $v_f = 40$, $v_o = 30$, and $g = 17.5$,

$$s = \underline{\hspace{2cm}}$$

82. Use a calculator for these. Round to tenths.

a) In $\boxed{d = \sqrt{\dfrac{I_A - I_B}{m}}}$, when $I_A = 395$, $I_B = 70$, and $m = 3$,

$$d = \underline{\hspace{2cm}}$$

b) In $\boxed{v_o = \sqrt{(v_f)^2 - 2gs}}$, when $v_f = 37.6$, $g = 32.2$, and

$$s = 14 , \quad v_o = \underline{\hspace{2cm}}$$

a) $I_1 = 5.4$

b) $s = 20$

a) $d = 10.4$ b) $v_o = 22.6$

4-8   FORMULAS IN PROPORTION FORM

In this section, we will discuss evaluations with formulas that are written in proportion form.

83. The formulas below are written in proportion form because they have only one fraction on each side.

$$\frac{P_1}{P_2} = \frac{V_2}{V_1} \qquad\qquad \frac{I_1}{I_2} = \frac{(d_2)^2}{(d_1)^2}$$

To perform evaluations with formulas of that type, we simply substitute and then solve the proportion. An example is shown. Use a calculator to complete the other evaluation.

Find "d_1" when $F_1 = 12$, $F_2 = 20$, and $d_2 = 40$.

Find V_1 when $V_2 = 120$, $T_1 = 12$, and $T_2 = 18$.

$$\boxed{\dfrac{F_1}{F_2} = \dfrac{d_1}{d_2}}$$

$$\frac{12}{20} = \frac{d_1}{40}$$

$$(12)(40) = (20)(d_1)$$

$$d_1 = \frac{(12)(40)}{20}$$

$$d_1 = 24$$

$$\boxed{\dfrac{V_1}{V_2} = \dfrac{T_1}{T_2}}$$

84. We performed one evaluation with the formula below. Use a calculator to complete the other evaluation.

Find T_2 when $P_1 = 10$, $V_1 = 16$, $T_1 = 4$, $P_2 = 20$, and $V_2 = 11$.

$$\boxed{\frac{P_1 V_1}{T_1} = \frac{P_2 V_2}{T_2}}$$

$$\frac{(10)(16)}{4} = \frac{(20)(11)}{T_2}$$

$$(10)(16)(T_2) = (4)(20)(11)$$

$$T_2 = \frac{(4)(20)(11)}{(10)(16)}$$

$$T_2 = 5.5$$

Find P_2 when $P_1 = 12$, $V_1 = 25$, $T_1 = 3$, $V_2 = 50$, and $T_2 = 7$.

$$\boxed{\frac{P_1 V_1}{T_1} = \frac{P_2 V_2}{T_2}}$$

$V_1 = 80$, from:

$$\frac{(120)(12)}{18}$$

85. We performed one evaluation with the formula below. Use a calculator to complete the other evaluation.

Find I_1 when $I_2 = 10$, $d_2 = 4$, and $d_1 = 8$.

$$\boxed{\frac{I_1}{I_2} = \frac{(d_2)^2}{(d_1)^2}}$$

$$\frac{I_1}{10} = \frac{(4)^2}{(8)^2}$$

$$(I_1)(8^2) = (10)(4^2)$$

$$I_1 = \frac{(10)(4^2)}{8^2}$$

$$I_1 = 2.5$$

Find I_2 when $I_1 = 18$, $d_2 = 6$, and $d_1 = 3$.

$$\boxed{\frac{I_1}{I_2} = \frac{(d_2)^2}{(d_1)^2}}$$

$P_2 = 14$, from:

$$P_2 = \frac{(12)(25)(7)}{(3)(50)}$$

86. Use a calculator for these. Round to tenths in (b).

a) Find d_2 when $F_1 = 14.8$, $F_2 = 40.7$, and $d_1 = 21.6$.

$$\boxed{\frac{F_1}{F_2} = \frac{d_1}{d_2}}$$

b) Find V_2 when $V_1 = 87.5$, $T_1 = 65$, and $T_2 = 45$.

$$\boxed{\frac{V_1}{V_2} = \frac{T_1}{T_2}}$$

$I_2 = 4.5$, from:

$$\frac{(18)(3^2)}{6^2}$$

87.   Use a calculator for these. Round to tenths.

a) Find V_1 when $P_1 = 10.7$,
$T_1 = 55$, $P_2 = 25.8$, $V_2 = 63.9$,
and $T_2 = 90$.

$$\boxed{\dfrac{P_1 V_1}{T_1} = \dfrac{P_2 V_2}{T_2}}$$

b) Find I_2 when $I_1 = 25.6$,
$d_2 = 7.98$, and $d_1 = 5.74$.

$$\boxed{\dfrac{I_1}{I_2} = \dfrac{(d_2)^2}{(d_1)^2}}$$

a) $d_2 = 59.4$

b) $V_2 = 60.6$

a) $V_1 = 94.2$ b) $I_2 = 13.2$

4-9   EVALUATIONS INVOLVING NUMBERS IN POWER-OF-TEN FORM

In this section, we will discuss formula evaluations involving numbers in power-of-ten form.

88.   As we saw earlier, a number in scientific notation or some other power-of-ten form is entered on a calculator by entering the first factor, pressing
$\boxed{\text{EE}}$, and then entering the exponent. For example:

To enter 5.78×10^{12} , we press 5.78 , $\boxed{\text{EE}}$, and 12.

To enter 370×10^{-9} , we press 370 , $\boxed{\text{EE}}$, 9 , and $\boxed{+/-}$.

The calculator steps for the evaluation below are shown.

In $\boxed{Q = CE}$, find Q when $C = 8.55 \times 10^{-12}$ and $E = 60$.

Enter	Press	Display	
8.55	$\boxed{\text{EE}}$	8.55 00	
12	$\boxed{+/-}$ $\boxed{\text{x}}$	8.55 −12	
60	$\boxed{=}$	5.13 −10	$(Q = \underline{5.13 \times 10^{-10}})$

Use a calculator for this one. Round the first factor of the answer to two decimal places.

In $\boxed{E = hf}$, when $h = 6.63 \times 10^{-34}$ and $f = 3.25 \times 10^{18}$,

$$E = \underline{\hspace{3cm}}$$

$E = 2.15 \times 10^{-15}$

89.   To perform the evaluations below, we enter the value for "f" and then press the reciprocal key $\boxed{1/x}$. Round the first factor of the answer to hundredths.

In $\boxed{\lambda = \dfrac{1}{f}}$:   a) find λ when f = 386 x 10^6. λ = _____

b) find λ when f = 275 x 10^6. λ = _____

---

90.   To square a number in power-of-ten form, we enter it in the usual way and then press $\boxed{x^2}$. For example:

To square 47×10^{12} , we press 47 , $\boxed{EE}$, 12 , and $\boxed{x^2}$.

The calculator steps for the evaluation below are shown.

In $\boxed{E = mc^2}$, find E when m = 50 and c = 3 x 10^8 .

Enter	Press	Display	
50	$\boxed{x}$	50.	
3	$\boxed{EE}$	3. 00	
8	$\boxed{x^2}$ $\boxed{=}$	4.5 18	(E = $\underline{4.5 \times 10^{18}}$)

Use a calculator for this one.

In $\boxed{E = mc^2}$, when m = 690 and c = 3 x 10^8 ,   E = _____

Answers:
a) $\lambda = 2.59 \times 10^{-9}$

b) $\lambda = 3.64 \times 10^{-9}$

---

91.   Use a calculator for these. Round the first factor of the answer to two decimal places.

a) In $\boxed{M = \dfrac{gs^2}{G}}$, when g = 9.80 , s = 6.37 x 10^6 , and G = 6.67 x 10^{-11} , M = _____

b) In $\boxed{F = \dfrac{kQ_1 Q_2}{s^2}}$, when k = 9 x 10^9 , Q_1 = 3 x 10^{-4} , Q_2 = 5 x 10^{-5} , and s = .5 , F = _____

Answer:
E = 6.21 x 10^{19}

a) M = 5.96 x 10^{24}

b) F = 5.4 x 10^2 (or 540)

92.   The calculator steps for the evaluation below are shown. Notice that we
pressed $\boxed{=}$ to complete the operation within the radical before pressing
$\boxed{\sqrt{x}}$.

In $\boxed{Z = \sqrt{R^2 + X^2}}$, find Z when $R = 8.5 \times 10^6$ and $X = 3.4 \times 10^6$.

Enter	Press	Display
8.5	$\boxed{EE}$	8.5 00
6	$\boxed{x^2}$ $\boxed{+}$	7.225 13
3.4	$\boxed{EE}$	3.4 00
6	$\boxed{x^2}$ $\boxed{=}$ $\boxed{\sqrt{x}}$	9.1548 06 ($Z = \underline{9.1548 \times 10^6}$)

Using the same steps, complete this one. Round the first factor of the answer
to two decimal places.

In $\boxed{Z = \sqrt{R^2 + X^2}}$, when $R = 9.5 \times 10^6$ and $X = 4.8 \times 10^6$,

Z = _____

$Z = 1.06 \times 10^7$

4-10   EVALUATING "xy" EQUATIONS

To graph an "xy" equation, we have to prepare a table of solutions. To do so, we have to evaluate the
equation. We will do some evaluations of that type on a calculator in this section.

93.   Evaluations with "xy" equations are done in the usual way on a calculator.
Find "y" when x = 5 in each equation below.

a) $y = 1.5x$ _____   c) $y = 2.5x^2$ _____

b) $y = 0.8x + 7$ _____   d) $y = 10 - 1.2x$ _____

94.   Find "y" when x = 12 in each equation below.

a) $y = 3x^2 + 2x$ _____   c) $y = x^2 - 5x - 10$ _____

b) $y = \dfrac{240}{x}$ _____   d) $y = 100 - 7.5x$ _____

a) 7.5   c) 62.5
b) 11   d) 4

95.   Find "y" when x = 2.5 in each equation below. (Note: Remember that
you have to press $\boxed{+/-}$ to enter a negative value on a calculator.)

a) $y = -2x + 15$ _____   b) $y = -3x^2$ _____

a) 456   c) 74
b) 20   d) 10

a) 10   b) -18.75

96.   Find "y" when x = -1.5 in each equation.

   a) y = 25 - 7x _____   b) y = $\frac{60}{x}$ _____

---

97.   Find "y" when x = -7 in each equation below.

   a) y = -2x - 9 _____   b) y = $-5x^2 + 6x - 7$ _____

a) 35.5   b) -40

---

98.   The equations below contain a "-x" and a "$-x^2$". To enter positive values
        for "x", we enter the number, square it if necessary, and then press
        $\boxed{+/-}$. An example is shown.

a) 5   b) -294

   In $\boxed{y = -x^2 + 5x}$, find "y" when x = 20 .

Enter	Press	Display
20	$\boxed{x^2}$ $\boxed{+/-}$ $\boxed{+}$	-400.
5	$\boxed{x}$	5.
20	$\boxed{=}$	-300. (y = -300)

   Find "y" when x = 10 in each equation below.

   a) y = -x + 25 _____   b) y = $-x^2 - 3x + 50$ _____

---

99.   To enter negative values for "x" in the same type of equations, we enter
        the negative number (which includes pressing $\boxed{+/-}$), square it if necessary,
        and then press $\boxed{+/-}$ again. An example is shown.

a) 15   b) -80

   In $\boxed{y = -x^2 + 7x}$, find "y" when x = -3.5 .

Enter	Press	Display
3.5	$\boxed{+/-}$ $\boxed{x^2}$ $\boxed{+/-}$ $\boxed{+}$	-12.25
7	$\boxed{x}$	7
3.5	$\boxed{+/-}$ $\boxed{=}$	-36.75 (y = -36.75)

   Find "y" when x = -4 in each equation below.

   a) y = -x - 2.5 _____   b) y = $-x^2 + 5x - 30$ _____

---

100.   To enter fractional values for "x", we convert them to decimal numbers.
        For example:

a) 1.5   c) -66

   $\frac{1}{2}$ is converted to 0.5 $-2\frac{1}{2}$ is converted to -2.5

Continued on following page.

100.   Continued

Do these on a calculator.

a) In $\boxed{y = x^2}$, when $x = 1\frac{1}{2}$, y = _____

b) In $\boxed{y = 2x^2 - 5x}$, when $x = \frac{1}{2}$, y = _____

c) In $\boxed{y = \dfrac{100}{x}}$, when $x = -2\frac{1}{2}$, y = _____

---

101.   When a term contains a fractional coefficient, we can write the term as a division. For example:

$\frac{1}{2}x$ can be written $\frac{x}{2}$ $\frac{1}{3}x^2$ can be written $\frac{x^2}{3}$

Do these on a calculator.

a) In $\boxed{y = \dfrac{1}{2}x + 10}$, when $x = 1\frac{1}{2}$, y = _____

b) In $\boxed{y = \dfrac{1}{3}x^2 + 6x}$, when $x = -9$, y = _____

a) 2.25

b) -2

c) -40

---

102.   In $\boxed{y = \sqrt{r^2 - x^2}}$:   a) when $r = 15$ and $x = 12$, y = _____

b) when $r = 13$ and $x = -5$, y = _____

a) 10.75

b) -27

a) 9   b) 12

SELF-TEST 12 (pages 143-153)

1. Find "h_1". Round to hundredths.

$\boxed{h_1 = \dfrac{P - p_1}{w} - h}$

P = 91.2
p_1 = 15.9
w = 4.88
h = 7.56

_____

2. Find F. Round to the nearest whole number.

$\boxed{F = \dfrac{m_1 m_2}{kd^2}}$

m_1 = 3,790
m_2 = 2,140
k = 0.168
d = 422

_____

3. Find P_1. Round to tenths.

$\boxed{\dfrac{P_2}{P_1} = \dfrac{V_1}{V_2}}$

P_2 = 41.7
V_1 = 520
V_2 = 370

_____

4. Find I_1. Round to two decimal places.

$\boxed{\dfrac{I_1}{I_2} = \dfrac{(d_2)^2}{(d_1)^2}}$

I_2 = 4.05
d_2 = 15.3
d_1 = 11.7

_____

Continued on following page.

Self-Test 12 - Continued

In Problems 5 and 6, report each answer in scientific notation with the first factor rounded to two decimal places.

5. Find "t" when $f = 51.2 \times 10^8$.

$$t = \frac{1}{f}$$

6. Find Q when: $C = 157 \times 10^{-12}$
$E = 2.25 \times 10^3$

$$Q = CE$$

7. Find "y" when $x = -7.5$.

$$y = -x + 18$$

8. Find "y" when $x = 2.5$.

$$y = 2x - x^2$$

ANSWERS:

1. $h_1 = 7.87$ 3. $P_1 = 29.7$ 5. $t = 1.95 \times 10^{-10}$ 7. $y = 25.5$

2. $F = 271$ 4. $I_1 = 6.93$ 6. $Q = 3.53 \times 10^{-7}$ 8. $y = -1.25$

4-11   FORMULAS CONTAINING PARENTHESES (CALCULATORS WITH PARENTHESES)

In this section, we will show how the parentheses symbols $\boxed{(}$ and $\boxed{)}$ on a calculator can be used for evaluations with formulas containing parentheses.

Note: 1) If your calculator has parentheses symbols, do this section and section 4-13.

2) If your calculator does not have parentheses symbols, skip this section and do sections 4-12 and 4-14.

---

103.   In the evaluation below, we performed the addition within the parentheses first.

In $\boxed{D = P(Q + R)}$, find D when $P = 5$, $Q = 4$, and $R = 3$.

$$D = P(Q + R) = 5(4 + 3) = 5(7) = 35$$

To do the same evaluation on a calculator, the parentheses symbols $\boxed{(}$ and $\boxed{)}$ can be used. The steps are shown below. Notice that we pressed $\boxed{x}$ after entering 5.

Enter	Press	Display
5	$\boxed{x}$ $\boxed{(}$	5.
4	$\boxed{+}$	4.
3	$\boxed{)}$ $\boxed{=}$	35.

Use the same method for these. Don't forget to press $\boxed{x}$ before $\boxed{(}$.

a) In $\boxed{M = P(L - X)}$, when $P = 20$, $L = 44.5$, and $X = 29.7$, $M = \underline{\hspace{1cm}}$

b) In $\boxed{Q = C(T_1 - T_2)}$, when $C = 2.5$, $T_1 = 90$, and $T_2 = 55$, $Q = \underline{\hspace{1cm}}$

104. In the evaluation below, we performed the subtraction within the parentheses first.

    In $\boxed{H = ms(t_2 - t_1)}$, find H when m = 4, s = 3, $t_2 = 10$, and $t_1 = 5$.

    $$H = ms(t_2 - t_1) = (4)(3)(10 - 5) = (4)(3)(5)$$
    $$= 12 \ (5) = 60$$

    The calculator steps for the same evaluation are shown below. Notice that we simply press $\boxed{(}$ and $\boxed{)}$ where they appear in the formula.

Enter	Press	Display
4	$\boxed{x}$	4.
3	$\boxed{x}$ $\boxed{(}$	12.
10	$\boxed{-}$	10.
5	$\boxed{)}$ $\boxed{=}$	60.

    Use a calculator for these. Don't forget to press $\boxed{x}$ before $\boxed{(}$.

    a) In $\boxed{K = mt(x - r)}$, when m = 25.7 , t = 40 , x = 19.7 , and r = 12.2 , K = _____

    b) In $\boxed{V = SR(b_1 + b_2)}$, when S = 3.7 , R = 4.2 , $b_1 = 25$, and $b_2 = 75$, V = _____

a) M = 296

b) Q = 87.5

---

105. In the evaluation below, we began by squaring 4 and 3. Then we performed the subtraction within the parentheses.

    In $\boxed{d = a(b^2 - c^2)}$, find "d" when a = 10 , b = 4 , and c = 3 .
    $$d = a(b^2 - c^2) = 10(4^2 - 3^2) = 10(16 - 9) = 10(7) = 70$$

    The calculator steps for the same evaluation are shown below. Notice that we simply press $\boxed{x^2}$ after entering 4 and 3 .

Enter	Press	Display
10	$\boxed{x}$ $\boxed{(}$	10.
4	$\boxed{x^2}$ $\boxed{-}$	16.
3	$\boxed{x^2}$ $\boxed{)}$ $\boxed{=}$	70.

    Use a calculator for these. Round to tens in (a) and to hundredths in (b).

    a) In $\boxed{A = 3.14(R^2 - r^2)}$, when R = 25.5 and r = 13.7 ,
    A = _____

    b) In $\boxed{A = 0.785(D^2 - d^2)}$, when D = 2.54 and d = 1.89 ,
    A = _____

a) K = 7,710

b) V = 1,554

106. The calculator steps for the evaluation below are shown. Notice that we began by pressing $\boxed{(}$. Also, we pressed $\boxed{x}$ after $\boxed{)}$.

In $\boxed{H = (1 - C^2)h}$, find H when $C = 0.6$ and $h = 50$.

Enter	Press	Display
	$\boxed{(}$	0.
1	$\boxed{-}$	1.
0.6	$\boxed{x^2}$ $\boxed{)}$ $\boxed{x}$	0.64
50	$\boxed{=}$	32. (H = 32)

Use a calculator for these. Round to hundredths.

a) In $\boxed{d = (1 - t^2)R}$, when $t = 0.75$ and $R = 15$, $d =$ _____

b) In $\boxed{m = (a^2 + b^2)T}$, when $a = 1.25$, $b = 2.46$, and $T = 3.04$,

m = _____

a) A = 1,450.

b) A = 2.26

---

107. The calculator steps for the evaluation below are shown. Notice that we simply evaluated the numerator and then divided by 2.

In $\boxed{A = \dfrac{h(b_1 + b_2)}{2}}$, find A when $h = 20$, $b_1 = 4$, and $b_2 = 10$.

Enter	Press	Display
20	$\boxed{x}$ $\boxed{(}$	20.
4	$\boxed{+}$	4.
10	$\boxed{)}$ $\boxed{÷}$	280.
2	$\boxed{=}$	140. (A = 140)

Use a calculator for these. Round to hundredths in (b).

a) In $\boxed{r = \dfrac{R(w - 2F)}{w}}$, when $R = 12$, $w = 18$, and $F = 6$,

r = _____

b) In $\boxed{H = \dfrac{3.14dR(F_1 - F_2)}{33,000}}$, when $d = 100$, $R = 15$, $F_1 = 35.8$,

and $F_2 = 27.4$, H = _____

a) d = 6.56

b) m = 23.15

a) r = 4

b) H = 1.20

108. The calculator steps for the evaluation below are shown. Notice that we evaluated the numerator and then divided by 2 and 3.

In $\boxed{a = \dfrac{c(d - f)}{2b}}$, find "a" when $c = 4$, $d = 20$, $f = 8$, and $b = 3$.

Enter	Press	Display
4	$\boxed{\text{x}}$ $\boxed{(}$	4.
20	$\boxed{-}$	20.
8	$\boxed{)}$ $\boxed{\div}$	48.
2	$\boxed{\div}$	24.
3	$\boxed{=}$	8. $(a = 8)$

Use the same steps for this one. Round to hundredths.

In $\boxed{F = \dfrac{W(R - r)}{2R}}$, when $W = 12.4$, $R = 75.1$, and $r = 57.3$,

$$F = \underline{\hspace{2cm}}$$

---

109. Use a calculator for these. Round to tenths in (b).

a) In $\boxed{E = \dfrac{(m_1 + m_2)v^2}{2}}$, when $m_1 = 35.5$, $m_2 = 55.5$, and

$v = 40$, $E = \underline{\hspace{2cm}}$

b) In $\boxed{H = \dfrac{M[(V_1)^2 - (V_2)^2]}{1,100gt}}$, when $M = 50$, $V_1 = 225$, $V_2 = 110$,

$g = 16$, and $t = 5$, $H = \underline{\hspace{2cm}}$

F = 1.47

---

110. To perform the evaluations below, use $\boxed{\text{EE}}$ to enter the numbers in scientific notation.

a) In $\boxed{\Delta L = \alpha L(T - T_0)}$, when $\alpha = 1.1 \times 10^{-5}$, $L = 60$, $T = 80$,

and $T_0 = 0$, $\Delta L = \underline{\hspace{2cm}}$

b) In $\boxed{\Delta V = \beta V(T - T_0)}$, when $\beta = 9.6 \times 10^{-4}$, $V = 100$, $T = 30$,

and $T_0 = 0$, $\Delta V = \underline{\hspace{2cm}}$

a) E = 72,800

b) H = 21.9

---

a) $\Delta L = 5.28 \times 10^{-2}$

b) $\Delta V = 2.88 \times 10^{0}$

or 2.88

111. To perform the evaluations below, we simply use parentheses after pressing $\boxed{-}$.

a) In $\boxed{A \;=\; 180 \;-\; (B + C)}$, when $B = 47$ and $C = 29$, $A = \underline{\hspace{2cm}}$

b) In $\boxed{A \;=\; 360 \;-\; (B + C + D)}$, when $B = 110$, $C = 55$, and $D = 71$, $A = \underline{\hspace{2cm}}$

---

112. Use parentheses when necessary for these.

a) In $\boxed{P \;=\; p_1 + w(h - h_1)}$, when $p_1 = 20$, $w = 25$, $h = 75$, and $h_1 = 40$, $P = \underline{\hspace{2cm}}$

b) In $\boxed{H \;=\; \alpha + t(P_1 - P_2)}$, when $\alpha = 75$, $t = 12$, $P_1 = 80$, and $P_2 = 35$, $H = \underline{\hspace{2cm}}$

a) $A = 104$

b) $A = 124$

---

113. Use parentheses when necessary for the evaluation below. Be sure to press $\boxed{=}$ to complete the subtraction in the numerator before dividing.

In $\boxed{I_B \;=\; \dfrac{I_C - (\beta + 1)I_{CO}}{\beta}}$, when $I_C = 87.5$, $\beta = 0.5$, and $I_{CO} = 42.6$, $I_B = \underline{\hspace{2cm}}$

a) $P = 895$

b) $H = 615$

---

114. The evaluation below contains two sets of parentheses.

In $\boxed{F \;=\; (a + 3)(b + 4)}$, find F when $a = 5$ and $b = 2$.

$$F \;=\; (a + 3)(b + 4) \;=\; (5 + 3)(2 + 4) \;=\; (8)(6) \;=\; 48$$

The calculator steps for the same evaluation are shown. Notice that we pressed $\boxed{x}$ between the two sets of parentheses.

Enter	Press	Display
	$\boxed{(}$	0.
5	$\boxed{+}$	5.
3	$\boxed{)}$ x $\boxed{(}$	8.
2	$\boxed{+}$	2.
4	$\boxed{)}$ $\boxed{=}$	48.

$I_B = 47.2$

Continued on following page.

114.   Continued

Use the same method for these.

a) In $\boxed{m = (C - D)(T - S)}$, when C = 35 ,   D = 25 ,   T = 87 ,

and S = 51 ,   m = _____

b) In $\boxed{V = 2t(b - t)(a - t_1)}$, when t = 15 ,   b = 45 ,   a = 37 ,

and $t_1 = 20$,   V = _____

---

a) m = 360   b) V = 15,300

4-12   FORMULAS CONTAINING PARENTHESES (CALCULATORS WITHOUT PARENTHESES)

In this section, we will show how to perform evaluations with formulas containing parentheses on a calculator that does not have parentheses symbols.

---

115.   In the evaluation below, we performed the addition within the parentheses first.

In $\boxed{D = P(Q + R)}$, find D when P = 5 ,   Q = 4 ,   and R = 3 .

D = P(Q + R) = 5(4 + 3) = 5(7) = 35

To do the same evaluation on a calculator without parentheses, we also perform the addition within parentheses first. The steps are shown below. Notice that we have to press $\boxed{=}$ to complete the addition before multiplying by 5.

Enter	Press	Display
4	$\boxed{+}$	4.
3	$\boxed{=}$ $\boxed{x}$	7.
5	$\boxed{=}$	35.

Use the same method for these. Don't forget to press $\boxed{=}$ to complete the subtraction before multiplying.

a) In $\boxed{M = P(L - X)}$, when P = 20 ,   L = 44.5 ,   and X = 29.7 ,

M = _____

b) In $\boxed{Q = C(T_1 - T_2)}$, when C = 2.5 , $T_1 = 90$,   and $T_2 = 55$,

Q = _____

---

a) M = 296   b) Q = 87.5

116. In the evaluation below, we performed the subtraction within the parentheses first.

In $\boxed{H = ms(t_2 - t_1)}$, find H when m = 4, s = 3, $t_2 = 10$, and $t_1 = 5$.

$$H = ms(t_2 - t_1) = (4)(3)(10 - 5) = \underline{(4)(3)}(5)$$
$$= 12 \ (5) = 60$$

The calculator steps for the same evaluation are:

Enter	Press	Display
10	$\boxed{-}$	10.
5	$\boxed{=}$ $\boxed{x}$	5.
4	$\boxed{x}$	20.
3	$\boxed{=}$	60.

Use a calculator for these. Don't forget to press $\boxed{=}$ to complete the subtraction or addition before multiplying.

a) In $\boxed{K = mt(x - r)}$, when m = 25.7 , t = 40 , x = 19.7 , and r = 12.2 , K = _____

b) In $\boxed{V = SR(b_1 + b_2)}$, when S = 3.7 , R = 4.2 , $b_1 = 25$, and $b_2 = 75$, V = _____

117. In the evaluation below, we began by squaring 4 and 3. Then we performed the subtraction within the parentheses.

In $\boxed{d = a(b^2 - c^2)}$, find "d" when a = 10 , b = 4 , and c = 3 .

$$d = a(b^2 - c^2) = 10(4^2 - 3^2) = 10(16 - 9) = 10(7) = 70$$

The calculator steps for the same evaluation are shown below. Notice that we press $\boxed{x^2}$ after entering 4 and 3 .

Enter	Press	Display
4	$\boxed{x^2}$ $\boxed{-}$	16.
3	$\boxed{x^2}$ $\boxed{=}$ $\boxed{x}$	7.
10	$\boxed{=}$	70.

Use a calculator for these. Round to tens in (a) and to hundredths in (b).

a) In $\boxed{A = 3.14(R^2 - r^2)}$, when R = 25.5 and r = 13.7 , A = _____

b) In $\boxed{A = 0.785(D^2 - d^2)}$, when D = 2.54 and d = 1.89 , A = _____

a) K = 7,710

b) V = 1,554

118.   The calculator steps for the evaluation below are shown.

In $\boxed{H = (1 - C^2)h}$, find H when $C = 0.6$ and $h = 50$.

Enter	Press	Display
1	$\boxed{-}$	1.
0.6	$\boxed{x^2}$ $\boxed{=}$ $\boxed{x}$	0.64
50	$\boxed{=}$	32. (H = 32)

Use a calculator for these. Round to hundredths.

a) In $\boxed{d = (1 - t^2)R}$, when $t = 0.75$ and $R = 15$, $d =$ _____

b) In $\boxed{m = (a^2 + b^2)T}$, when $a = 1.25$, $b = 2.46$, and $T = 3.04$,

m = _____

a) A = 1,450.

b) A = 2.26

---

119.   The calculator steps for the evaluation below are shown. Notice that we simply evaluated the numerator and then divided by 2.

In $\boxed{A = \dfrac{h(b_1 + b_2)}{2}}$, find A when $h = 20$, $b_1 = 4$, and $b_2 = 10$.

Enter	Press	Display
4	$\boxed{+}$	4.
10	$\boxed{=}$ $\boxed{x}$	14.
20	$\boxed{\div}$	280.
2	$\boxed{=}$	140. (A = 140)

Use a calculator for these. Round to hundredths in (b).

a) In $\boxed{r = \dfrac{R(w - 2F)}{w}}$, when $R = 12$, $w = 18$, and $F = 6$,

r = _____

b) In $\boxed{H = \dfrac{3.14dR(F_1 - F_2)}{33,000}}$, when $d = 100$, $R = 15$, $F_1 = 35.8$,

and $F_2 = 27.4$, H = _____

a) d = 6.56

b) m = 23.15

a) r = 4

b) H = 1.20

120.   The calculator steps for the evaluation below are shown. Notice that we evaluated the numerator and then divided by 2 and 3.

In $\boxed{a = \dfrac{c(d - f)}{2b}}$, find "a" when c = 4 , d = 20 , f = 8 , and b = 3 .

Enter	Press	Display
20	$\boxed{-}$	20.
8	$\boxed{=}$ $\boxed{\text{x}}$	12.
4	$\boxed{\div}$	48.
2	$\boxed{\div}$	24.
3	$\boxed{=}$	8. (a = 8)

Use the same steps for this one. Round to hundredths.

In $\boxed{F = \dfrac{W(R - r)}{2R}}$, when W = 12.4 , R = 75.1 , and r = 57.3 ,

$$F = \underline{\hspace{2cm}}$$

---

121.   Use a calculator for these. Round to tenths in (b).

a) In $\boxed{E = \dfrac{(m_1 + m_2)v^2}{2}}$, when $m_1 = 35.5$, $m_2 = 55.5$, and

v = 40 ,   E = \underline{\hspace{2cm}}

b) In $\boxed{H = \dfrac{M[(V_1)^2 - (V_2)^2]}{1,100gt}}$, when M = 50 , $V_1 = 225$, $V_2 = 110$,

g = 16 , and t = 5 ,   H = \underline{\hspace{2cm}}

F = 1.47

---

122.   To perform the evaluations below, use $\boxed{\text{EE}}$ to enter the numbers in scientific notation.

a) In $\boxed{\Delta L = \alpha L(T - T_0)}$, when $\alpha = 1.1 \times 10^{-5}$, L = 60 , T = 80 ,

and $T_0 = 0$, $\Delta L = \underline{\hspace{2.5cm}}$

b) In $\boxed{\Delta V = \beta V(T - T_0)}$, when $\beta = 9.6 \times 10^{-4}$, V = 100 , T = 30 ,

and $T_0 = 0$, $\Delta V = \underline{\hspace{2.5cm}}$

a) E = 72,800

b) H = 21.9

---

a) $\Delta L = 5.28 \times 10^{-2}$

b) $\Delta V = 2.88 \times 10^{0}$

or 2.88

123. To perform the evaluation below, we begin by evaluating $w(h - h_1)$ and then add the value of p_1 . The steps are shown.

In $\boxed{P = p_1 + w(h - h_1)}$, find P when $p_1 = 20$, $w = 25$, $h = 75$, and $h_1 = 40$.

Enter	Press	Display	
75	$\boxed{-}$	75.	
40	$\boxed{=}$ $\boxed{x}$	35.	
25	$\boxed{+}$	875.	
20	$\boxed{=}$	895.	(P = 895)

Using the same steps, do this one.

In $\boxed{H = \alpha + t(P_1 - P_2)}$, when $\alpha = 75$, $t = 12$, $P_1 = 80$, and $P_2 = 35$, H = _____

H = 615

124. For the evaluation below, we evaluate $(B + C)$ and put that value in storage by pressing $\boxed{STO}$. Then we enter 180 and subtract the stored value by pressing the recall key $\boxed{RCL}$. The steps are shown.

In $\boxed{A = 180 - (B + C)}$, find A when $B = 47$ and $C = 29$.

Enter	Press	Display	
47	$\boxed{+}$	47.	
29	$\boxed{=}$ $\boxed{STO}$	76.	
180	$\boxed{-}$ $\boxed{RCL}$ $\boxed{=}$	104.	(A = 104)

Using the same method, complete these.

a) In $\boxed{A = 180 - (B + C)}$, when $B = 87$ and $C = 14$, A = _____

b) In $\boxed{A = 360 - (B + C + D)}$, when $B = 110$, $C = 55$, and $D = 71$, A = _____

a) A = 79

b) A = 124

125. To evaluate the numerator below, we evaluate $(\beta + 1)I_{CO}$ and put that value in storage in order to subtract it from I_C. Then we press $\boxed{=}$ to complete the subtraction in the numerator before dividing. The steps are shown.

In $\boxed{I_B = \dfrac{I_C - (\beta + 1)I_{CO}}{\beta}}$, find I_B when $I_C = 87.5$, $\beta = 0.5$, and $I_{CO} = 42.6$.

Enter	Press	Display
0.5	$\boxed{+}$	0.5
1	$\boxed{=}$ $\boxed{\text{x}}$	1.5
42.6	$\boxed{=}$ $\boxed{\text{STO}}$	63.9
87.5	$\boxed{-}$ $\boxed{\text{RCL}}$ $\boxed{=}$ $\boxed{\div}$	23.6
0.5	$\boxed{=}$	47.2 $(I_B = 47.2)$

126. The evaluation below contains two sets of parentheses.

In $\boxed{F = (a + 3)(b + 4)}$, find F when $a = 5$ and $b = 2$.

$$F = (a + 3)(b + 4) = (5 + 3)(2 + 4) = (8)(6) = 48$$

The calculator steps for the same evaluations are shown. Notice that we evaluated $(5 + 3)$ and put that value in storage, then evaluated $(2 + 4)$ and multiplied that value by the stored value. The steps are shown.

Enter	Press	Display
5	$\boxed{+}$	5.
3	$\boxed{=}$ $\boxed{\text{STO}}$	8.
2	$\boxed{+}$	2.
4	$\boxed{=}$ $\boxed{\text{x}}$ $\boxed{\text{RCL}}$ $\boxed{=}$	48.

Use the same method for this one.

In $\boxed{m = (C - D)(T - S)}$, when $C = 35$, $D = 25$, $T = 87$, and $S = 51$, $m =$ _____

$m = 360$

127.   For the evaluation below, we evaluate $2t(b - t)$ and store that value. Then we evaluate $(a - t_1)$ and multiply that value by the stored value. The steps are shown.

In $\boxed{V = 2t(b - t)(a - t_1)}$, find V when $t = 15$, $b = 45$, $a = 37$,

and $t_1 = 20$.

Enter	Press	Display
45	$\boxed{-}$	45.
15	$\boxed{=}$ $\boxed{\text{x}}$	30.
2	$\boxed{\text{x}}$	60.
15	$\boxed{=}$ $\boxed{\text{STO}}$	900.
37	$\boxed{-}$	37.
20	$\boxed{=}$ $\boxed{\text{x}}$ $\boxed{\text{RCL}}$ $\boxed{=}$	15300. (V = 15,300)

4-13   DIVISIONS REQUIRING PARENTHESES (CALCULATORS WITH PARENTHESES)

In this section, we will show how the parentheses symbols on a calculator can be used for evaluations involving divisions in which the denominator contains an addition, subtraction, or set of parentheses.

128.   To perform the evaluation below, we simplified the denominator and then divided.

In $\boxed{I = \dfrac{E}{R_1 + R_2}}$, find I when $E = 100$, $R_1 = 15$, and $R_2 = 5$.

$$I = \frac{E}{R_1 + R_2} = \frac{100}{15 + 5} = \frac{100}{20} = 5$$

To do the same evaluation on a calculator, we can treat $R_1 + R_2$ as if there were parentheses around it. The steps are:

Enter	Press	Display
100	$\boxed{\div}$ $\boxed{(}$	100.
15	$\boxed{+}$	15.
5	$\boxed{)}$ $\boxed{=}$	5.

Use the same method for these. Round to hundredths.

a) In $\boxed{A = \dfrac{B}{B + 1}}$, when $B = 28.3$, $A = \underline{\hspace{1.5cm}}$

b) In $\boxed{P = \dfrac{W_s}{V_1 - V_2}}$, when $W_s = 9.75$, $V_1 = 8.64$, and $V_2 = 5.73$,

$P = \underline{\hspace{1.5cm}}$

129. Do these on a calculator by treating the denominator as if there were parentheses around it. Round to two decimal places.

a) In $R_T = \dfrac{R_1 R_2}{R_1 + R_2}$, when $R_1 = 15$ and $R_2 = 20$,

$$R_T = \underline{\hspace{2cm}}$$

b) In $W = \dfrac{2RF}{R - r}$, when $R = 37.5$, $F = 4.6$, and $r = 2.9$,

$$W = \underline{\hspace{2cm}}$$

a) A = 0.97

b) P = 3.35

130. Use parentheses around the denominator for these. Round to thousandths in (a) and to tenths in (b).

a) In $A = \dfrac{A_f}{1 + BA_f}$, when $A_f = 12.7$ and $B = 2.5$,

$$A = \underline{\hspace{2cm}}$$

b) In $R = \dfrac{wr}{w - 2F}$, when $w = 7.58$, $r = 10.5$, and $F = 2.14$,

$$R = \underline{\hspace{2cm}}$$

a) $R_T = 8.57$

b) W = 9.97

131. Use parentheses around the denominator for these. Round to tenths in (b).

a) In $I = \dfrac{E}{R_1 + R_2 + R_3}$, when $E = 200$, $R_1 = 15$, $R_2 = 25$, and $R_3 = 40$, $I = \underline{\hspace{2cm}}$

b) In $T = \dfrac{BV_1(P_2 - P_1)}{V_1 - V_2}$, when $B = 2.5$, $V_1 = 80$, $P_2 = 30$, $P_1 = 20$, and $V_2 = 50$, $T = \underline{\hspace{2cm}}$

a) A = 0.388

b) R = 24.1

132. Do these on a calculator.

a) In $t = \dfrac{h^2}{1 - h}$, when $h = 0.75$, $t = \underline{\hspace{2cm}}$

b) In $m = \dfrac{M}{F^2 - 1}$, when $M = 150$ and $F = 11$, $m = \underline{\hspace{2cm}}$

a) I = 2.5

b) T = 66.7

133. In the evaluations below, be sure to press $\boxed{=}$ to complete the subtraction in the numerator before using parentheses to divide.

    a) In $\boxed{m = \dfrac{y_2 - y_1}{x_2 - x_1}}$, when $y_2 = 20$, $y_1 = 8$, $x_2 = 10$,

                and $x_1 = 2$, $m =$ _____

    b) In $\boxed{v = \dfrac{s_2 - s_1}{t_2 - t_1}}$, when $s_2 = 100$, $s_1 = 10$, $t_2 = 60$,

                and $t_1 = 20$, $v =$ _____

a) $t = 2.25$

b) $m \neq 1.25$

---

134. Use a calculator for these. Round to hundredths.

    a) In $\boxed{w = \dfrac{P - p_1}{h - h_1}}$, when $P = 45.5$, $p_1 = 28.7$, $h = 40$,

                and $h_1 = 25$, $w =$ _____

    b) In $\boxed{I_{CO} = \dfrac{I_C - \beta I_B}{\beta + 1}}$, when $I_C = 100$, $\beta = 2.75$, and

                $I_B = 30$, $I_{CO} =$ _____

a) $m = 1.5$

b) $v = 2.25$

---

135. In the evaluation below, we must divide H by both "m" and $(t_2 - t_1)$. Parentheses can be used for the second division. The steps are shown.

    In $\boxed{s = \dfrac{H}{m(t_2 - t_1)}}$, find s when $H = 90$, $m = 1.5$, $t_2 = 50$, and $t_1 = 20$.

Enter	Press	Display	
90	$\boxed{\div}$	90.	
1.5	$\boxed{\div}$ $\boxed{(}$	60.	
50	$\boxed{-}$	50.	
20	$\boxed{)}$ $\boxed{=}$	2.	(s = 2)

Use the same steps for this one. Round to hundredths.

    In $\boxed{A = \dfrac{3T}{G(F - G)}}$, when $T = 50.5$, $G = 12.8$, and $F = 15.1$,

                $A =$ _____

a) $w = 1.12$

b) $I_{CO} = 4.67$

136. Do these. In (a), be sure to press $\boxed{=}$ to complete the subtraction in the numerator. Round to thousandths.

    a) In $\boxed{\alpha = \dfrac{L_2 - L_1}{L_1(t_2 - t_1)}}$, when $L_2 = 75$, $L_1 = 40$, $t_2 = 30$, and $t_1 = 15$, $\alpha =$ _____

    b) In $\boxed{\beta = \dfrac{T(V_1 - V_2)}{V_1(P_2 - P_1)}}$, when $T = 20$, $V_1 = 90$, $V_2 = 55$, $P_2 = 80$, and $P_1 = 65$, $\beta =$ _____

$A = 5.15$

---

137. To perform the evaluations below, we have to divide each numerator by the three factors in the denominator. Round to two decimal places.

    a) In $\boxed{T = \dfrac{HL}{AK(t_2 - t_1)}}$, when $H = 33.8$, $L = 200$, $A = 10$, $K = 22.5$, $t_2 = 30$, and $t_1 = 25$, $T =$ _____

    b) In $\boxed{R = \dfrac{33,000H}{3.14d(F_1 - F_2)}}$, when $H = 1.5$, $d = 25$, $F_1 = 150$, and $F_2 = 75$, $R =$ _____

a) $\alpha = 0.058$

b) $\beta = 0.519$

---

138. In the evaluation below, there are four factors in the denominator. Parentheses must be used for two of the divisions.

    In $\boxed{S = \dfrac{T}{2t(a - t)(b - t_1)}}$, when $T = 500$, $t = 25$, $a = 35$, $b = 40$, and $t_1 = 20$, $S =$ _____

a) $T = 6.01$

b) $R = 8.41$

$S = 0.05$

4-14   DIVISIONS REQUIRING STORAGE (CALCULATORS WITHOUT PARENTHESES)

In this section, we will show how storage on a calculator without parentheses symbols can be used for evaluations involving divisions in which the denominator contains an addition, subtraction, or set of parentheses.

---

139.   To perform the evaluation below, we simplified the denominator and then divided.

In $\boxed{I = \dfrac{E}{R_1 + R_2}}$, find I when $E = 100$, $R_1 = 15$, and $R_2 = 5$.

$$I = \frac{E}{R_1 + R_2} = \frac{100}{15 + 5} = \frac{100}{20} = 5$$

To do the same evaluation on a calculator without parentheses symbols, we evaluate the denominator first and put that value in storage; then we enter the value of E and divide by the stored value. The steps are:

Enter	Press	Display
15	$\boxed{+}$	15.
5	$\boxed{=}$ $\boxed{\text{STO}}$	20.
100	$\boxed{\div}$ $\boxed{\text{RCL}}$ $\boxed{=}$	5.

Use the same method for these. Round to hundredths.

a) In $\boxed{A = \dfrac{B}{B+1}}$, when $B = 28.3$, A = _____

b) In $\boxed{P = \dfrac{W_S}{V_1 - V_2}}$, when $W_S = 9.75$, $V_1 = 8.64$, and $V_2 = 5.73$,

P = _____

---

140.   Do these on a calculator by evaluating the denominator first and storing that value, then evaluating the numerator and dividing by the stored value. Round to hundredths.

a) In $\boxed{R_T = \dfrac{R_1 R_2}{R_1 + R_2}}$, when $R_1 = 15$ and $R_2 = 20$,

R_T = _____

b) In $\boxed{W = \dfrac{2RF}{R - r}}$, when $R = 37.5$, $F = 4.6$, and $r = 2.9$,

W = _____

a) A = 0.97

b) P = 3.35

---

a) $R_T = 8.57$ b) W = 9.97

141.   Use storage for these.   Round to thousandths in (a) and to tenths in (b).

a)   In $A = \dfrac{A_f}{1 + BA_f}$,   when $A_f = 12.7$ and $B = 2.5$,

$$A = \underline{\hspace{2cm}}$$

b)   In $R = \dfrac{wr}{w - 2F}$,   when $w = 7.58$, $r = 10.5$,   and $F = 2.14$,

$$R = \underline{\hspace{2cm}}$$

---

142.   Use storage for these.   In (b), be sure to evaluate $(P_2 - P_1)$ first and press $\boxed{=}$ to complete that subtraction when evaluating the numerator. Round to tenths in (b).

a)   In $I = \dfrac{E}{R_1 + R_2 + R_3}$,   when $E = 200$, $R_1 = 15$, $R_2 = 25$,

and $R_3 = 40$, $I = \underline{\hspace{2cm}}$

b)   In $T = \dfrac{BV_1(P_2 - P_1)}{V_1 - V_2}$,   when $B = 2.5$, $V_1 = 80$, $P_2 = 30$,

$P_1 = 20$,   and $V_2 = 50$, $T = \underline{\hspace{2cm}}$

a) A = 0.388

b) R = 24.1

---

143.   Use storage to do these on a calculator.

a)   In $t = \dfrac{h^2}{1 - h}$,   when $h = 0.75$, $t = \underline{\hspace{2cm}}$

b)   In $m = \dfrac{M}{F^2 - 1}$,   when $M = 150$ and $F = 11$, $m = \underline{\hspace{2cm}}$

a) I = 2.5

b) T = 66.7

---

144.   In the evaluations below, be sure to press $\boxed{=}$ to complete the subtraction in the numerator before dividing by the stored value.

a)   In $m = \dfrac{y_2 - y_1}{x_2 - x_1}$,   when $y_2 = 20$, $y_1 = 8$, $x_2 = 10$,

and $x_1 = 2$, $m = \underline{\hspace{2cm}}$

b)   In $v = \dfrac{s_2 - s_1}{t_2 - t_1}$,   when $s_2 = 100$, $s_1 = 10$, $t_2 = 60$,

and $t_1 = 20$, $v = \underline{\hspace{2cm}}$

a) t = 2.25

b) m = 1.25

145. Use a calculator for these. Round to hundredths.

a) In $w = \dfrac{P - p_1}{h - h_1}$, when $P = 45.5$, $p_1 = 28.7$, $h = 40$,

and $h_1 = 25$, $w =$ _____

b) In $I_{CO} = \dfrac{I_C - \beta I_B}{\beta + 1}$, when $I_C = 100$, $\beta = 2.75$, and

$I_B = 30$, $I_{CO} =$ _____

a) $m = 1.5$

b) $v = 2.25$

---

146. In the evaluation below, we must evaluate the denominator first and store that value. To evaluate the denominator, we evaluate $(t_2 - t_1)$ first and press $\boxed{=}$ to complete the subtraction. The steps are shown.

In $s = \dfrac{H}{m(t_2 - t_1)}$, find s when $H = 90$, $m = 1.5$, $t_2 = 50$, and $t_1 = 20$.

Enter	Press		Display	
50	$\boxed{-}$		50.	
20	$\boxed{=}$ $\boxed{x}$		30.	
1.5	$\boxed{=}$ $\boxed{STO}$		45.	
90	$\boxed{\div}$ $\boxed{RCL}$ $\boxed{=}$		2.	(s = 2)

Use the same steps for this one. Round to hundredths.

In $A = \dfrac{3T}{G(F - G)}$, when $T = 50.5$, $G = 12.8$, and $F = 15.1$,

$A =$ _____

a) $w = 1.12$

b) $I_{CO} = 4.67$

---

147. Do these. In (a), be sure to press $\boxed{=}$ to complete the subtraction in the numerator before dividing by the stored value. Round to thousandths.

a) In $\alpha = \dfrac{L_2 - L_1}{L_1(t_2 - t_1)}$, when $L_2 = 75$, $L_1 = 40$, $t_2 = 30$,

and $t_1 = 15$, $\alpha =$ _____

b) In $\beta = \dfrac{T(V_1 - V_2)}{V_1(P_2 - P_1)}$, when $T = 20$, $V_1 = 90$, $V_2 = 55$,

$P_2 = 80$, and $P_1 = 65$, $\beta =$ _____

$A = 5.15$

148. To evaluate the denominators below, we begin by performing the subtraction in the parentheses and press $\boxed{=}$ to complete that subtraction. Round to two decimal places.

a) In $\boxed{T = \dfrac{HL}{AK(t_2 - t_1)}}$, when $H = 33.8$, $L = 200$, $A = 10$,
$K = 22.5$, $t_2 = 30$, and $t_1 = 25$,

$$T = \underline{\hspace{2cm}}$$

b) In $\boxed{R = \dfrac{33,000H}{3.14d(F_1 - F_2)}}$, when $H = 1.5$, $d = 25$, $F_1 = 150$,
and $F_2 = 75$, $R = \underline{\hspace{2cm}}$

a) $\alpha = 0.058$

b) $\beta = 0.519$

---

149. To evaluate the denominator below, we must evaluate $2t(a - t)$ and store that value, then evaluate $(b - t_1)$ and multiply by the stored value. After placing the denominator in storage, we divide T by the value of the denominator. The steps are shown.

In $\boxed{S = \dfrac{T}{2t(a - t)(b - t_1)}}$, find S when $T = 500$, $t = 25$, $a = 35$,
$b = 40$, and $t_1 = 20$.

Enter	Press	Display
35	$\boxed{-}$	35.
25	$\boxed{=}$ $\boxed{x}$	10.
2	$\boxed{x}$	20.
25	$\boxed{=}$ $\boxed{\text{STO}}$	500.
40	$\boxed{-}$	40.
20	$\boxed{=}$ $\boxed{x}$ $\boxed{\text{RCL}}$ $\boxed{=}$ $\boxed{\text{STO}}$	10000.
500	$\boxed{\div}$ $\boxed{\text{RCL}}$ $\boxed{=}$	0.05 (S = 0.05)

a) $T = 6.01$

b) $R = 8.41$

SELF-TEST 13 (pages 153-172)

1. Find F. Round to the nearest whole number.

$$F = \frac{W(R - r)}{2R}$$

  $W = 456$

  $R = 27.3$

  $r = 13.9$

2. Find A. Round to hundredths.

$$A = 0.7854(D^2 - d^2)$$

  $D = 8.17$

  $d = 7.82$

3. Find H when: $F = 109$
                    $G = 27$

$$H = 180 - (F + G)$$

4. Find R_t when: $R_1 = 150$
                        $R_2 = 350$

$$R_t = \frac{R_1 R_2}{R_1 + R_2}$$

5. Find "m". Round to tenths.

$$m = \frac{y_2 - y_1}{x_2 - x_1}$$

  $y_2 = 35$

  $y_1 = 18$

  $x_2 = 22$

  $x_1 = 13$

6. Find "s". Round to thousandths.

$$s = \frac{H}{m(t_2 - t_1)}$$

  $H = 76.4$

  $m = 23.9$

  $t_2 = 8.55$

  $t_1 = 3.37$

ANSWERS: 1. $F = 112$ 3. $H = 44$ 5. $m = 1.9$

                  2. $A = 4.40$ 4. $R_t = 105$ 6. $s = 0.617$

SUPPLEMENTARY PROBLEMS - CHAPTER 4

Assignment 10

1. In $\boxed{R = V - T}$, find R when V = 512.7 and T = 179.8 .

2. In $\boxed{V = LWH}$, find V when L = 188 , W = 96 , and H = 0.355 . Round to hundreds.

3. In $\boxed{Q = br - hs}$, find Q when b = 6.29, r = 31.3, h = 2.46, and s = 57.8 . Round to tenths.

4. In $\boxed{L = \dfrac{A}{W}}$, find L when A = 9,410 and W = 58 . Round to the nearest whole number.

5. In $\boxed{t = \dfrac{1}{f}}$, find "t" when f = 576 . Round to five decimal places.

6. In $\boxed{v = \dfrac{Ftg}{m}}$, find "v" when F = 450, t = 0.836, g = 32.2, m = 1,790 . Round to hundredths.

7. In $\boxed{X = L - \dfrac{M}{P}}$, find X when L = 58.7 , M = 263 , and P = 9.14 . Round to tenths.

8. In $\boxed{P = \dfrac{H - E}{V}}$, find P when H = 850 , E = 570 , and V = 390 . Round to thousandths.

9. In $\boxed{C = \dfrac{5F - 160}{9}}$, find C when F = -40 .

10. In $\boxed{T = \dfrac{a - b}{c} + h}$, find T when a = 12.47 , b = 8.93 , c = 0.0152 , and h = 68.3 .
    Round to the nearest whole number.

11. In $\boxed{H = \dfrac{B}{VT}}$, find H when B = 138 , V = 0.515 , and T = 143 . Round to two decimal places.

12. In $\boxed{m = \dfrac{1}{4ap}}$, find "m" when a = 22.4 and p = 0.128 . Round to four decimal places.

13. In $\boxed{E = \dfrac{PL}{Ae}}$, find E when P = 2,000 , L = 15 , A = 0.0205 , and e = 0.047 .
    Round to millions.

14. In $\boxed{F = D - \dfrac{H}{ms}}$, find F when D = 20.7 , H = 590 , m = 8.22 , and s = 11.6 .
    Round to one decimal place.

Assignment 11

1. In $\boxed{P = I^2R}$, find P when I = 4.8 and R = 35 . Round to the nearest whole number.

2. In $\boxed{A = 0.7854d^2}$, find A when d = 0.942 . Round to three decimal places.

3. In $\boxed{R = \dfrac{pL}{d^2}}$, find R when p = 10.4 , L = 750 , and d = 64.1 . Round to hundredths.

Continued on following page.

4. In $\boxed{s = \dfrac{v^2}{2g}}$, find "s" when $v = 120$ and $g = 32.2$. Round to the nearest whole number.

5. In $\boxed{a = (b - c)^2}$, find "a" when $b = 5.42$ and $c = 1.86$. Round to one decimal place.

6. In $\boxed{p = m + r\sqrt{t}}$, find "p" when $m = 2.94$, $r = 1.37$, and $t = 20.6$.
Round to two decimal places.

7. In $\boxed{E = \sqrt{PR}}$, find E when $P = 80$ and $R = 180$.

8. In $\boxed{b = \sqrt{c^2 - a^2}}$, find "b" when $c = 510$ and $a = 170$. Round to tens.

9. In $\boxed{w = 2g\sqrt{\dfrac{a}{r}}}$, find "w" when $g = 47$, $a = 53$, and $r = 79$. Round to the nearest whole number.

10. In $\boxed{F = \dfrac{W}{\sqrt{ht}}}$, find F when $W = 4,860$, $h = 12.6$, and $t = 250$. Round to tenths.

11. In $\boxed{t = \dfrac{1}{b\sqrt{m}}}$, find "t" when $b = 8$ and $m = 5.32$. Round to thousandths.

12. In $\boxed{C = \sqrt{1 - \dfrac{H}{h}}}$, find C when $H = 75.3$ and $h = 91.8$. Round to three decimal places.

13. In $\boxed{V = s^3}$, find V when $s = 89.2$. Round to thousands.

14. In $\boxed{s = \sqrt[3]{V}}$, find "s" when $V = 260$. Round to hundredths.

Assignment 12

1. In $\boxed{v_1 = v_2 - at}$, find "v_1" when $v_2 = 950$, $a = 140$, and $t = 5.5$.

2. In $\boxed{e = ir_1 + ir_2}$, find "e" when $i = 0.325$, $r_1 = 65$, and $r_2 = 35$.

3. In $\boxed{I_a = \dfrac{E_1 - E_o}{R_a}}$, find I_a when $E_1 = 67.5$, $E_o = 22.5$, $R_a = 84.2$. Round to thousandths.

4. In $\boxed{F = \dfrac{m_1 m_2}{rd^2}}$, find F when $m_1 = 180$, $m_2 = 400$, $r = 0.0615$, and $d = 720$.
Round to hundredths.

5. In $\boxed{d = \sqrt{\dfrac{I_A - I_B}{m}}}$, find "d" when $I_A = 76.2$, $I_B = 49.6$, and $m = 15.8$.
Round to two decimal places.

6. In $\boxed{\dfrac{V_1}{V_2} = \dfrac{T_1}{T_2}}$, find V_1 when $V_2 = 1,500$, $T_1 = 300$, and $T_2 = 273$. Round to tens.

7. In $\boxed{\dfrac{F_1}{F_2} = \dfrac{d_2}{d_1}}$, find F_2 when $F_1 = 480$, $d_2 = 18.7$, and $d_1 = 32.6$.
Round to the nearest whole number.

8. In $\boxed{\dfrac{P_1 V_1}{T_1} = \dfrac{P_2 V_2}{T_2}}$, find P_1 when $V_1 = 120$, $T_1 = 500$, $P_2 = 90$, $V_2 = 60$, and $T_2 = 300$.

Continued on following page.

In Problems 9 to 11, report each answer in scientific notation with the first factor rounded to two decimal places.

9. In $\boxed{Q = CE}$, find Q when $C = 175 \times 10^{-12}$ and $E = 4.55 \times 10^3$.

10. In $\boxed{t = \dfrac{1}{f}}$, find "t" when $f = 1.88 \times 10^9$.

11. In $\boxed{Z = \sqrt{R^2 + X^2}}$, find Z when $R = 57.2 \times 10^6$ and $X = 39.6 \times 10^6$.

In Problems 12 to 14, find "y" in each equation when $x = 12.5$.

12. $\boxed{y = -5x + 8}$ 13. $\boxed{y = \dfrac{200}{x}}$ 14. $\boxed{y = 3x^2 - 7x}$

Assignment 13

1. In $\boxed{M = P(L - K)}$, find M when $P = 160$, $L = 48.3$, and $K = 17.5$. Round to hundreds.

2. In $\boxed{A = 0.7854(D^2 - d^2)}$, find A when $D = 3.28$ and $d = 2.93$. Round to hundredths.

3. In $\boxed{r = \dfrac{R(w - 2F)}{w}}$, find "r" when $R = 59.2$, $w = 435$, and $F = 127$. Round to tenths.

4. In $\boxed{d = (1 - t^2)K}$, find "d" when $t = 0.28$ and $K = 6,100$. Round to hundreds.

5. In $\boxed{E = \dfrac{(m_1 + m_2)v^2}{2}}$, find E when $m_1 = 375$, $m_2 = 540$, and $v = 81.9$.
   Round to hundred-thousands.

6. In $\boxed{A = 180 - (B + C)}$, find A when $B = 39$ and $C = 48$.

7. In $\boxed{\Delta L = \alpha L(T - T_0)}$, find ΔL when $\alpha = 1.3 \times 10^{-5}$, $L = 60$, $T = 100$, and $T_0 = 20$.

8. In $\boxed{I = \dfrac{E}{R_1 + R_2}}$, find I when $E = 12$, $R_1 = 220$, and $R_2 = 470$. Round to thousandths.

9. In $\boxed{R = \dfrac{wr}{w - 2F}}$, find R when $w = 783$, $r = 2.08$, and $F = 195$.
   Round to two decimal places.

10. In $\boxed{v = \dfrac{s_2 - s_1}{t_2 - t_1}}$, find "v" when $s_2 = 81.4$, $s_1 = 28.7$, $t_2 = 3.26$, and $t_1 = 1.19$.
    Round to one decimal place.

11. In $\boxed{m = \dfrac{M}{F^2 - 1}}$, find "m" when $M = 32,600$ and $F = 1.82$. Round to hundreds.

12. In $\boxed{R = \dfrac{33,000H}{3.14d(F_1 - F_2)}}$, find R when $H = 46.3$, $d = 5.72$, $F_1 = 6,180$, and $F_2 = 4,930$.
    Round to tenths.

Chapter 5

MEASUREMENT SYSTEMS

In this chapter, we will discuss the units of length, weight, time, liquid measures, and temperature in both the English System and the Metric System. Conversions of units within each system and between the two systems are discussed. The substitution and unity-fraction methods are shown for all conversions. The decimal-point-shift method is also shown for conversions within the Metric System. The metric prefixes are related to powers of ten.

5-1   LENGTH - ENGLISH SYSTEM

In this section, we will discuss the common units of length in the English System. The procedures for converting from one unit to another are shown.

> Note: The English System is sometimes called the "British" System or "Customary" System. Though it is called the "English" System, it is presently used primarily in the United States.

---

1.   The common equivalent English units of length are:

```
┌─────────────────────────────────────┐
│   1 foot (ft)  =  12 inches (in)     │
│   1 yard (yd)  =  3 ft               │
│   1 mile (mi)  =  5,280 ft or 1,760 yd│
└─────────────────────────────────────┘
```

To convert from a larger unit to a smaller unit, we use the conversion facts above to substitute equivalent measures. That is:

To convert from feet to inches, we substitute "12 in" for "1 ft".

   4 ft = 4(1 ft) = 4(12 in) = 48 in

To convert from yards to feet, we substitute "3 ft" for "1 yd".

   9 yd = 9(1 yd) = 9(3 ft) = 27 ft

To convert from miles to yards, we substitute "1,760 yd" for "1 mi".

   5 mi = 5(1 mi) = 5(1,760 yd) = 8,800 yd

Continued on following page.

1.   Continued

Following the examples, complete these conversions. Use a calculator when necessary.

a) 15 ft = _____ in   c) 10 mi = _____ ft

b) 20 yd = _____ ft   d) 12 mi = _____ yd

---

2.   The same method can be used to convert decimal-number measures. For example, we converted 2.5 yd to feet below.

$$2.5 \text{ yd} = 2.5(1 \text{ yd}) = 2.5(3 \text{ ft}) = 7.5 \text{ ft}$$

Complete these conversions.

a) 6.4 mi = _____ ft   b) 7.35 ft = _____ in

a) 180 in (from 15 x 12)

b) 60 ft (from 20 x 3)

c) 52,800 ft
       (from 10 x 5,280)

d) 21,120 yd
       (from 12 x 1,760)

---

3.   To convert yards to inches, we can use a two-step process: first converting yards to feet and then converting feet to inches. For example, we converted 5 yd to inches in two steps below.

   Step 1: Converting 5 yd to feet
            $$5 \text{ yd} = 5(1 \text{ yd}) = 5(3 \text{ ft}) = 15 \text{ ft}$$

   Step 2: Converting 15 ft to inches
            $$15 \text{ ft} = 15(1 \text{ ft}) = 15(12 \text{ in}) = 180 \text{ in}$$

The same conversion can be made in one step by substituting "36 in" for "1 yd".
            $$5 \text{ yd} = 5(1 \text{ yd}) = 5(36 \text{ in}) = 180 \text{ in}$$

Using either method, complete these conversions.

a) 3.5 yd = _____ in   b) 10 yd = _____ in

a) 33,792 ft

b) 88.2 in

---

4.   To convert from a smaller unit to a larger unit, we derive other conversion facts from the basic conversion facts. For example, we derived

$1 \text{ in} = \dfrac{1}{12} \text{ ft}$ from the basic fact below by multiplying both sides by $\dfrac{1}{12}$.

Notice that doing so is the same as dividing the "1" on the right side by "12".

$$12 \text{ in} = 1 \text{ ft}$$

$$\frac{1}{12}(12 \text{ in}) = \frac{1}{12}(1 \text{ ft})$$

$$1 \text{ in} = \frac{1}{12} \text{ ft}$$

Continued on following page.

a) 126 in

b) 360 in

4.   Continued

   Using the same method, complete these:

   a) Since 3 ft = 1 yd, 1 ft = _____ yd

   b) Since 5,280 ft = 1 mile, 1 ft = _____ mi

---

5.   Using the derived facts, we also convert from a smaller unit to a larger unit by substituting equivalent measures. That is:

   To convert from inches to feet, we substitute "$\frac{1}{12}$ ft" for "1 in".

   $$72 \text{ in } = 72(1 \text{ in}) = 72\left(\frac{1}{12} \text{ ft}\right) = \frac{72}{12} \text{ ft } = 6 \text{ ft}$$

   To convert from feet to yards, we substitute "$\frac{1}{3}$ yd" for "1 ft".

   $$27 \text{ ft } = 27(1 \text{ ft}) = 27\left(\frac{1}{3} \text{ yd}\right) = \frac{27}{3} \text{ yd } = 9 \text{ yd}$$

   To convert from feet to miles, we substitute "$\frac{1}{5,280}$ mi" for "1 ft".

   $$15,840 \text{ ft } = 15,840(1 \text{ ft}) = 15,840\left(\frac{1}{5,280} \text{ mi}\right) = \frac{15,840}{5,280} \text{ mi } = 3 \text{ mi}$$

   Following the examples, complete these conversions. Use a calculator when necessary.

   a) 264 in = _____ ft   c) 26,400 ft = _____ mi

   b) 81 ft = _____ yd   d) 12,320 yd = _____ mi

---

6.   When converting 49 ft to yards below, we rounded to the nearest tenth.

   $$49 \text{ ft } = 49(1 \text{ ft}) = 49\left(\frac{1}{3} \text{ yd}\right) = \frac{49}{3} \text{ yd } = 16.3 \text{ yd}$$

   Complete these conversions. Round to the nearest tenth.

   a) 21,890 yd = _____ mi   b) 74.9 in = _____ ft

Answer column (right side):

a) $\frac{1}{3}$ yd

b) $\frac{1}{5,280}$ mi

a) 22 ft $\left(\text{from } \frac{264}{12}\right)$

b) 27 yd $\left(\text{from } \frac{81}{3}\right)$

c) 5 mi $\left(\text{from } \frac{26,400}{5,280}\right)$

d) 7 mi $\left(\text{from } \frac{12,320}{1,760}\right)$

a) 12.4 mi

b) 6.2 ft

7. To convert inches to yards, we can use a two-step process: first converting inches to feet and then converting feet to yards. For example, we converted 288 in to yards in two steps below.

Step 1: Converting 288 in to feet

$$288 \text{ in} = 288(1 \text{ in}) = 288\left(\frac{1}{12} \text{ ft}\right) = \frac{288}{12} \text{ ft} = 24 \text{ ft}$$

Step 2: Converting 24 ft to yards

$$24 \text{ ft} = 24(1 \text{ ft}) = 24\left(\frac{1}{3} \text{ yd}\right) = \frac{24}{3} \text{ yd} = 8 \text{ yd}$$

The same conversion can be made in one step by substituting "$\frac{1}{36}$ yd" for "1 in".

$$288 \text{ in} = 288(1 \text{ in}) = 288\left(\frac{1}{36} \text{ yd}\right) = \frac{288}{36} \text{ yd} = 8 \text{ yd}$$

Using either method, complete these conversions.

a) 216 in = _____ yd   b) 900 in = _____ yd

---

8. When converting from a larger unit to a smaller unit, we multiply. Therefore we get more of the smaller unit. For example:

$$5 \text{ ft} = 5(12 \text{ in}) = 60 \text{ in} \qquad (60 \text{ is more than } 5)$$

When converting from a smaller unit to a larger unit, we divide. Therefore we get less of the larger unit. For example:

$$30 \text{ ft} = \frac{30}{3} \text{ yd} = 10 \text{ yd} \qquad (10 \text{ is less than } 30)$$

Using the facts above, complete these:

a) If we convert 5 miles to yards, will we get "more than 5" or "less than 5" yards? _____

b) If we convert 108 inches to yards, will we get "more than 108" or "less than 108" yards? _____

a) 6 yd   b) 25 yd

a) more than 5, since we multiply by 1,760

b) less than 108, since we divide by 36

5-2   LENGTH - METRIC SYSTEM

In this section, we will discuss the units of length in the <u>Metric System</u>. The procedures for English-Metric conversions and for conversions within the Metric System are discussed.

Note: The Metric System is the only system of measurement used in all major countries other than the United States. Its use in the United States is increasing.

---

9.   The basic unit of length in the metric system is the "<u>meter</u>". A meter is defined below in terms of inches.

$$1 \text{ meter } = 39.37 \text{ inches}$$

Since 1 yard is 36 inches, you can see that 1 meter is a little longer than 1 yard or 3 feet.

We converted 1 meter to the nearest hundredth of a yard below.

$$1 \text{ meter } = 39.37 \text{ inches } = \frac{39.37}{36} \text{ yard } = 1.09 \text{ yard}$$

Convert 1 meter to the nearest hundredth of a foot.

$$1 \text{ meter } = 39.37 \text{ inches } = \frac{39.37}{12} \text{ feet } = \underline{\hspace{2cm}} \text{ feet}$$

---

10.   Another commonly used unit of length in the metric system is the "<u>centimeter</u>". 1 centimeter equals $\frac{1}{100}$ meter. From the fact below, you can see that there are a little more than $2\frac{1}{2}$ centimeters in an inch.

$$1 \text{ inch } = 2.54 \text{ centimeters}$$

Since 1 inch is about $2\frac{1}{2}$ centimeters, 1 foot is about $2\frac{1}{2}$ x 12 or 30 centimeters and 1 yard is about $2\frac{1}{2}$ x 36 or 90 centimeters. Complete the exact conversions below.

a) 1 foot = 12 inches = 12(2.54 centimeters) = _____ centimeters

b) 1 yard = 36 inches = 36(2.54 centimeters) = _____ centimeters

Answer (margin): 3.28 feet $\left(\text{from } \frac{39.37}{12} \right)$

---

11.   To convert from feet to meters or from centimeters to inches, we can use the derived facts at the right below.

Since 1 meter = 3.28 feet , 1 foot = $\frac{1}{3.28}$ meter

Since 1 inch = 2.54 centimeters , 1 centimeter = $\frac{1}{2.54}$ inch

Continued on following page.

Answer (margin):
a) 30.48 centimeters
b) 91.44 centimeters

11. Continued

Using the derived facts, complete these. Round to hundredths.

a) 10 feet = _____ meters   b) 50 centimeters = _____ inches

---

12. A third commonly used unit of length in the metric system is the
"kilometer". 1 kilometer equals 1,000 meters. As you can see from
the facts below, a kilometer is about six-tenths of a mile and there are
a little more than $1\frac{1}{2}$ kilometers in a mile.

> 1 kilometer = 0.621 mile
>
> 1 mile = 1.61 kilometers

Using the facts above, complete these.

a) 100 kilometers = _____ miles   b) 100 miles = _____ kilometers

a) 3.05 meters
$\left(\text{from } \dfrac{10}{3.28}\right)$

b) 19.69 inches
$\left(\text{from } \dfrac{50}{2.54}\right)$

---

13. The other units of length in the metric system are defined in terms of a
meter. The definitions are:

> 1 **kilo**meter (km) = 1,000 meters (m)
>
> 1 **hecto**meter (hm) = 100 meters (m)
>
> 1 **deka**meter (dam) = 10 meters (m)
>
> 1 **deci**meter (dm) = $\dfrac{1}{10}$ meter (m)
>
> 1 **centi**meter (cm) = $\dfrac{1}{100}$ meter (m)
>
> 1 **milli**meter (mm) = $\dfrac{1}{1,000}$ meter (m)

The names for the other units contain a prefix followed by the word "meter".
Since the same prefixes and conversion factors are used in all types of
metric measures, they should be memorized. For example:

kilo- means 1,000

deka- means 10

centi- means $\dfrac{1}{100}$

Write the conversion factor that goes with each prefix.

a) hecto- _____   b) deci- _____   c) milli- _____

a) 62.1 miles

b) 161 kilometers

a) 100   b) $\dfrac{1}{10}$ c) $\dfrac{1}{1,000}$

14.   Write the prefix that goes with each conversion factor.

a) 1,000 _____   b) 10 _____   c) $\frac{1}{100}$ _____

---

15.   The abbreviations for units of length in the metric system are given in the table in Frame 13. They should also be memorized. For example:

The abbreviation for "meter" is "m".

The abbreviation for "hectometer" is "hm".

The abbreviation for "millimeter" is "mm".

Write the abbreviation for each unit of length.

a) kilometer _____   c) dekameter _____

b) centimeter _____   d) decimeter _____

a) kilo-

b) deka-

c) centi-

---

16.   Write the unit of length that goes with each abbreviation.

a) dam _____   c) hm _____

b) dm _____   d) mm _____

a) km   c) dam

b) cm   d) dm

---

17.   To convert dekameters, hectometers, and kilometers to meters, we multiply by 10, 100, or 1,000. Doing so is the same as shifting the decimal point one, two, or three places to the right. For example:

7 hm = 7(1 hm) = 7(100 m) = 7.00 or 700 m

Use the same method for these.

a) 19 dam = _____ m   b) 250 km = _____ m

a) dekameter

b) decimeter

c) hectometer

d) millimeter

---

18.   The decimal-point-shift method was used to convert 4.75 dam to meters below.

4.75 dam = 4.75(1 dam) = 4.75(10 m) = 4.75 or 47.5 m

Use the same method for these.

a) 12.5 hm = _____ m   b) .83 km = _____ m

a) 190 m (from 19(10))

b) 250,000 m
   (from 250(1,000))

---

a) 1,250 m

b) 830 m

19. To convert decimeters, centimeters, and millimeters to meters, we
     multiply by $\frac{1}{10}$, $\frac{1}{100}$, or $\frac{1}{1,000}$. Doing so is the same as dividing by 10,
     100, or 1,000, which can be done by shifting the decimal point one, two,
     or three places <u>to the left</u>. For example:

$$87 \text{ cm} = 87(1 \text{ cm}) = 87\left(\frac{1}{100}\text{ m}\right) = \frac{87}{100}\text{ m} = .87, \text{ or } .87 \text{ m}$$

     Use the same method for these.

     a) 9 dm = _____ m   b) 146 mm = _____ m

---

20. The decimal-point-shift method was used to convert 3.9 cm to meters below.

$$3.9 \text{ cm} = 3.9(1 \text{ cm}) = 3.9\left(\frac{1}{100}\text{ m}\right) = \frac{3.9}{100}\text{ m} = 03.9 \text{ or } .039 \text{ m}$$

     Use the same method for these.

     a) 1.44 dm = _____ m   b) 26.7 mm = _____ m

a) .9 m $\left(\text{from } \frac{9}{10}\right)$

b) .146 m $\left(\text{from } \frac{146}{1,000}\right)$

---

21. To convert from meters to dekameters, hectometers, or kilometers, we
     derive other conversion facts from the basic conversion facts. For
     example, we derived $1 \text{ m} = \frac{1}{100}\text{ hm}$ below by multiplying both sides by $\frac{1}{100}$
     Doing so is the same as dividing the "1" on the right side by 100.

$$100 \text{ m} = 1 \text{ hm}$$

$$\frac{1}{100}(100 \text{ m}) = \frac{1}{100}(1 \text{ hm})$$

$$1 \text{ m} = \frac{1}{100}\text{ hm}$$

     Using the same method, complete these:

     a) Since 10 m = 1 dam, 1 m = _____ dam

     b) Since 1,000 m = 1 km, 1 m = _____ km

a) .144 m

b) .0267 m

a) $\frac{1}{10}$ dam

b) $\frac{1}{1,000}$ km

22.   Using the derived facts, we can convert meters to dekameters, hectometers, and kilometers. For example, we converted 200 m to hectometers below by substituting "$\frac{1}{100}$ hm" for "1 m".

$$200 \text{ m} = 200(1 \text{ m}) = 200\left(\frac{1}{100} \text{ hm}\right) = \frac{200}{100} \text{ hm} = 2.00. \text{ or 2 hm}$$

Use the same method for these.

   a) 400 m = _____ dam   b) 99,000 m = _____ km

---

23.   We converted 36 m to dekameters below by substituting "$\frac{1}{10}$ dam" for "1 m".

$$36 \text{ m} = 36(1 \text{ m}) = 36\left(\frac{1}{10} \text{ dam}\right) = \frac{36}{10} \text{ dam} = 3.6. \text{ or 3.6 dam}$$

Use the same method to complete these.

   a) 925 m = _____ hm   b) 70 m = _____ km

a) 40 dam $\left(\text{from } \frac{400}{10}\right)$

b) 99 km $\left(\text{from } \frac{99,000}{1,000}\right)$

---

24.   To convert from meters to decimeters, centimeters, or millimeters, we also derive other conversion facts from the basic conversion facts. For example, we derived 1 m = 100 cm below by multiplying both sides by 100.

$$\frac{1}{100} \text{ m} = 1 \text{ cm}$$

$$100\left(\frac{1}{100} \text{ m}\right) = 100(1 \text{ cm})$$

$$1 \text{ m} = 100 \text{ cm}$$

Using the same method, complete these.

   a) Since $\frac{1}{10}$ m = 1 dm,   1 m = _____ dm

   b) Since $\frac{1}{1,000}$ m = 1 mm,   1 m = _____ mm

a) 9.25 hm

b) .07 km

---

25.   Using the derived facts, we can convert meters to decimeters, centimeters, or millimeters. For example, we converted 15 m to centimeters below by substituting "100 cm" for "1 m".

$$15 \text{ m} = 15(1 \text{ m}) = 15(100 \text{ cm}) = 15.00. \text{ or 1,500 cm}$$

Use the same method for these.

   a) .7 m = _____ dm   b) .69 m = _____ mm

a) 10 dm

b) 1,000 mm

26.   Complete each conversion.

    a) .056 m = _____ cm   b) .00715 m = _____ mm

a) 7 dm $\left(\text{from } .7(10)\right)$

b) 690 mm
$\left(\text{from } .69(1,000)\right)$

---

a) 5.6 cm   b) 7.15 mm

5-3   UNITY FRACTIONS AND LENGTH CONVERSIONS

In this section, we will show how unity fractions can be used for length conversions within the English System, between the English System and the Metric System, and within the Metric System.

27.   Using the fact that 1 foot = 12 inches, we set up two fractions below. Each fraction equals "1" because the numerator and denominator of each are equal. Because the fractions equal "1", they are called "<u>unity</u>" fractions.

$$\frac{1 \text{ ft}}{12 \text{ in}} = 1 \qquad\qquad \frac{12 \text{ in}}{1 \text{ ft}} = 1$$

Using the fact that 1 mile = 1,760 yards, set up two unity fractions below.

---

28.   Two unity fractions can be set up from any conversion fact. For example:

    1 inch = 2.54 centimeters   1 kilometer = 1,000 meters

$$\frac{1 \text{ in}}{2.54 \text{ cm}} \qquad \frac{2.54 \text{ cm}}{1 \text{ in}} \qquad\qquad \frac{1 \text{ km}}{1,000 \text{ m}} \qquad \frac{1,000 \text{ m}}{1 \text{ km}}$$

Set up two unity fractions from each conversion fact below.

    a) 1 kilometer = 0.621 mile   b) 1 meter = 100 centimeters

$\dfrac{1 \text{ mi}}{1,760 \text{ yd}}$ and $\dfrac{1,760 \text{ yd}}{1 \text{ mi}}$

---

29.   Using a unity fraction set up from 1 yard = 3 feet, we converted 5 yards to feet below. Notice that we cancelled the "1 yd's" as if they were numbers.

$$5 \text{ yd} = 5 \text{ yd}\left(\frac{3 \text{ ft}}{1 \text{ yd}}\right) = 5(1 \text{ yd})\left(\frac{3 \text{ ft}}{1 \text{ yd}}\right) = 5(3 \text{ ft}) = 15 \text{ ft}$$

The same conversion can be made without writing 5 yd explicitly as 5(1 yd). That is:

$$5 \text{ yd} = 5 \text{ yd}\left(\frac{3 \text{ ft}}{1 \text{ yd}}\right) = 5(3 \text{ ft}) = 15 \text{ ft}$$

a) $\dfrac{1 \text{ km}}{0.621 \text{ mi}}$ and $\dfrac{0.621 \text{ mi}}{1 \text{ km}}$

b) $\dfrac{1 \text{ m}}{100 \text{ cm}}$ and $\dfrac{100 \text{ cm}}{1 \text{ m}}$

Continued on following page.

29.   Continued

a) Complete the following conversion of 4 mi to kilometers.

$$4 \text{ mi } = 4 \text{ mi}\left(\frac{1.61 \text{ km}}{1 \text{ mi}}\right) = \underline{\hspace{2cm}} \text{ km}$$

b) Complete the following conversion of 3.5 m to centimeters.

$$3.5 \text{ m } = 3.5 \text{ m}\left(\frac{100 \text{ cm}}{1 \text{ m}}\right) = \underline{\hspace{2cm}} \text{ cm}$$

---

30.   Using a unity fraction set up from 12 inches = 1 foot, we converted 84 in to feet below.

$$84 \text{ in } = 84 \text{ in}\left(\frac{1 \text{ ft}}{12 \text{ in}}\right) = \frac{84}{12} \text{ ft } = 7 \text{ ft}$$

a) Complete the following conversion of 12.7 cm to inches.

$$12.7 \text{ cm } = 12.7 \text{ cm}\left(\frac{1 \text{ in}}{2.54 \text{ cm}}\right) = \underline{\hspace{2cm}} \text{ in}$$

b) Complete the following conversion of 45 m to kilometers.

$$45 \text{ m } = 45 \text{ m}\left(\frac{1 \text{ km}}{1,000 \text{ m}}\right) = \underline{\hspace{2cm}} \text{ km}$$

---

31.   The steps in the unity-fraction method for the conversion below are described.

$$3.5 \text{ mi } = \underline{\hspace{2cm}} \text{ ft}$$

1) Identify the conversion fact relating miles and feet.
It is: 1 mi = 5,280 ft

2) Of the two possible unity fractions, use the one with feet in the numerator and miles in the denominator so that miles can be cancelled.

Therefore: $3.5 \text{ mi } = 3.5 \text{ mi}\left(\dfrac{5,280 \text{ ft}}{1 \text{ mi}}\right) = 3.5(5,280 \text{ ft}) = 18,480 \text{ ft}$

Use the unity-fraction method for each conversion below.

a) 6 in = \underline{\hspace{2cm}} cm   b) .55 hm = \underline{\hspace{2cm}} m

---

Answers (right column):

a) 6.44 km,
   from: 4(1.61 km)

b) 350 cm,
   from: 3.5(100 cm)

a) 5 in, from: $\left(\dfrac{12.7}{2.54}\right)$in

b) .045 km,

   from: $\left(\dfrac{45}{1,000}\right)$km

a) 15.24 cm, from:

   $6 \text{ in}\left(\dfrac{2.54 \text{ cm}}{1 \text{ in}}\right)$

b) 55 m, from:

   $.55 \text{ hm}\left(\dfrac{100 \text{ m}}{1 \text{ hm}}\right)$

32. The steps in the unity-fraction method for the conversion below are described.

$$647 \text{ cm} = \underline{\hspace{2cm}} \text{ m}$$

1) Identify the conversion fact relating centimeters and meters. It is: 1 m = 100 cm

2) Of the two possible unity fractions, use the one with meters in the numerator and centimeters in the denominator <u>so that</u> centimeters <u>can be cancelled</u>.

Therefore: $647 \text{ cm} = 647 \text{ cm}\left(\dfrac{1 \text{ m}}{100 \text{ cm}}\right) = \left(\dfrac{647}{100}\right)\text{m} = 6.47 \text{ m}$

Use the unity-fraction method for each conversion below.

a) 21,120 yd = $\underline{\hspace{2cm}}$ mi   b) 566 m = $\underline{\hspace{2cm}}$ km

---

33. Use the unity-fraction method for each conversion. Round the answer for (b) to hundredths.

a) 3 m = $\underline{\hspace{2cm}}$ in   b) 50,000 ft = $\underline{\hspace{2cm}}$ mi

a) 12 mi, from:

$21,120 \text{ yd}\left(\dfrac{1 \text{ mi}}{1,760 \text{ yd}}\right)$

b) .566 km, from:

$566 \text{ m}\left(\dfrac{1 \text{ km}}{1,000 \text{ m}}\right)$

---

34. When using the unity-fraction method for conversions within the metric system, use conversion facts that do not involve fractions. For example:

Instead of $1 \text{ mm} = \dfrac{1}{1,000} \text{ m}$, use 1 m = 1,000 mm

Instead of $1 \text{ m} = \dfrac{1}{100} \text{ hm}$, use 1 hm = 100 m

Use the unity-fraction method for these.

a) 47 cm = $\underline{\hspace{2cm}}$ m   b) 2,500 m = $\underline{\hspace{2cm}}$ km

a) 118.11 in, from:

$3 \text{ m}\left(\dfrac{39.37 \text{ in}}{1 \text{ m}}\right)$

b) 9.47 mi, from:

$50,000 \text{ ft}\left(\dfrac{1 \text{ mi}}{5,280 \text{ ft}}\right)$

---

a) .47 m, from: $47 \text{ cm}\left(\dfrac{1 \text{ m}}{100 \text{ cm}}\right)$

b) 2.5 km, from: $2,500 \text{ m}\left(\dfrac{1 \text{ km}}{1,000 \text{ m}}\right)$

5-4   TWO-STEP CONVERSIONS IN THE METRIC SYSTEM

A two-step process can be used to convert from one non-basic unit to another non-basic unit in the metric system. We will discuss that two-step process in this section.

---

35.   To convert 3.7 km to dekameters, we can use a two-step process that involves converting to the basic unit "meter" first.

    1) <u>Converting 3.7 km to meters</u>.

$$3.7 \text{ km} = 3.7(1,000 \text{ m}) = 3,700 \text{ m}$$

    2) <u>Then converting 3,700 m to dekameters</u>.

$$3,700 \text{ m} = 3,700\left(\frac{1}{10} \text{ dam}\right) = \frac{3,700}{10} \text{ dam} = 370 \text{ dam}$$

Use the two-step process below to complete: 85 hm = _____ km

85 hm = _____ m   and   _____ m = _____ km

---

36.   To convert 375 mm to centimeters, we used the same two-step process below.

    1) <u>Converting 375 mm to meters</u>.

$$375 \text{ mm} = 375\left(\frac{1}{1,000} \text{ m}\right) = \frac{375}{1,000} \text{ m} = .375 \text{ m}$$

    2) <u>Then converting .375 m to centimeters</u>.

$$.375 \text{ m} = .375(100 \text{ cm}) = 37.5 \text{ cm}$$

Use the two-step process below to complete: 6.9 cm = _____ mm

6.9 cm = _____ m   and   _____ m = _____ mm

---

37.   To convert 503 dm to dekameters, we used the same two-step process below.

    1) <u>Converting 503 dm to meters</u>.

$$503 \text{ dm} = 503\left(\frac{1}{10} \text{ m}\right) = \frac{503}{10} \text{ m} = 50.3 \text{ m}$$

    2) <u>Then converting 50.3 m to dekameters</u>.

$$50.3 \text{ m} = 50.3\left(\frac{1}{10} \text{ dam}\right) = \frac{50.3}{10} \text{ dam} = 5.03 \text{ dam}$$

Continued on following page.

Answer column:

85 hm = 8,500 m

and

8,500 m = <u>8.5 km</u>

6.9 cm = .069 m

and

.069 m = <u>69 mm</u>

37.   Continued

Use the two-step process below to complete: 1.9 hm = _____ dm

1.9 hm = _____ m   and   _____ m = _____ dm

---

38.   We used the unity-fraction method to convert 99 dam to hectometers in a
two-step process below.

1) Converting 99 dam to meters.

$$99 \text{ dam} = 99 \text{ dam}\left(\frac{10 \text{ m}}{1 \text{ dam}}\right) = 990 \text{ m}$$

2) Then converting 990 m to hectometers.

$$990 \text{ m} = 990 \text{ m}\left(\frac{1 \text{ hm}}{100 \text{ m}}\right) = \frac{990}{100} \text{ hm} = 9.9 \text{ hm}$$

When the unity-fraction method is used, the conversion above can be done
in one step by multiplying by both unity fractions at the same time. That is:

$$99 \text{ dam} = 99 \text{ dam}\left(\frac{10 \text{ m}}{1 \text{ dam}}\right)\left(\frac{1 \text{ hm}}{100 \text{ m}}\right) = \frac{990}{100} \text{ hm} = 9.9 \text{ hm}$$

Use the unity-fraction method to complete: .63 km = _____ dam
Try to do it in one step.

---

39.   We used the unity-fraction method to convert 83 cm to millimeters in a
two-step process below.

1) Converting 83 cm to meters.

$$83 \text{ cm} = 83 \text{ cm}\left(\frac{1 \text{ m}}{100 \text{ cm}}\right) = \frac{83}{100} \text{ m} = .83 \text{ m}$$

2) Then converting .83 m to millimeters.

$$.83 \text{ m} = .83 \text{ m}\left(\frac{1,000 \text{ mm}}{1 \text{ m}}\right) = 830 \text{ mm}$$

The same conversion was done in one step below by multiplying by both
unity fractions at the same time.

$$83 \text{ cm} = 83 \text{ cm}\left(\frac{1 \text{ m}}{100 \text{ cm}}\right)\left(\frac{1,000 \text{ mm}}{1 \text{ m}}\right) = \frac{83,000}{100} \text{ mm} = 830 \text{ mm}$$

Use the unity-fraction method to complete: 295 mm = _____ dm
Try to do it in one step.

Answer column:

1.9 hm = 190 m
and
190 m = 1,900 dm

63 dam, from:

$$.63 \text{ km}\left(\frac{1,000 \text{ m}}{1 \text{ km}}\right)\left(\frac{1 \text{ dam}}{10 \text{ m}}\right)$$

40.   We used the unity-fraction method to convert 648 cm to dekameters in a two-step process below.

   1) Converting 648 cm to meters.

$$648 \text{ cm} = 648 \text{ cm}\left(\frac{1 \text{ m}}{100 \text{ cm}}\right) = \frac{648}{100} \text{ m} = 6.48 \text{ m}$$

   2) Then converting 6.48 m to dekameters.

$$6.48 \text{ m} = 6.48 \text{ m}\left(\frac{1 \text{ dam}}{10 \text{ m}}\right) = \frac{6.48}{10} \text{ dam} = .648 \text{ dam}$$

The same conversion was done in one step below by multiplying by both unity fractions at the same time.

$$648 \text{ cm} = 648 \text{ cm}\left(\frac{1 \text{ m}}{100 \text{ cm}}\right)\left(\frac{1 \text{ dam}}{10 \text{ m}}\right) = \frac{648}{1,000} \text{ dam} = .648 \text{ dam}$$

Use the unity-fraction method to complete: 7 hm = _____ dm
Try to do it in one step.

2.95 dm, from:

$$295 \text{mm}\left(\frac{1 \text{ m}}{1,000 \text{mm}}\right)\left(\frac{10 \text{dm}}{1 \text{m}}\right)$$

7,000 dm, from: $7 \text{ hm}\left(\frac{100 \text{ m}}{1 \text{ hm}}\right)\left(\frac{10 \text{ dm}}{1 \text{ m}}\right)$

5-5   THE DECIMAL-POINT-SHIFT METHOD FOR METRIC CONVERSIONS

Since the metric system is based on 10, unit conversions in the metric system involve only a shift of the decimal point. We will discuss the decimal-point-shift method for metric conversions in this section.

---

41.   The units of length in the metric system are diagrammed below with the larger units at the left. As you can see, the units are similar to the places in the decimal number system.

1,000	100	10	1	$\frac{1}{10}$	$\frac{1}{100}$	$\frac{1}{1,000}$
km	hm	dam	m	dm	cm	mm

To convert from a larger unit to a smaller unit, we can simply shift the decimal point to the right. The number of places shifted depends on the number of unit-places in the conversion. For example:

   To convert from dekameters to meters, we shift 1 place to the right because "m" is 1 unit-place to the right of "dam".

$$2.5 \text{ dam} = 2.5 \text{ or } 25 \text{ m}$$

Continued on following page.

41. Continued

To convert from meters to centimeters, we shift 2 places to the right because "cm" is 2 unit-places to the right of "m".

$$37 \text{ m} = 37.\underset{\frown}{00} \text{ or } 3,700 \text{ cm}$$

To convert from hectometers to decimeters, we shift 3 places to the right because "dm" is 3 unit-places to the right of "hm".

$$.28 \text{ hm} = \underline{\hspace{2cm}} \text{ dm}$$

---

42. How many places to the right would we shift the decimal point for each conversion below?

a) kilometers to meters _____

b) meters to decimeters _____

c) dekameters to centimeters _____

d) centimeters to millimeters _____

.280 or 280 dm

---

43. Use the decimal-point-shift method for each conversion.

a) .65 km = _____ m

b) 1.85 m = _____ cm

c) 7.7 dam = _____ dm

d) 49 cm = _____ mm

a) 3 places   c) 3 places

b) 1 place   d) 1 place

---

44. To convert from a smaller unit to a larger unit, we can simply shift the decimal point to the left. The number of places shifted depends on the number of unit-places in the conversion. For example:

To convert from decimeters to meters, we shift 1 place to the left because "m" is 1 unit-place to the left of "dm".

$$75 \text{ dm} = 7\underset{\frown}{5}. \text{ or } 7.5 \text{ m}$$

To convert from meters to hectometers, we shift 2 places to the left because "hm" is 2 unit-places to the left of "m".

$$8.3 \text{ m} = \underset{\frown}{} 08.3 \text{ or } .083 \text{ hm}$$

To convert from centimeters to dekameters, we shift 3 places to the left because "dam" is 3 unit-places to the left of "cm".

$$249 \text{ cm} = \underline{\hspace{2cm}} \text{ dam}$$

a) 650 m   c) 770 dm

b) 185 cm   d) 490 mm

---

45. How many places to the left would we shift the decimal point for each conversion below?

a) centimeters to meters _____

b) meters to kilometers _____

c) decimeters to dekameters _____

d) millimeters to centimeters _____

.249 or .249 dam

46. Use the decimal-point-shift method for these.

    a) 775 mm = _____ m   c) 940 cm = _____ dm

    b) 43 m = _____ dam   d) 19 dm = _____ hm

---

47. When using the decimal-point-shift method, remember these facts:

    To convert from a larger to a smaller unit, we shift to the right.
    To convert from a smaller to a larger unit, we shift to the left.

    Use the decimal-point-shift method for these.

    a) .48 km = _____ m   c) 1.8 m = _____ cm

    b) 27 m = _____ dam   d) 67 mm = _____ m

---

48. Use the decimal-point-shift method for these.

    a) 285 mm = _____ cm   c) 95 dam = _____ km

    b) 6.5 dm = _____ mm   d) 149 km = _____ hm

---

49. When converting from a larger unit to a smaller unit, we shift the decimal point to the right. Therefore, we get more of the smaller unit. For example:

$$3 \text{ km} = 3,000 \text{ m} \qquad (3,000 \text{ is more than } 3)$$
$$5 \text{ m} = 500 \text{ cm} \qquad (500 \text{ is more than } 5)$$

When converting from a smaller unit to a larger unit, we shift the decimal point to the left. Therefore, we get less of the larger unit. For example:

$$39 \text{ m} = .039 \text{ km} \qquad (.039 \text{ is less than } 39)$$
$$644 \text{ cm} = 6.44 \text{ m} \qquad (6.44 \text{ is less than } 644)$$

a) If we convert 16 km to dekameters, will we get "more than 16" or "less than 16" dekameters? _____

b) If we convert 75 cm to decimeters, will we get "more than 75" or "less than 75" decimeters? _____

SELF-TEST 14 (pages 176-193)

Do the following length conversions within the English system.

1.   198 in = _____ ft

2.   Round to tenths.

   38,000 ft = _____ mi

Do the following English-to-metric length conversions.

3.   15 in = _____ cm

4.   Round to tenths.

   300 ft = _____ m

Do the following metric-to-English length conversions.

5.   50 km = _____ mi

6.   Round to hundredths.

   3.8 cm = _____ in

Do these conversions involving meters.

7.   5,000 m = _____ km

8.   6.4 m = _____ cm

9.   83.9 mm = _____ m

Do the following length conversions within the metric system.

10.   98 cm = _____ mm

11.   42.7 mm = _____ dm

12.   .82 km = _____ hm

ANSWERS:

1. 16.5 ft	5. 31.05 mi	9. .0839 m
2. 7.2 mi	6. 1.50 in	10. 980 mm
3. 38.1 cm	7. 5 km	11. .427 dm
4. 91.4 m	8. 640 cm	12. 8.2 hm

5-6   WEIGHT (OR MASS)

In this section, we will discuss conversions of weight (or mass) units within the English System, between the English System and the Metric System, and within the Metric System.

50.   The common equivalent English units of weight (or mass) are:

> 1 pound (lb) = 16 ounces (oz)
>
> 1 ton (T) = 2,000 pounds

To convert from a larger unit to a smaller unit, we use the facts above. Complete these:

a) 10 lb = _____ oz   b) 5 T = _____ lb

---

51.   To convert from a smaller unit to a larger unit, we can use the derived facts below.

$$1 \text{ oz} = \frac{1}{16} \text{ lb} \qquad 1 \text{ lb} = \frac{1}{2,000} \text{ T}$$

Use the facts above to complete these.

a) 48 oz = _____ lb   b) 16,000 lb = _____ T

a) 160 oz   b) 10,000 lb

---

52.   Unity fractions can also be used for weight conversions. For example:

$$5 \text{ lb} = 5 \text{ lb}\left(\frac{16 \text{ oz}}{1 \text{ lb}}\right) = 5(16 \text{ oz}) = 80 \text{ oz}$$

$$24,000 = 24,000 \text{ lb}\left(\frac{1 \text{ T}}{2,000 \text{ lb}}\right) = \left(\frac{24,000}{2,000}\right)\text{T} = \underline{\hspace{2cm}}$$

a) 3 lb   b) 8 T

---

53.   Use either method for these.

a) 72 oz = _____ lb   b) 6.75 T = _____ lb

12 T

---

54.   The basic unit of weight (or mass) in the metric system is the "gram". Its abbreviation is "g". From the fact below, you can see that a gram is a very small measure because there are more than 28 grams in 1 ounce.

> 1 oz = 28.35 g

Using the fact above, complete these.

a) 8 oz = _____ grams   b) 1 lb = _____ grams

a) 4.5 lb

b) 13,500 lb

55. A kilogram (kg) is 1,000 grams (g). As you can see from the fact below, there are a little more than 2 pounds in 1 kilogram.

$$1 \text{ kg} = 2.20 \text{ lb}$$

Use the fact above to complete these.

a) 5 kg = _____ lb   b) 100 kg = _____ lb

a) 226.8 grams

b) 453.6 grams

---

56. To convert from grams to ounces or from pounds to kilograms, we can use the derived facts below.

$$1 \text{ g} = \frac{1}{28.35} \text{ oz} \qquad 1 \text{ lb} = \frac{1}{2.20} \text{ kg}$$

Use the derived facts above for these. Round to tenths.

a) 100 g = _____ oz   b) 100 lb = _____ kg

a) 11 lb   b) 220 lb

---

57. Unity fractions can also be used for English-Metric weight conversions. Complete these. In (b), round to tenths.

a) $12 \text{ oz} = 12 \cancel{oz}\left(\dfrac{28.35 \text{ g}}{1 \cancel{oz}}\right) = 12(28.35 \text{ g}) = $ _____ g

b) $10 \text{ lb} = 10 \cancel{lb}\left(\dfrac{1 \text{ kg}}{2.20 \cancel{lb}}\right) = \left(\dfrac{10}{2.20}\right)\text{kg} = $ _____ kg

a) 3.5 oz   b) 45.5 kg

---

58. Use either method for these. In (b), round to hundredths.

a) 7.5 kg = _____ lb   b) 150 g = _____ oz

a) 340.2 g

b) 4.5 kg

---

59. Except for a metric ton which is defined in terms of kilograms, other units of weight (or mass) in the metric system are defined in terms of a gram. The definitions are:

> 1 **metric ton** (MT or t) = 1,000 kilograms (kg)
>
> 1 **kilogram** (kg) = 1,000 grams (g)
>
> 1 **hectogram** (hg) = 100 grams (g)
>
> 1 **dekagram** (dag) = 10 grams (g)
>
> 1 **decigram** (dg) = $\dfrac{1}{10}$ gram (g)
>
> 1 **centigram** (cg) = $\dfrac{1}{100}$ gram (g)
>
> 1 **milligram** (mg) = $\dfrac{1}{1,000}$ gram (g)

a) 16.5 lb b) 5.29 oz

Continued on following page.

59.   Continued

With the exception of a metric ton, you can see these facts:

1. The same prefixes and conversion factors that were used for length are used.

2. The same abbreviations are used with "g" substituted for "m".

Using the facts above, complete these:

a) 3 kg = _____ g   c) 25 dg = _____ g

b) 5 MT = _____ kg   d) 17.9 mg = _____ g

---

60.   To convert from a gram to the other units, we derive facts like those below.

$$\text{Since} \quad 1 \text{ kg} = 1,000 \text{ g}, \quad 1 \text{ g} = \frac{1}{1,000} \text{ kg}$$

$$\text{Since} \quad 1 \text{ cg} = \frac{1}{100} \text{ g}, \quad 1 \text{ g} = 100 \text{ cg}$$

Using derived facts like those above, complete these.

a) 75 g = _____ hg   c) 3.88 g = _____ mg

b) 2.6 g = _____ dg   d) 5,000 kg = _____ MT

a) 3,000 g

b) 5,000 kg

c) 2.5 g

d) .0179 g

---

61.   Unity fractions can be used for weight conversions in the metric system. For example:

a) $3.5 \text{ dag} = 3.5 \text{ dag}\left(\frac{10 \text{ g}}{1 \text{ dag}}\right) = 3.5(10 \text{ g}) = $ _____ g

b) $168 \text{ cg} = 168 \text{ cg}\left(\frac{1 \text{ g}}{100 \text{ cg}}\right) = \frac{168}{100} \text{ g} = $ _____ g

a) .75 hg

b) 26 dg

c) 3,880 mg

d) 5 MT

---

62.   A two-step process can be used to convert from one non-basic unit to another non-basic unit in the metric system. For example, to convert 78 dg to centigrams below, we used two steps.

1) <u>Converting 78 dg to the basic unit "grams"</u>.

$$78 \text{ dg} = 78\left(\frac{1}{10} \text{ g}\right) = \frac{78}{10} \text{ g} = 7.8 \text{ g}$$

2) <u>Then converting 7.8 g to centigrams</u>.

$$7.8 \text{ g} = 7.8(100 \text{ cg}) = 780 \text{ cg}$$

Use the two-step process below to complete: 3.9 hg = _____ kg

3.9 hg = _____ g and _____ g = _____ kg

a) 35 g

b) 1.68 g

63. We used unity fractions in the two-step process below to convert 8.5 mg to decigrams.

1) Converting 8.5 mg to the basic unit "grams".

$$8.5 \text{ mg} = 8.5 \text{ mg}\left(\frac{1 \text{ g}}{1,000 \text{ mg}}\right) = \left(\frac{8.5}{1,000}\right)\text{g} = .0085 \text{ g}$$

2) Then converting .0085 g to decigrams.

$$.0085 \text{ g} = .0085 \text{ g}\left(\frac{10 \text{ dg}}{1 \text{ g}}\right) = .0085(10 \text{ dg}) = .085 \text{ dg}$$

When using unity fractions, the same conversion can be made in one step by multiplying by both unity fractions at the same time. That is:

$$8.5 \text{ mg} = 8.5 \text{ mg}\left(\frac{1 \text{ g}}{1,000 \text{ mg}}\right)\left(\frac{10 \text{ dg}}{1 \text{ g}}\right) = \left(\frac{85}{1,000}\right)\text{dg} = .085 \text{ dg}$$

Use the unity fraction method to complete: 1.9 kg = _____ dag
Try to do it in one step.

3.9 hg = 390 g

and

390 g = .39 kg

---

64. The decimal-point-shift method is the most efficient method for metric conversions. To use that method for weight conversions, the following table can be used.

1,000	100	10	1	$\frac{1}{10}$	$\frac{1}{100}$	$\frac{1}{1,000}$
kg	hg	dag	g	dg	cg	mg

If we are converting from a larger unit to a smaller unit, we shift the decimal point to the right. For example:

6.8 kg = 6.800 or 6,800 g   (Shifted 3 places)

42 cg =   42.0 or 420 mg   (Shifted 1 place)

If we are converting from a smaller unit to a larger unit, we shift the decimal point to the left. For example:

95 g = 95. or .95 hg   (Shifted 2 places)

62 hg = 62. or 6.2 kg   (Shifted 1 place)

Use the decimal-point-shift method for each conversion.

a) 6.48 dag = _____ g   c) 750 mg = _____ g

b) .26 g = _____ cg   d) 88 g = _____ dag

190 dag, from:

$$1.9 \text{ kg}\left(\frac{1,000 \text{ g}}{1 \text{ kg}}\right)\left(\frac{1 \text{ dag}}{10 \text{ g}}\right)$$

65.   Use the decimal-point-shift method for these.

    a) 27 kg = _____ dag   c) 45 dg = _____ dag

    b) 5.1 cg = _____ mg   d) 185 cg = _____ dg

a) 64.8 g	c) .75 g
b) 26 cg	d) 8.8 dag

a) 2,700 dag	b) 51 mg	c) .45 dag	d) 18.5 dg

5-7   TIME

The same units of time are used in both the English System and the Metric System. We will discuss those units and conversions between them in this section.

66.   The following units of time are used in both the English system and the metric system.

$$1 \text{ minute (min)} = 60 \text{ seconds (sec)}$$
$$1 \text{ hour (hr)} = 60 \text{ minutes (min)}$$

Use the facts above to complete these:

    a) 10 min = _____ sec   b) 5.25 hr = _____ min

67.   To convert from seconds to minutes or from minutes to hours, we can use the derived facts below.

$$1 \text{ sec} = \frac{1}{60} \text{ min} \qquad 1 \text{ min} = \frac{1}{60} \text{ hr}$$

Using the derived facts, complete these.

    a) 750 sec = _____ min   b) 645 min = _____ hr

a) 600 sec

b) 315 min

68.   Unity fractions can also be used for conversions of time units. For example:

    a) $15 \text{ min} = 15 \text{ min}\left(\frac{60 \text{ sec}}{1 \text{ min}}\right) = 15(60 \text{ sec}) = $ _____ sec

    b) $1,200 \text{ min} = 1,200 \text{ min}\left(\frac{1 \text{ hr}}{60 \text{ min}}\right) = \left(\frac{1,200}{60}\right) \text{ hr} = $ _____ hr

a) 12.5 min

b) 10.75 hr

a) 900 sec

b) 20 hr

69.   Since there are 60 minutes in 1 hour and 60 seconds in 1 minute, there are 60 x 60 or 3,600 seconds in 1 hour. That is:

$$1 \text{ hr} = 3,600 \text{ sec}$$

Using the fact above, complete these.

a) 2.5 hr = _____ sec   b) 1,800 sec = _____ hr

---

70.   To get a unit of time smaller than a second, the unit "millisecond" is used in the metric system. The usual conversion factor and abbreviation for "milli-" are used. That is:

$$1 \text{ millisecond (msec)} = \frac{1}{1,000} \text{ second (sec)}$$

Using the fact above, complete these.

a) 479 msec = _____ sec   b) 1.8 sec = _____ msec

a) 9,000 sec

b) .5 hr

a) .479 sec   b) 1,800 msec

5-8   LIQUID MEASURES

In this section, we will discuss the units that are used for liquid measures in both the English and Metric Systems.

---

71.   The common equivalent units of liquid measure in the English System are:

$$
\begin{aligned}
1 \text{ pint (pt)} &= 16 \text{ ounces (oz)} \\
1 \text{ quart (qt)} &= 2 \text{ pints} = 32 \text{ ounces} \\
1 \text{ gallon (gal)} &= 4 \text{ quarts} = 8 \text{ pints}
\end{aligned}
$$

Note: The unit "ounce" that is used for liquid measures is not the same as the unit "ounce" that is used as a measure of weight (or mass).

Using the facts above, complete these.

a) 3 pt = _____ oz   c) 5 gal = _____ qt

b) 4 qt = _____ pt   d) 5 qt = _____ oz

a) 48 oz   c) 20 qt

b) 8 pt   d) 160 oz

72.   Complete: a) 32 oz = _____ pt   c) 8 qt = _____ gal

       b) 10 pt = _____ qt   d) 40 pt = _____ gal

---

73.   Unity fractions can be used for conversions of liquid measures. For
       example:

         a) 10 gal = 10 gal$\left(\dfrac{4 \text{ qt}}{1 \text{ gal}}\right)$ = 10(4 qt) = _____ qt

         b) 320 oz = 320 oz$\left(\dfrac{1 \text{ pt}}{16 \text{ oz}}\right)$ = $\dfrac{320}{16}$ pt = _____ pt

a) 2 pt   c) 2 gal

b) 5 qt   d) 5 gal

---

74.   The basic unit of liquid measure in the metric system is the "liter".
       Its abbreviation is "ℓ". As you can see from the fact below, a liter
       is a little more than 1 quart.

       ┌───┐
       │   1 liter (ℓ) = 1.057 quart (qt)   │
       └───┘

       Using the fact above, complete these conversions.

         a) 10 ℓ = _____ qt   b) 1 ℓ = 1.057 qt = _____ pt

a) 40 qt

b) 20 pt

---

75.   Using the derived fact 1 qt = $\dfrac{1}{1.057}$ ℓ , complete these. Round to
       hundredths.

         a) 3 qt = _____ ℓ b) 10 qt = _____ ℓ

a) 10.57 qt

b) 2.114 pt

---

76.   The following fact can be used to convert from gallons to liters.

       ┌───┐
       │   1 gallon (gal) = 3.79 liters (ℓ)   │
       └───┘

       Using the fact above, complete these. Where necessary, round to
       hundredths.

         a) 10 gal = _____ ℓ c) 10 ℓ = _____ gal

         b) 25 gal = _____ ℓ d) 25 ℓ = _____ gal

a) 2.84 ℓ

b) 9.46 ℓ

---

77.   Another unit of liquid measure that is used in the metric system is the
       "milliliter". The usual conversion factor and abbreviation for "milli-"
       are used. That is:

       ┌───┐
       │   1 milliliter (mℓ) = $\dfrac{1}{1,000}$ liter (ℓ)   │
       └───┘

       Therefore: a) 275 mℓ = _____ ℓ b) 1.4 ℓ = _____ mℓ

a) 37.9 ℓ

b) 94.75 ℓ

c) 2.64 gal

d) 6.60 gal

78.   Complete:

      a) 25 mℓ = _____ ℓ b) .5 ℓ = _____ mℓ

    a) .275 ℓ

    b) 1,400 mℓ

---

    a) .025 ℓ b) 500 mℓ

5-9   RATE

In this section, we will discuss measures of rate in both the English System and the Metric System.

---

79.   Both velocity (or speed) and flow rate are rate measures. Each is stated as a ratio of a quantity to 1 unit of time. Two examples are discussed.

    If a car travels 150 miles in 3 hours, its velocity (or speed) is the ratio of miles traveled to hours traveled. The ratio is reduced so that the velocity (or speed) is stated in terms of 1 hour. That is:

$$\frac{150 \text{ miles}}{3 \text{ hours}} \text{ is reduced to } \frac{50 \text{ miles}}{1 \text{ hour}}$$

    If 60 liters of water flow into a tank in 10 minutes, the flow rate is the ratio of liters to minutes. The ratio is also reduced so that the flow rate is stated in terms of 1 minute. That is:

$$\frac{60 \text{ liters}}{10 \text{ minutes}} \text{ is reduced to } \frac{6 \text{ liters}}{1 \text{ minute}}$$

Ordinarily a slanted line is used to express a rate. When a slanted line is used, the "1" is not explicitly written in front of the time unit. That is:

$$\frac{50 \text{ miles}}{1 \text{ hour}} \text{ is written } 50 \text{ miles/hour}$$

$$\frac{6 \text{ liters}}{1 \text{ minute}} \text{ is written } \underline{\hspace{4cm}}$$

6 liters/minute

---

80.   Abbreviations are ordinarily used when writing rates, and the word "per" is used in their word names. For example:

    The word name for 50 mi/hr is "50 miles per hour".

    The word name for 6 ℓ/min is "_____".

6 liters per minute

81.   To convert miles per hour to kilometers per hour or liters per minute to gallons per minute, we simply convert miles to kilometers or liters to gallons. We use the facts: 1 mi = 1.61 km and 1 gal = 3.79 ℓ

Since   50 mi = 50(1.61) = 80.5 km, 50 mi/hr = 80.5 km/hr

Since   6 ℓ = $6\left(\dfrac{1}{3.79}\right)$ = 1.58 gal,   6 ℓ/min = 1.58 gal/min

Using the same facts, complete these conversions.

a) 10 gal/sec = _____ ℓ/sec   b) 100 km/hr = _____ mi/hr

---

82.   To convert meters per second to feet per second or vice versa, we use the fact: 1 m = 3.28 ft. The unity-fraction method can be used for the conversions.   For example:

$$50 \text{ m/sec} = \frac{50 \text{ m}}{1 \text{ sec}}\left(\frac{3.28 \text{ ft}}{1 \text{ m}}\right) = \frac{50(3.28 \text{ ft})}{1 \text{ sec}} = \frac{164 \text{ ft}}{1 \text{ sec}} = 164 \text{ ft/sec}$$

$$75 \text{ ft/sec} = \frac{75 \text{ ft}}{1 \text{ sec}}\left(\frac{1 \text{ m}}{3.28 \text{ ft}}\right) = \frac{75 \text{ m}}{3.28 \text{ sec}} = \frac{22.9 \text{ m}}{1 \text{ sec}} = 22.9 \text{ m/sec}$$

Use the unity-fraction method for these. In (b), round to tenths.

a) 100 m/sec = _____ ft/sec

b) 50 ℓ/min = _____ gal/min

a) 37.9 ℓ/sec

b) 62.1 mi/hr

---

83.   To convert gal/min to gal/sec or m/sec to m/min, we simply convert minutes to seconds or seconds to minutes. For example:

$$300 \text{ gal/min} = \frac{300 \text{ gal}}{1 \text{ min}} = \frac{300 \text{ gal}}{60 \text{ sec}} = \frac{5 \text{ gal}}{1 \text{ sec}} = 5 \text{ gal/sec}$$

$$2.5 \text{ m/sec} = \frac{2.5 \text{ m}}{1 \text{ sec}} = \frac{2.5 \text{ m}}{\frac{1}{60} \text{ min}} = \frac{(2.5 \text{ m})(60)}{1 \text{ min}} = 150 \text{ m/min}$$

Note: $\dfrac{2.5 \text{ m}}{\frac{1}{60}}$ = (2.5 m)$\left(\text{the reciprocal of } \dfrac{1}{60}\right)$ = (2.5 m)(60) = 150 m

Using the same method, complete these conversions.

a) 630 mi/hr = _____ mi/min

b) 7.5 cm/sec = _____ cm/min

a) 328 ft/sec, from:

$$\frac{100 \text{ m}}{1 \text{ sec}}\left(\frac{3.28 \text{ ft}}{1 \text{ m}}\right)$$

b) 13.2 gal/min, from:

$$\frac{50 \ell}{1 \text{ min}}\left(\frac{1 \text{ gal}}{3.79 \ell}\right)$$

84.   The unity-fraction method can also be used to change the time unit. For example:

$$225 \, \ell/\text{min} = \left(\frac{225 \, \ell}{1 \, \cancel{\text{min}}}\right)\left(\frac{1 \, \cancel{\text{min}}}{60 \, \text{sec}}\right) = \frac{225 \, \ell}{60 \, \text{sec}} = \frac{3.75 \, \ell}{1 \, \text{sec}} = 3.75 \, \ell/\text{sec}$$

$$1.5 \, \text{mi}/\text{min} = \left(\frac{1.5 \, \text{mi}}{1 \, \cancel{\text{min}}}\right)\left(\frac{60 \, \cancel{\text{min}}}{1 \, \text{hr}}\right) = \frac{(1.5 \, \text{mi})(60)}{1 \, \text{hr}} = \frac{90 \, \text{mi}}{1 \, \text{hr}} = 90 \, \text{mi}/\text{hr}$$

Use the unity-fraction method for these.

a) 495 mi/hr = _____ mi/min

b) 3.4 ℓ/sec = _____ ℓ/min

---

85.   To convert 40 mi/hr to ft/sec, we must convert both the length unit and the time unit. A two-step process can be used. (<u>Note</u>: 1 hr = 3,600 sec)

1) Converting 40 mi to ft.

$$\frac{40 \, \text{mi}}{1 \, \text{hr}} = \frac{40(5,280 \, \text{ft})}{1 \, \text{hr}} = \frac{211,200 \, \text{ft}}{1 \, \text{hr}}$$

2) Then converting 211,200 ft/hr to ft/sec.

$$\frac{211,200 \, \text{ft}}{1 \, \text{hr}} = \frac{211,200 \, \text{ft}}{3,600 \, \text{sec}} = \frac{58.7 \, \text{ft}}{1 \, \text{sec}} = 58.7 \, \text{ft}/\text{sec}$$

Use the two-step process to convert 84 km/hr to m/sec below. Round to tenths.

a) Convert 84 km/hr to m/hr.

b) Then convert m/hr to m/sec.

Right column answers:

a) 10.5 mi/min, from:

$$\frac{630 \, \text{mi}}{60 \, \text{sec}}$$

b) 450 cm/min, from:

$$\frac{7.5 \, \text{cm}}{\frac{1}{60} \, \text{min}} = \frac{(7.5 \, \text{cm})(60)}{1 \, \text{min}}$$

a) 8.25 mi/min, from:

$$\left(\frac{495 \, \text{mi}}{1 \, \text{hr}}\right)\left(\frac{1 \, \text{hr}}{60 \, \text{min}}\right)$$

b) 204 ℓ/min, from:

$$\left(\frac{3.4 \, \ell}{1 \, \text{sec}}\right)\left(\frac{60 \, \text{sec}}{1 \, \text{min}}\right)$$

a) 84 km/hr = 84,000 m/hr

b) 23.3 m/sec, from:

$$\frac{84,000 \, \text{m}}{1 \, \text{hr}} = \frac{84,000 \, \text{m}}{3,600 \, \text{sec}}$$

86.   Unity fractions can be used in the two-step process. For example, we
       converted 1.4 ℓ/sec to gal/min below. (Note: 1 gal = 3.79 ℓ)

   1) Converting 1.4 ℓ/sec to gal/sec.

$$\frac{1.4\ \ell}{1\ \text{sec}} = \left(\frac{1.4\ \ell}{1\ \text{sec}}\right)\left(\frac{1\ \text{gal}}{3.79\ \ell}\right) = \frac{1.4\ \text{gal}}{3.79\ \text{sec}} = \frac{.37\ \text{gal}}{1\ \text{sec}}$$

   2) Then converting .37 gal/sec to gal/min.

$$\frac{.37\ \text{gal}}{1\ \text{sec}} = \left(\frac{.37\ \text{gal}}{1\ \text{sec}}\right)\left(\frac{60\ \text{sec}}{1\ \text{min}}\right) = \frac{(60)(.37\ \text{gal})}{1\ \text{min}} = \frac{22.2\ \text{gal}}{1\ \text{min}} = 22.2\ \text{gal/min}$$

When the unity-fraction method is used, we can do the conversion above in
one step by multiplying by both unity fractions at the same time. That is:

$$\frac{1.4\ \ell}{1\ \text{sec}} = \left(\frac{1.4\ \ell}{1\ \text{sec}}\right)\left(\frac{1\ \text{gal}}{3.79\ \ell}\right)\left(\frac{60\ \text{sec}}{1\ \text{min}}\right) = \frac{(1.4)(60)(1\ \text{gal})}{3.79(1\ \text{min})} = \frac{84\ \text{gal}}{3.79\ \text{min}} = \frac{22.2\ \text{gal}}{1\ \text{min}} = 22.2\ \text{gal/min}$$

Use unity-fractions for this conversion. Try to do it in one step.

   35 cm/sec = _____ m/min

---

87.   We used unity fractions in a one-step process to convert 100 ft/sec to
       mi/hr below.

$$\frac{100\ \text{ft}}{1\ \text{sec}} = \left(\frac{100\ \text{ft}}{1\ \text{sec}}\right)\left(\frac{1\ \text{mi}}{5,280\ \text{ft}}\right)\left(\frac{3,600\ \text{sec}}{1\ \text{hr}}\right) = \frac{360,000\ \text{mi}}{5,280\ \text{hr}} = \frac{68.2\ \text{mi}}{1\ \text{hr}} = 68.2\ \text{mi/hr}$$

Use the one step process for these. Round (b) to hundredths.

   a) 50 m/sec = _____ km/hr

   b) 40 gal/min = _____ ℓ/sec

---

21 m/min, from:

$$\left(\frac{35\ \text{cm}}{1\ \text{sec}}\right)\left(\frac{1\ \text{m}}{100\ \text{cm}}\right)\left(\frac{60\ \text{sec}}{1\ \text{min}}\right)$$

a) 180 km/hr, from:

$$\left(\frac{50\ \text{m}}{1\ \text{sec}}\right)\left(\frac{1\ \text{km}}{1,000\ \text{m}}\right)\left(\frac{3,600\ \text{sec}}{1\ \text{hr}}\right)$$

b) 2.53 ℓ/sec, from:

$$\left(\frac{40\ \text{gal}}{1\ \text{min}}\right)\left(\frac{3.79\ \ell}{1\ \text{gal}}\right)\left(\frac{1\ \text{min}}{60\ \text{sec}}\right)$$

5-10   REPEATED CONVERSIONS

In this section, we will show how repeated conversions of the same type can be done on a calculator by either using "memory" or the constant key ⎾ K ⏌ (if the calculator has that key).

88.   To convert feet to inches, we multiply by 12 since 1 ft = 12 in . If we have to convert many feet measures to inches, we can use a calculator's "memory" capability. That is, we can use the "storage" key $\boxed{STO}$ to store 12 and then we can use the "recall" key $\boxed{RCL}$ to recall 12 when necessary. Following these steps, convert 3 ft, 10 ft, 20 ft, and 75 ft to inches.

Enter	Press	Display
3	$\boxed{x}$	3.
12	$\boxed{STO}$ $\boxed{=}$	36.
10	$\boxed{x}$ $\boxed{RCL}$ $\boxed{=}$	120.
20	$\boxed{x}$ $\boxed{RCL}$ $\boxed{=}$	240.
75	$\boxed{x}$ $\boxed{RCL}$ $\boxed{=}$	900.

Note: We did not have to use $\boxed{RCL}$ for the first multiplication.

89.   Using 1 inch = 2.54 centimeters and the steps in the last frame, complete these on a calculator.

a) 5 in = _____ cm   c) 17.5 in = _____ cm

b) 10 in = _____ cm   d) 48 in = _____ cm

90.   To convert seconds to minutes, we divide by 60 since 1 sec = $\frac{1}{60}$ min .

Use the steps below to convert 300 sec, 450 sec, 915 sec, and 1,230 sec to minutes on a calculator.

Enter	Press	Display
300	$\boxed{\div}$	300.
60	$\boxed{STO}$ $\boxed{=}$	5.
450	$\boxed{\div}$ $\boxed{RCL}$ $\boxed{=}$	7.5
915	$\boxed{\div}$ $\boxed{RCL}$ $\boxed{=}$	15.25
1,230	$\boxed{\div}$ $\boxed{RCL}$ $\boxed{=}$	20.5

Note: We did not have to use $\boxed{RCL}$ for the first division.

a) 12.7 cm

b) 25.4 cm

c) 44.45 cm

d) 121.92 cm

91.   Using the steps in the last frame, complete these on a calculator. Divide
        by 1.057 since 1 quart $= \dfrac{1}{1.057}$ ℓ . Round to two decimal places.

        a) 2 qt = _____ ℓ c) 5.5 qt = _____ ℓ

        b) 4 qt = _____ ℓ d) 10 qt = _____ ℓ

---

92.   If your calculator has a "constant" key $\boxed{\text{ K }}$, you can use that key for
        repeated conversions. For example, to convert 2 mi, 3.5 mi, 12 mi,
        and 15.75 mi to feet below, we used $\boxed{\text{ K }}$ to multiply by 5,280. Notice
        that $\boxed{\text{ x }}$ is pressed before $\boxed{\text{ K }}$ and therefore $\boxed{\text{ x }}$ does not have to
        be repeated.

Enter	Press	Display
5,280	$\boxed{\text{x}}$ $\boxed{\text{K}}$	5,280.
2	$\boxed{=}$	10,560.
3.5	$\boxed{=}$	18,480.
12	$\boxed{=}$	63,360.
15.75	$\boxed{=}$	83,160.

        Using the same steps, complete these by multiplying by 28.35 , since
        1 oz = 28.35 g .

        a) 2 oz = _____ g   c) 10 oz = _____ g

        b) 4 oz = _____ g   d) 16 oz = _____ g

a) 1.89 ℓ

b) 3.78 ℓ

c) 5.20 ℓ

d) 9.46 ℓ

---

93.   To convert 32 oz, 48 oz, 160 oz, and 248 oz to pounds, we divide by 16
        since 1 oz $= \dfrac{1}{16}$ lb . Following the steps below, we can use the $\boxed{\text{ K }}$
        key to do so. Notice that we entered 16 and pressed $\boxed{\div}$ before
        pressing $\boxed{\text{ K }}$.

Enter	Press	Display
16	$\boxed{\div}$ $\boxed{\text{K}}$	16.
32	$\boxed{=}$	2.
48	$\boxed{=}$	3.
160	$\boxed{=}$	10.
248	$\boxed{=}$	15.5

        Using the same steps, complete these by dividing by 2.20 since
        1 lb $= \dfrac{1}{2.20}$ kg . Round to tenths.

        a) 5 lb = _____ kg   c) 17.9 lb = _____ kg

        b) 10 lb = _____ kg   d) 24.3 lb = _____ kg

a) 56.7 g

b) 113.4 g

c) 283.5 g

d) 453.6 g

---

a) 2.3 kg   b) 4.5 kg   c) 8.1 kg   d) 11.0 kg

SELF-TEST 15 (pages 194-207)

Do the following English-to-metric weight conversions.

1. Round to the nearest gram.

    32 oz = _____ g

2. Round to hundredths.

    15 lb = _____ kg

Do the following metric-to-English weight conversions.

3. Round to the nearest pound.

    4.6 kg = _____ lb

4. Round to tenths.

    250 g = _____ oz

Do the following weight conversions within the metric system.

5. .83 kg = _____ g

6. 60 mg = _____ g

7. 4.7 cg = _____ mg

Do the following time conversions.

8. 204 sec = _____ min

9. .09 sec = _____ msec

Do these liquid measure conversions.

10. 750 ml = _____ l

11. Round to tenths.   20 l = _____ gal

12. Convert 60 miles per hour to kilometers per hour. That is:

    60 mi/hr = _____ km/hr

13. Convert 2 liters per second to liters per minute. That is:

    2 l/sec = _____ l/min

14. Using the | STO | or | K | calculator key, convert the following to grams by multiplying each by 28.35 .

    a) 8 oz = _____ g   b) 16 oz = _____ g   c) 60 oz = _____ g

ANSWERS:

1. 907 g	5. 830 g	9. 90 msec	13. 120 l/min
2. 6.82 kg	6. .06 g	10. .75 l	14. a) 226.8 g
3. 10 lb	7. 47 mg	11. 5.3 gal	b) 453.6 g
4. 8.8 oz	8. 3.4 min	12. 96.6 km/hr	c) 1,701 g

5-11   TEMPERATURE

In this section, we will discuss measures of temperature in both the English and the Metric Systems. The Fahrenheit scale, the Celsius scale, and the Kelvin scale are discussed.

94.   The <u>Fahrenheit</u> scale is used to measure temperature in the English system.
The <u>Celsius</u> scale is used to measure temperature in the metric system.
The relationship between the two scales is outlined below.

From the outline, you can see
these facts:

Water boils at 212°
Fahrenheit and
100° Celsius.

Room temperature is
68° Fahrenheit and
20° Celsius.

Water freezes at
32° Fahrenheit and
0° Celsius.

When it is 0°
Fahrenheit, it is
_____° Celsius.

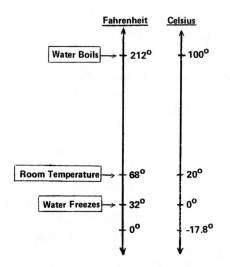

95.   The following formula is used to convert from degrees-Fahrenheit (F) to
degrees-Celsius (C).

$$C = \frac{5}{9}(F - 32°)$$

Therefore: If F = 86°,   C = $\frac{5}{9}$(86° - 32°) = $\frac{5}{9}$($\overset{6}{\cancel{54}}$°) = 30°

If F = 14°,   C = $\frac{5}{9}$(14° - 32°) = $\frac{5}{9}$($\overset{2}{-\cancel{18}}$°) = -10°

Using the above formula, complete these conversions.

a) If F = 104°, C = _____   b) If F = -13°, C = _____

-17.8° Celsius

a) C = 40°

b) C = -25°

96. The formula for converting degrees-Fahrenheit to degrees-Centigrade can be written in various forms. For example:

$$C = \frac{5}{9}(F - 32°) \qquad C = \frac{5(F - 32°)}{9} \qquad \boxed{C = \frac{5F - 160°}{9}}$$

Using the boxed form at the right, we converted 59°F to 15°C below.

$$C = \frac{5F - 160°}{9} = \frac{5(59°) - 160°}{9} = \frac{295° - 160°}{9} = \frac{135°}{9} = 15°$$

A calculator can be used for the same conversion. When doing so, it is easier to use the boxed form of the above formula. The steps are shown below.

Enter	Press	Display
5	$\boxed{\text{x}}$	5.
59	$\boxed{-}$	295.
160	$\boxed{=}$ $\boxed{÷}$	135.
9	$\boxed{=}$	15.

Using $C = \frac{5F - 160°}{9}$ and the same steps, do these on a calculator.

  a) If F = 122°, C = _____   b) If F = 41°, C = _____

97. We used $C = \frac{5F - 160°}{9}$ to convert -22°F to -30°C below.

$$C = \frac{5F - 160°}{9} = \frac{5(-22°) - 160°}{9} = \frac{-110° - 160°}{9} = \frac{-270°}{9} = -30°$$

Using the steps below, we can do the same conversion on a calculator.

Enter	Press	Display
5	$\boxed{\text{x}}$	5.
22	$\boxed{+/-}$ $\boxed{-}$	-110.
160	$\boxed{=}$ $\boxed{÷}$	-270.
9	$\boxed{=}$	-30.

Do these on a calculator.

  a) If F = -40°, C = _____   b) If F = 23°, C = _____

a) C = 50°

b) C = 5°

a) C = -40°

b) C = -5°

98.   Use a calculator and the same form of the formula for these. Round to tenths.

   a) If F = 196°,   C = _____   b) If F = -25°,   C = _____

---

99.   The following formula is used to convert from degrees-Celsius (C) to degrees-Fahrenheit (F).

$$F = \frac{9C}{5} + 32°$$

   Therefore: If C = 25°, $F = \frac{9(25°)}{5} + 32° = \frac{225°}{5} + 32° = 45° + 32° = 77°$

   If C = -20°, $F = \frac{9(-20°)}{5} + 32° = \frac{-180°}{5} + 32° = -36° + 32° = -4°$

Using the above formula, complete these conversions.

   a) If C = 10°,   F = _____   b) If C = -10°,   F = _____

a) C = 91.1°

b) C = -31.7°

---

100.   We used the same formula to convert 45°C to 113°F below.

$$F = \frac{9C}{5} + 32° = \frac{9(45°)}{5} + 32° = \frac{405°}{5} + 32° = 81° + 32° = 113°$$

A calculator can be used for the same conversion. The steps are:

Enter	Press	Display
9	x	9.
45	÷	405.
5	+	81.
32	=	113.

Use the same formula and a calculator for these.

   a) If C = 60°,   F = _____   b) If C = 5°,   F = _____

a) F = 50°

b) F = 14°

a) F = 140°

b) F = 41°

101.   We used the same formula to convert -15°C to 5°F below.

$$F = \frac{9C}{5} + 32° = \frac{9(-15°)}{5} + 32° = \frac{-135°}{5} + 32° = -27° + 32° = 5°$$

Using the steps below, we can do the same conversion on a calculator.

Enter	Press	Display
9	x	9.
15	+/- ÷	-135.
5	+	-27.
32	=	5.

Do these on a calculator.

a) If C = -30°, F = _____   b) If C = -50°, F = _____

102.   Use a calculator for these.

a) If C = 93°, F = _____   b) If C = -7°, F = _____

a) F = -22°
b) F = -58°

103.   Very high temperatures are used for processes like melting metals.
When converting high temperatures, the two formulas below are still used.

$$C = \frac{5F - 160°}{9} \qquad\qquad F = \frac{9C}{5} + 32°$$

Use a calculator for these conversions. Round to the nearest whole
number when necessary.

a) 900°F = _____ °C   c) 750°C = _____ °F

b) 1,500°F = _____ °C   d) 1,253°C = _____ °F

a) F = 199.4°
b) F = 19.4°

a) 482°C
b) 816°C
c) 1,382°F
d) 2,287°F

104.   The <u>Kelvin</u> scale is also used to measure temperature in the metric system. The relationship between the Fahrenheit, Celsius, and Kelvin scales is outlined below.

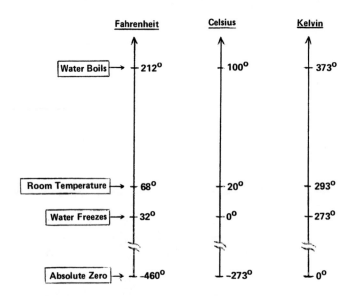

The advantage of the Kelvin scale is the fact that it has 0° at "absolute zero". Therefore, there are no negative values.

   Note: Temperature is related to the movement of molecules in substances. At "absolute zero", all movement of molecules ceases. Therefore, there are no temperatures below "absolute zero".

Though "absolute zero" is 0° Kelvin, it is _____ ° Fahrenheit and _____ ° Celsius.

---

105.   You can see from the outline in the last frame that the following formula relates degrees-Kelvin (K) and degrees-Celsius (C).

$$K = C + 273°$$

   Therefore: If C = 50°,   K = 50° + 273° = 323°

               If C = -100°,   K = -100° + 273° = 173°

Using the same formula, complete these.

   a) 125°C = _____ °K   b) -29°C = _____ °F

-460°F and -273°C

a) 398°K   b) 244°K

106. The following formula can be used to convert from degrees-Kelvin to degrees-Celsius.

$$C = K - 273°$$

Therefore: If K = 200°, C = 200° - 273° = -73°

If K = 500°, C = 500° - 273° = 227°

Using the same formula, complete these.

a) 189°K = _____ °C   b) 450°K = _____ °C

---

107. Complete these conversions.

a) 1,200°C = _____ °K   b) 2,000°K = _____ °C

a) -84°C   b) 177°C

---

108. To convert 323°K to degrees-Fahrenheit, we must convert to degrees-Celsius first. Therefore, two steps are needed.

1) <u>Converting 323°K to degrees-Celsius.</u>

C = K - 273° = 323° - 273° = 50°

2) <u>Converting 50°C to degrees-Fahrenheit.</u>

$$F = \frac{9C}{5} + 32° = \frac{9(50°)}{5} + 32° = 122°$$

Using the same two-step process, convert 253°K to degrees-Fahrenheit.

253°K = _____ °C = _____ °F

a) 1,473°K

b) 1,727°C

---

109. To convert 95°F to degrees-Kelvin, we must also convert to degrees-Celsius first. Therefore, a two-step process is needed.

1) <u>Converting 95°F to degrees-Celsius.</u>

$$C = \frac{5F - 160°}{9} = \frac{5(95°) - 160°}{9} = 35°$$

2) <u>Converting 35°C to degrees-Kelvin.</u>

K = C + 273° = 35° + 273° = 308°

Using the same two-step process, convert 5°F to degrees-Kelvin.

5°F = _____ °C = _____ °K

-20°C = -4°F

-15°C = 258°K

5-12   POWERS OF TEN AND METRIC PREFIXES

The conversion factors for metric prefixes are usually stated as powers of ten. In this section, we will discuss that use of powers of ten and extend the metric system to include prefixes with larger and smaller conversion factors.

110.   The conversion factors for the metrix prefixes discussed up to this point are expressed as powers of ten in the table below.

Prefix	Abbreviation	Conversion Factor	Place-Name
kilo-	k	$1000 = 10^3$	thousands
hecto-	h	$100 = 10^2$	hundreds
deka-	da	$10 = 10^1$	tens
Basic Unit $\longrightarrow$		10^0	ones
deci-	d	$\frac{1}{10} = 0.1 = 10^{-1}$	tenths
centi-	c	$\frac{1}{100} = 0.01 = 10^{-2}$	hundredths
milli-	m	$\frac{1}{1000} = 0.001 = 10^{-3}$	thousandths

Write the power-of-ten conversion factor for each prefix.

a) kilo- _____   b) deka- _____   c) centi- _____

---

111.   Write the <u>metric prefix</u> for each conversion factor.

a) 10^2 _____   b) 10^{-1} _____   c) 10^{-3} _____

a) 10^3 b) 10^1 c) 10^{-2}

---

112.   Write the <u>place-name</u> related to each prefix.

a) centi- _____   b) kilo- _____

a) hecto-

b) deci-

c) milli-

---

113.   Write the <u>prefix</u> related to each place-name.

a) thousandths _____   b) hundreds _____

a) hundredths

b) thousands

---

a) milli-

b) hecto-

114. In the table below, we have extended the metric system to include prefixes with larger and smaller conversion factors. Because of their size, the conversion factors are expressed only as powers of ten.

*Prefix	Abbreviation	Conversion Factor	Place-Name
tera-	T	10^{12}	trillions
giga-	G	10^{9}	billions
mega-	M	10^{6}	millions
kilo-	k	10^{3}	thousands
hecto-	h	10^{2}	hundreds
deka-	da	10^{1}	tens
Basic Unit →		10^{0}	ones
deci-	d	10^{-1}	tenths
centi-	c	10^{-2}	hundredths
milli-	m	10^{-3}	thousandths
micro-	μ	10^{-6}	millionths
nano-	n	10^{-9}	billionths
pico-	p	10^{-12}	trillionths

Note: 1) Beyond kilo- and milli-, prefixes are only added for every third place.

2) The abbreviation for micro- is "μ", which is the Greek letter for "m".

Write the power-of-ten conversion factor for each prefix.

a) giga- _____   b) micro- _____   c) pico- _____

115. Write the prefix for each conversion factor.

a) 10^{12} _____   b) 10^{6} _____   c) 10^{-9} _____

a) 10^{9}

b) 10^{-6}

c) 10^{-12}

116. Write the place-name related to each prefix.

a) mega- _____   c) micro- _____

b) giga- _____   d) pico- _____

a) tera-

b) mega-

c) nano-

*Note these pronunciations:

tera- (ter' ah)   nano- (nan' oh)
giga- (jig' ah)   pico- (peek' oh)

a) millions   c) millionths

b) billions   d) trillionths

117.   Write the prefix related to each place-name.

a) millions _____   c) millionths _____

b) trillions _____   d) billionths _____

---

118.   We wrote the abbreviations for some metric units below.

megameters = Mm	microseconds = μsec
gigagrams = Gg	picometers = pm

Write the abbreviation for each unit.

a) teragrams = _____   c) micrograms = _____

b) gigameters = _____   d) nanoseconds = _____

a) mega-   c) micro-

b) tera-   d) nano-

---

119.   Write the unit that goes with each abbreviation.

a) Mm = _____   c) Tg = _____

b) μg = _____   d) psec = _____

a) Tg   c) μg

b) Gm   d) nsec

---

120.   To avoid confusion with the measuring instrument called a "micrometer", the word "micron" is used instead of "micrometer" in the metric system. That is:

μm = micron

Write the unit that goes with each abbreviation.

a) Gm = _____   c) μm = _____

b) μsec = _____   d) ng = _____

a) megameter

b) microgram

c) teragram

d) picosecond

---

121.   The metric system can be extended to prefixes beyond "tera-" and "pico-". Some additional units and their power-of-ten conversion factors are shown below.

Prefix	Abbreviation	Conversion Factor
exa-	E	10^{18}
peta-	P	10^{15}
femto-	f	10^{-15}
atto-	a	10^{-18}

a) gigameter

b) microsecond

c) micron

d) nanogram

5-13   USING POWERS OF TEN FOR CONVERSIONS INVOLVING A BASIC UNIT

In this section, we will show how powers of ten can be used to convert from a non-basic unit to a basic unit and from a basic unit to a non-basic unit in the metric system.

---

122. The power-of-ten conversion factor can be used to express a non-basic unit as a basic unit. For example:

$$2.58 \text{ kilometers} = 2.58 \times 10^3 \text{ meters}$$

$$47 \text{ microseconds} = 47 \times 10^{-6} \text{ seconds}$$

Following the examples, complete these:

a) 6.75 megameters = _____ meters

b) 110 nanoseconds = _____ second

c) 14.6 teragrams = _____ grams

d) .59 picometer = _____ meter

---

123. We used the same method to express each non-basic unit below as a basic unit.

$$4.69 \text{ Mg} = 4.69 \times 10^6 \text{ g}$$

$$275 \ \mu\text{m} = 275 \times 10^{-6} \text{ m}$$

Following the examples, complete these.

a) 14.3 Tm = _____ m   c) ₁ .66 Gg = _____ g

b) 165 nsec = _____ sec   d) 1.22 psec = _____ sec

> a) 6.75×10^6 meters
>
> b) 110×10^{-9} second
>
> c) 14.6×10^{12} grams
>
> d) $.59 \times 10^{-12}$ meter

---

124. When using power-of-ten factors to express units with the prefixes "kilo-" and "milli-" as a basic unit, we frequently convert to an ordinary number. To do so, we simply multiply by the power of ten. For example:

$$14.8 \text{ km} = 14.8 \times 10^3 \text{ m} = 14,800 \text{ m}$$

$$2.66 \text{ mg} = 2.66 \times 10^{-3} \text{ g} = .00266 \text{ g}$$

Following the examples, complete these.

a) 854 kg = _____ g = _____ g

b) .95 msec = _____ sec = _____ sec

> a) 14.3×10^{12} m
>
> b) 165×10^{-9} sec
>
> c) $.66 \times 10^9$ g
>
> d) 1.22×10^{-12} sec

> a) 854×10^3 g = 854,000 g
>
> b) $.95 \times 10^{-3}$ sec = .00095 sec

125. When using power-of-ten factors to express units with prefixes beyond "kilo-" and "milli-" as a basic unit, we do not ordinarily convert to an ordinary number because the number would be very large or very small. For example:

$$2.89 \text{ Gm} \text{ would be expressed } 2.89 \times 10^9 \text{ m}$$

(Note: $2.89 \times 10^9 = 2,890,000,000$)

$$3.25 \text{ } \mu\text{sec} \text{ would be expressed } 3.25 \times 10^{-6} \text{ sec}$$

(Note: $3.25 \times 10^{-6} = .00000325$)

Complete these. Do not convert to an ordinary number.

a) 45.5 Mm = _____ m   c) .63 Tg = _____ g

b) 275 nsec = _____ sec   d) 1.99 psec = _____ sec

---

126. When a basic unit is expressed with a power-of-ten factor, we can use that factor to convert directly to a non-basic unit. For example:

$$17.5 \times 10^6 \text{ meters} = 17.5 \text{ megameters}$$
$$6.38 \times 10^{-9} \text{ seconds} = 6.38 \text{ nanoseconds}$$

Following the examples, complete these.

a) 47×10^3 grams = 47 _____

b) 8.88×10^9 meters = 8.88 _____

c) 640×10^{-3} second = 640 _____

d) 12.5×10^{-6} meters = 12.5 _____

a) 45.5×10^6 m

b) 275×10^{-9} sec

c) $.63 \times 10^{12}$ g

d) 1.99×10^{-12} sec

---

127. We used the same method to convert each basic unit below to a non-basic unit.

$$22 \times 10^{12} \text{ m} = 22 \text{ Tm}$$
$$.37 \times 10^{-12} \text{ sec} = .37 \text{ psec}$$

Following the examples, complete these.

a) 7.14×10^6 g = 7.14 _____   c) 804×10^{-6} m = 804 _____

b) 55×10^{-9} sec = 55 _____   d) $.44 \times 10^9$ m = .44 _____

a) 47 kilograms

b) 8.88 gigameters

c) 640 milliseconds

d) 12.5 microns

---

a) 7.14 Mg

b) 55 nsec

c) 804 μm

d) .44 Gm

128. When a basic unit is expressed as an ordinary number, we can convert to "kilo-" by first converting to a power-of-ten expression with 10^3. Some examples are shown.

$$2,750 \text{ m} = 2.75 \times 10^3 \text{ m} = 2.75 \text{ km}$$
$$64,800 \text{ g} = 64.8 \times 10^3 \text{ g} = 64.8 \text{ kg}$$

Following the examples, complete these.

a) 7,700 m = _____ x 10^3 m = _____ km

b) 125,000 g = _____ x 10^3 g = _____ kg

---

129. When a basic unit is expressed as an ordinary number, we can convert to "milli-" by first converting to a power-of-ten expression with 10^{-3}. Some examples are shown.

$$.00278 \text{ m} = 2.78 \times 10^{-3} \text{ m} = 2.78 \text{ mm}$$
$$.0199 \text{ g} = 19.9 \times 10^{-3} \text{ g} = 19.9 \text{ mg}$$

Following the examples, complete these.

a) .0085 g = _____ x 10^{-3} g = _____ mg

b) .036 m = _____ x 10^{-3} m = _____ mm

a) 7.7×10^3 m = 7.7 km

b) 125×10^3 g = 125 kg

a) 8.5×10^{-3} g = 8.5 mg   b) 36×10^{-3} m = 36 mm

5-14   THE DECIMAL-POINT-SHIFT METHOD FOR CONVERSIONS INVOLVING LARGER AND SMALLER UNITS

In this section, we will show how the decimal-point-shift method can be used for conversions involving units larger than <u>kilo-</u> and smaller than <u>milli-</u>.

---

130. The metric prefixes are shown below with the larger ones to the left. The prefixes are similar to the places in the decimal number system.

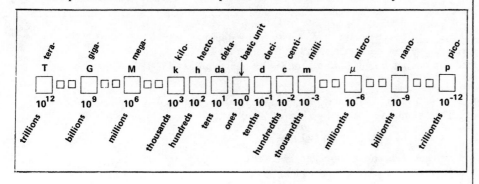

Continued on following page.

130. Continued

You can see the following facts from the diagram.

  giga- is larger than mega-

  (since 10^9 or billions is larger than 10^6 or millions)

  milli- is larger than micro-

  (since 10^{-3} or thousandths is larger than 10^{-6} or millionths)

Identify the larger prefix in each pair.

  a) tera- or giga-   c) micro- or nano-

  b) kilo- or mega-   d) pico- or nano-

---

131. From the diagram in the last frame, you can also see these facts.

  kilo- is smaller than mega-

  (since 10^3 or thousands is smaller than 10^6 or millions)

  pico- is smaller than nano-

  (since 10^{-12} or trillionths is smaller than 10^{-9} or billionths)

Identify the smaller prefix in each pair.

  a) mega- or giga-   c) nano- or micro-

  b) tera- or giga-   d) milli- or micro-

a) tera-   c) micro-

b) mega-   d) nano-

---

132. Identify the larger measure in each pair.

  a) km or Mm   b) ng or pg   c) Tm or Mm

a) mega-   c) nano-

b) giga-   d) micro-

---

133. Identify the smaller measure in each pair.

  a) μsec or nsec   b) Mg or Gg   c) nsec or msec

a) Mm   b) ng   c) Tm

---

134. To convert from a larger unit to a smaller unit, we shift the decimal point to the right and get more of the smaller unit. The number of places shifted depends on the number of unit-places in the conversion. For example:

  To convert from megameters to kilometers, we shift 3 places to the right because "km" is 3 unit-places to the right of "Mm".

$$1.5 \text{ Mm} = 1.500 \text{ or } 1,500 \text{ km}$$

  To convert from milliseconds to nanoseconds, we shift 6 places to the right because "nsec" is 6 unit-places to the right of "msec".

$$38 \text{ msec} = \underline{\hspace{3cm}} \text{ nsec}$$

a) nsec

b) Mg

c) nsec

135. How many places to the right would we shift the decimal point for each conversion below?

    a) Tg to Gg _____   b) nsec to psec _____   c) Gm to km _____

38.000000, or

38,000,000 nsec

---

136. Use the decimal-point-shift method for each conversion.

    a) 9.7 Mm = _____ km   c) .58 Tg = _____ Mg

    b) 14.6 μsec = _____ nsec   d) 2.3 mm = _____ nm

a) 3 places

b) 3 places

c) 6 places

---

137. To convert <u>from</u> a <u>smaller</u> <u>unit</u> <u>to</u> a <u>larger</u> <u>unit</u>, we shift the decimal point <u>to the left</u> and get <u>less</u> of the larger unit. The number of places shifted depends on the number of unit-places in the conversion. For example:

    To convert from megameters to gigameters, we shift <u>3 places</u> <u>to the left</u> because "Gm" is 3 unit-places to the left of "Mm".

$$8.4 \text{ Mm} = {}_{\wedge}008.4 \text{ or } .0084 \text{ Gm}$$

    To convert from picoseconds to microseconds, we shift <u>6 places</u> <u>to the left</u> because "μsec" is 6 unit-places to the left of "psec".

$$475 \text{ psec} = \text{_____} \text{ μsec}$$

a) 9,700 km

b) 14,600 nsec

c) 580,000 Mg

d) 2,300,000 nm

---

138. How many places to the left would we shift the decimal point for each conversion below?

    a) km to Mm _____   b) nsec to μsec _____   c) Mg to Tg _____

${}_{\wedge}000475.$ or

.000475 μsec

---

139. Use the decimal-point-shift method for each conversion.

    a) 255 Mm = _____ Gm   c) 1,400 kg = _____ Gg

    b) 88 μsec = _____ msec   d) 980 pm = _____ μm

a) 3 places

b) 3 places

c) 6 places

---

140. When using the decimal-point-shift method, remember these facts.

    1) To convert <u>from</u> a <u>larger</u> <u>to</u> a <u>smaller</u> <u>unit</u>, we shift <u>to the right</u>.

    2) To convert <u>from</u> a <u>smaller</u> <u>to</u> a <u>larger</u> <u>unit</u>, we shift <u>to the left</u>.

Use the decimal-point-shift method for these.

    a) 7.5 nsec = _____ psec   c) 66 ng = _____ mg

    b) 480 km = _____ Mm   d) .029 Gm = _____ km

a) .255 Gm

b) .088 msec

c) .0014 Gg

d) .00098 μm

---

a) 7,500 psec   c) .000066 mg

b) .48 Mm   d) 29,000 km

141.   The decimal-point-shift method can also be used for conversions involving a basic unit. For example:

To convert from kilograms to grams, we shift 3 places to the right.

4.8 kg = 4.800 or 4,800 g

To convert from seconds to microseconds, we shift 6 places to the right.

75 sec = _____ or _____ μsec

---

142.   Two more examples of the same method are given below.

To convert from millimeters to meters, we shift 3 places to the left.

405 mm = 405. or .405 m

To convert from grams to megagrams, we shift 6 places to the left.

73 g = _____ or _____ Mg

---

143.   Use the decimal-point-shift method for these conversions.

a) .63 Mm = _____ m   c) 45 sec = _____ msec

b) 2.9 g = _____ kg   d) 3,500 μsec = _____ sec

75.000000 or

75,000,000 μsec

000073. or

.000073 Mg

---

a) 630,000 m

b) .0029 kg

c) 45,000 msec

d) .0035 sec

SELF-TEST 16 (pages 208-223)

Do the following temperature conversions.

1. 14°F = _____°C | 2. 1,500°C = _____°F | 3. -53°C = _____°K

4. Write the power of ten associated with the metric prefix "nano-". _____

5. Write the metric prefix associated with 10^6. _____

Do the following metric conversions using the power-of-ten method. Leave your answers in power-of-ten form.

6. 2.9 mg = _____ g | 7. 73.8 km = _____ m | 8. 525 μsec = _____ sec

Do the following metric conversions by any method.

9. 98 x 10^{-9} sec = _____ nsec | 10. 5,720 g = _____ kg | 11. .46 m = _____ cm

Use the decimal-point-shift method for the following conversions.

12. 640 μsec = _____ msec | 13. 8.16 Mm = _____ km | 14. 95 ng = _____ μg

ANSWERS:
1. -10°C	5. mega-	9. 98 nsec	13. 8,160 km
2. 2,732°F	6. 2.9 x 10^{-3} g	10. 5.72 kg	14. .095 μg
3. 220°K	7. 73.8 x 10^3 m	11. 46 cm	
4. 10^{-9}	8. 525 x 10^{-6} sec	12. .64 msec	

SUPPLEMENTARY PROBLEMS - CHAPTER 5

Assignment 14

Do these conversions involving lengths in the English system.

1. 8.5 ft = _____ in   2. 3 mi = _____ ft   3. 144 in = _____ yd   4. 12,320 yd = _____ mi

Do these conversions involving lengths in the English and metric systems.

5. 50 in = _____ cm   6. Round to hundredths.   7. Round to tenths.   8. 50 mi = _____ km

10 yd = _____ m   15 ft = _____ m

9. Round to hundredths. 10. Round to nearest mile. 11. Round to nearest foot. 12. Round to tenths.

20 cm = _____ in   200 km = _____ mi   100 m = _____ ft   98 cm = _____ ft

Convert meters to other metric units in these problems.

13. 4.2 m = _____ mm   14. 6,500 m = _____ km 15. 84 m = _____ hm   16. 2.25 m = _____ cm

17. .59 m = _____ cm   18. 6 m = _____ dm   19. 337 m = _____ km   20. .07 m = _____ mm

Convert other metric units to meters in these problems.

21. 8.14 km = _____ m   22. 3 dam = _____ m   23. 480 cm = _____ m   24. 500 mm = _____ m

25. 63.3 dm = _____ m   26. 7.5 cm = _____ m   27. 10 hm = _____ m   28. .026 km = _____ m

Do these conversions involving lengths in the metric system.

29. 15 cm = _____ mm   30. 63 mm = _____ cm   31. 2.4 dm = _____ mm 32. 9 cm = _____ dm

33. 8.2 km = _____ hm   34. 10 hm = _____ km   35. .3 dam = _____ km   36. 750 hm = _____ dam

Assignment 15

Do these conversions involving weights in the English system.

1. 2.75 lb = _____ oz   2. 400 oz = _____ lb   3. 7,000 lb = _____ T   4. .5T = _____ lb

Do these conversions involving weights in the English and metric systems.

5. Round to tenths.   6. Round to hundredths.   7. Round to nearest pound.   8. Round to tenths.

12.5 oz = _____ g   50 g = _____ oz   8.17 kg = _____ lb   140 lb = _____ kg

Do these conversions involving weights in the metric system.

9. 3.2 kg = _____ g   10. 57 cg = _____ g   11. .43 g = _____ mg   12. 914 g = _____ kg

13. 4 cg = _____ mg   14. 28 mg = _____ cg   15. 6.1 kg = _____ dag 16. 1.9 hg = _____ dg

Do the following time conversions.

17. 6.5 min = _____ sec   18. 48 sec = _____ min   19. 135 min = _____ hr

20. 1.25 hr = _____ sec   21. 70 msec = _____ sec   22. 3.86 sec = _____ msec

Continued on following page.

Do the following conversions of liquid measures.

23. 14 qt = _____ gal   24. 264 oz = _____ qt   25. 3,190 ml = _____ ℓ

26. .8 ℓ = _____ ml   27. Round to <u>nearest liter</u>.   28. Round to <u>tenths</u>.

5.8 gal = _____ ℓ   9.73 ℓ = _____ qt

Do the following rate conversions.

29. 18,000 mi/hr = ____ mi/sec 30. 4.7 gal/sec = ____ gal/min 31. 100 km/hr = _____ mi/hr

32. 8.4 m/min = _____ cm/sec 33. 9 m/sec = _____ km/hr   34. Round to <u>nearest gallon</u>.

900 ℓ/min = _____ gal/sec

Using the $\boxed{\text{STO}}$ or $\boxed{\text{K}}$ calculator key, convert the following to kilometers by multiplying each by 1.61 .

35. 30 mi = _____ km   36. 220 mi = _____ km   37. 400 mi = _____ km

<u>Assignment 16</u>

Do the following temperature conversions.

1. 68°F = _____°C   2. 1,832°F = _____°C 3. -15°C = _____°F   4. 100°C = _____°F

5. -140°C = _____°F 6. 23°F = _____°C   7. 300°K = _____°C   8. -210°C = _____°K

Write the <u>power of ten</u> associated with each metric prefix.

9. mega- _____   10. pico- _____   11. tera- _____   12. milli- _____

Write the <u>metric prefix</u> associated with each power of ten.

13. 10^{-6} _____   14. 10^3 _____   15. 10^{-9} _____   16. 10^9 _____

Use the power-of-ten method for these metric conversions. Leave your answers in power-of-ten form.

17. 18 msec = _____ sec 18. 5.6 km = _____ m 19. 240 μg = _____ g

Do these metric conversions by any method.

20. 5.2×10^6 m = _____ Mm   21. 19×10^{-3} g = _____ mg 22. 376×10^{-12} sec = _____ psec

23. 8,700 g = _____ kg   24. .0045 m = _____ mm 25. .0000294 sec = _____ μsec

Use the decimal-point-shift method for these conversions.

26. 3.4 cm = _____ mm   27. 600 ng = _____ μg   28. .75 mm = _____ μm

29. 1,200 μsec = _____ msec 30. 5.1 Mm = _____ km   31. .04 hm = _____ m

32. .0082 dag = _____ mg   33. 7.07 nsec = _____ psec 34. 93,800 dm = _____ hm

Chapter 6

AREAS AND VOLUMES

In this chapter, we will review the common formulas for area, volume, and surface area. The basic units for area and volume in both the English System and the Metric System are discussed. A brief discussion of density and weight is included.

6-1   RECTANGLES AND SQUARES

In this section, we will review the formulas for the perimeter and area of a rectangle and a square.

---

1. In a <u>rectangle</u>, there are two pairs of equal sides and four right angles (90° angles).

    The <u>longer</u> sides are called the "length" (L).

    The <u>shorter</u> sides are called the "width" (W).

    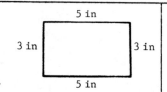

    The <u>perimeter</u> (P) of a rectangle is the distance around it. We can find the perimeter by adding the lengths of the four sides (L + W + L + W). Or we can double its length and double its width and add. That is:

    $$\boxed{P = 2L + 2W}$$

    When using the above formula, we treat the units as if they were letters. That is, for the rectangle above:

    $$P = 2(5 \text{ in}) + 2(3 \text{ in})$$
    $$= 10 \text{ in} + 6 \text{ in}$$
    $$= (10 + 6) \text{ in}$$
    $$= \underline{\qquad} \text{ in}$$

---

2. The area of a geometric figure is the number of "unit squares" needed to fill it. Some "unit squares" are shown below.

    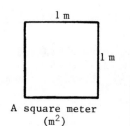

    A square inch
    (in^2)

    A square centimeter
    (cm^2)

    A square meter
    (m^2)

    Notice that we use the exponent "2" in the abbreviations for "unit squares".

    Therefore: The abbreviation for "a square foot" is "ft^2".

    The abbreviation for "a square kilometer" is $\underline{\qquad}$.

16 in

226

3. The word names for two area measures are given below.

    10 ft^2 is called "10 square feet".

    35 m^2 is called "35 square meters".

Write the abbreviation for each word name.

a) 25 square yards = _____   b) 84 square centimeters = _____

km^2

---

4. It takes 6 square centimeters to fill up the rectangle at the right. Therefore, its area (A) is 6 cm^2. Notice that we can find its area by multiplying its length and width. That is:

$$A = LW$$

When using the above formula, we also treat the units as if they were letters. That is, for the rectangle above:

A = (3 cm)(2 cm)

= (3)(2)(cm)(cm)

= _____ cm^2

(figure: rectangle, 3 cm wide, 2 cm high)

a) 25 yd^2

b) 84 cm^2

---

5. Use a calculator to find the perimeter and area of each rectangle. Don't forget to use square units when reporting an area.

a) (rectangle 4.6 yd by 1.8 yd)

b) (rectangle 3.4 m by 2.7 m)

P = _____   P = _____

A = _____   A = _____

6 cm^2

---

6. The symbols " and ' are also used as abbreviations for "inches" and "feet". For example:

9" means 9 inches   6' means 6 feet

a) 18.3' means 18.3 _____   b) 9.61" means 9.61 _____

a) P = 12.8 yd
   A = 8.28 yd^2

b) P = 12.2 m
   A = 9.18 m^2

---

7. Use a calculator to find the area of the rectangles with the following dimensions. Round to tenths.

a) L = 8.26', W = 2.77'   A = _____

b) L = 17.4", W = 3.51"   A = _____

a) feet

b) inches

8.   The formula A = LW can be rearranged to solve for either L or W. We get:

$$L = \frac{A}{W} \qquad\qquad W = \frac{A}{L}$$

The above formulas can be used to find L or W when the area and the other dimension are given. Use them to complete these. Round to two decimal places.

a) A = 108 cm², L = 25.5 cm   W = _____

b) A = 32.6 m², W = 4.75 m   L = _____

a) 22.9 ft²

b) 61.1 in²

---

9.   A <u>square</u> is a rectangle with four equal sides (s). Therefore:

The <u>perimeter</u> of a square is s + s + s + s. That is:

$$\boxed{P = 4s}$$

The <u>area</u> of a square is (s)(s). That is:

$$\boxed{A = s^2}$$

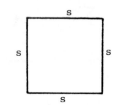

Using the formulas, find the perimeter and area of each square below. Round to the indicated place.

a)

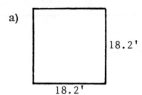

18.2'

18.2'

<u>Round to tenths</u>

P = _____

A = _____

b)

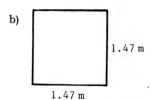

1.47 m

1.47 m

<u>Round to hundredths.</u>

P = _____

A = _____

a) W = 4.24 cm

b) L = 6.86 m

---

10.   The formula A = s² can be rearranged to solve for "s". We get:

$$s = \sqrt{A}$$

The new formula can be used to find the side of a square when its area is given. Using the formula, complete these. Round to tenths.

a) If A = 285 yd², s = _____

b) If A = 4.8 km², s = _____

a) P = 72.8 ft
   A = 331.2 ft²

b) P = 5.88 m
   A = 2.16 m²

a) s = 16.9 yd   b) s = 2.2 km

6-2   UNITS OF AREA

In this section, we will discuss the conversion facts for the basic units of area in both the English System and the Metric System.

11. Since 12 in = 1 ft, the figure at the right is 1 square foot. We converted 1 ft² to square inches below by substituting "12 in" for "1 ft".

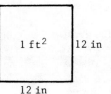

$$1 \text{ ft}^2 = (1 \text{ ft})(1 \text{ ft})$$
$$= (12 \text{ in})(12 \text{ in})$$
$$= 144 \text{ in}^2$$

Since 3 ft = 1 yd, the area of the square at the right is 1 square yard. Complete the conversion of 1 yd² to square feet below.

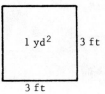

$$1 \text{ yd}^2 = (1 \text{ yd})(1 \text{ yd})$$
$$= (3 \text{ ft})(3 \text{ ft})$$
$$= \underline{\hspace{1cm}} \text{ ft}^2$$

12. The basic conversion facts for units of area in the English System are:

1 square foot (ft²) = 144 square inches (in²)
1 square yard (yd²) = 9 square feet (ft²)
1 square mile (mi²) = 640 acres

Using the facts above, complete these conversions.

a) 10 ft² = _____ in²   b) 10 yd² = _____ ft²

13. Using the same conversion facts, complete these. Round to tenths.

a) 500 in² = _____ ft²   b) 1,000 acres = _____ mi²

---

9 ft²

a) 1,440 in²

b) 90 ft²

a) 3.5 ft²

b) 1.6 mi²

14.   Since 1,000 m = 1 km, the figure at the right is 1 square kilometer. We converted 1 km² to square meters below by substituting "1,000 m" for "1 km".

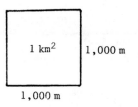

$$1 \text{ km}^2 = (1 \text{ km})(1 \text{ km})$$
$$= (1,000 \text{ m})(1,000 \text{ m})$$
$$= 1,000,000 \text{ m}^2$$

Since 100 cm = 1 m, the area of the figure at the right is 1 square meter. Complete the conversion of 1 m² to square centimeters below.

$$1 \text{ m}^2 = (1 \text{ m})(1 \text{ m})$$
$$= (100 \text{ cm})(100 \text{ cm})$$
$$= \underline{\hspace{1.5cm}} \text{ cm}^2$$

---

15.   The basic conversion facts for units of area in the metric system are:

1 square meter (m²) = 10,000 square centimeters (cm²)
1 square kilometer (km²) = 1,000,000 square meters (m²)
1 square kilometer (km²) = 100 hectares

Note: A "hectare" (pronounced "hect-air") is similar to an "acre" in the English system.

Using the facts above, complete these.

a) 5m² = _____ cm²   b) 9 km² = _____ hectares

10,000 cm²

---

16.   Using the same facts, complete these.

a) 5,000 cm² = _____ m²   b) 750 hectares = _____ km²

a) 50,000 cm²

b) 900 hectares

---

a) .5 m²

b) 7.5 km²

17. The basic conversion facts relating the English and metric systems are:

> $1 \text{ in}^2 = 6.45 \text{ cm}^2$
>
> $1 \text{ yd}^2 = 0.836 \text{ m}^2$
>
> $1 \text{ hectare (ha)} = 2.47 \text{ acres}$

Using the facts above, complete these:

a) $100 \text{ in}^2 = \underline{\hspace{1.5cm}} \text{ cm}^2$ b) $4 \text{ ha} = \underline{\hspace{1.5cm}} \text{ acres}$

---

18. Using the same facts, complete these. Round to hundredths.

a) $5 \text{ m}^2 = \underline{\hspace{1.5cm}} \text{ yd}^2$ b) $100 \text{ acres} = \underline{\hspace{1.5cm}} \text{ ha}$

a) 645 cm^2

b) 9.88 acres

---

a) 5.98 yd^2 b) 40.49 ha

6-3   PARALLELOGRAMS AND TRIANGLES

In this section, we will review the formulas for the area of a parallelogram and the area of a triangle.

---

19. A parallelogram is a four-sided figure <u>whose</u> <u>opposite</u> <u>sides</u> <u>are</u> <u>both</u> <u>parallel</u> <u>and</u> <u>equal</u>.

Identify the two pairs of parallel and equal sides in the parallelogram at the right.

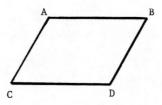

_____ and _____

_____ and _____

AB and CD

AC and BD

20.   Two lines are "__perpendicular__" to each other if they form right angles (90°) with each other. For example, lines "a", "b", and "c" are perpendicular to line "d" below.

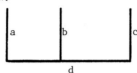

The "__height__" (or "altitude") of a parallelogram is the distance between two opposite sides. It is found by measuring the length of a perpendicular between the opposite sides. The perpendicular can be drawn at any point. For example, we drew three heights (h_1, h_2, and h_3) in the parallelogram below. The side to which they are drawn is called their "__base__" (b).

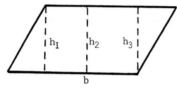

Parallel lines are equal distances apart at all points. Therefore, if

$h_1 = 5$ in, how long are h_2 and h_3 ? _____

---

21.   The area of a parallelogram is equal to its base (b) times its height (h). To show that fact, we can cut off part of the parallelogram and use it to form a rectangle as we have done below.

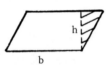 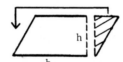

Since the area of the parallelogram equals the area of the rectangle, we can use the following formula for the area of a parallelogram.

$$A = bh$$

Using the formula, find the area of each parallelogram. Round to tenths.

a)

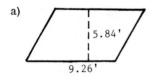

b)

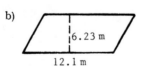

A = _____   A = _____

5 in

a) A = 54.1 ft^2

b) A = 75.4 m^2

22.   A triangle is a closed figure with three straight sides. A "height" of a triangle is the length of a line drawn from an angle <u>perpendicular</u> to the opposite side. The side to which the height is drawn is called its "<u>base</u>".

In the triangle at the right, we drew the height from angle B to side AC. Since the height is drawn to side AC we call side AC the _____ of that height.

---

23.   Since a height can be drawn from any of the three angles in a triangle to the opposite side, there are three <u>height</u>-<u>base</u> pairs in any triangle. For example, we drew the three heights for the triangle below.

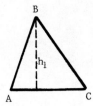 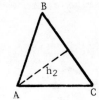 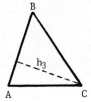

The base for height "h_1" is side AC.

a) The base for height "h_2" is side _____ .

b) The base for height "h_3" is side _____ .

base

---

24.   The area of a triangle equals one-half of any base times the height drawn to that base. That is:

$$A = \frac{1}{2}bh \qquad \text{or} \qquad A = \frac{bh}{2}$$

As you can see from the figure at the right, the formula makes sense because a triangle is simply half a parallelogram. Since the area of the parallelogram is "bh", the area of the triangle is half of "bh".

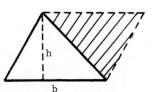

Using the form of the formula at the right above, find the area of these triangles. Round to tenths.

a)

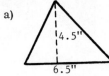

6.5"

A = _____

b)

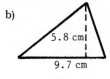

9.7 cm

A = _____

a) BC

b) AB

a) A = 14.6 in^2

b) A = 28.1 cm^2

25.   Find the area of each triangle.   Round to tenths.

a)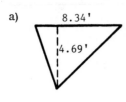

8.34'

4.69'

A = _____

b)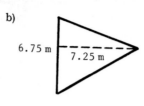

6.75 m

7.25 m

A = _____

---

26.   Sometimes we have to extend a side (or base) in order to draw the height to it.   For example, we extended side AC below so that a perpendicular could be drawn to it at D.   When a side is extended in order to draw a height to it, <u>the original side and not the extended side</u> is the base of that height.

Which of the following is the correct formula for the area of the triangle? _____

   a) $A = \frac{1}{2}(AC)(BD)$

   b) $A = \frac{1}{2}(AD)(BD)$

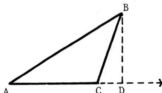

a) A = 19.6 ft^2

b) A = 24.5 m^2

---

27.   In the triangle at the right:

   a) The height is _____.

   b) The base is _____.

   c) The area is _____.

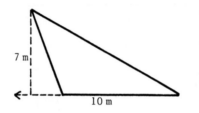

7 m

10 m

a) $A = \frac{1}{2}(AC)(BD)$

---

28.   In the triangle at the right:

   a) The height is _____.

   b) The base is _____.

   c) The area is _____
        (Round to thousandths.)

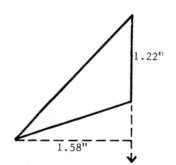

1.22"

1.58"

a) 7 m

b) 10 m

c) 35 m^2

a) 1.58 in

b) 1.22 in

c) 0.964 in^2

29. A right triangle is a triangle that contains
     a right angle (90°). The right angle is
     usually marked with a small square. In a
     right triangle:

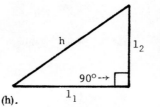

     1) The longest side (the one opposite the
         right angle) is called the "<u>hypotenuse</u>" (h).

     2) The two shorter sides are called the
         "<u>legs</u>" (l_1 and l_2).

a) In the triangle below,   b) In the triangle below,
    the hypotenuse is _____ .   the legs are _____ and _____ .

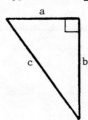

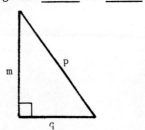

30. In a right triangle, each leg is the "<u>height</u>" to the other leg.   For   a) "c"
     example, in the triangle below:

                                                                                           b) "m" and "q"

         If a height is drawn from angle T
         to leg "t", it is identical to leg "a'.

         If a height is drawn from angle A
         to leg "a", it is identical to leg "t".

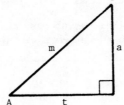

     In the triangle at the right:

     a) The height to leg "r" is leg _____ .

     b) The height to leg "v" is leg _____ .

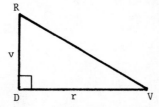

31. In the right triangle at the right: a) leg "v"

         1) leg 1 (or l_1) is the "<u>base</u>".   b) leg "r"

         2) leg 2 (or l_2) is the "<u>height</u>".

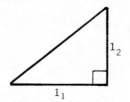

     Since the area of a triangle is half the base times the height, we use the
     following formula to find the area of a right triangle:

$$A = \frac{1}{2}(\text{leg } 1)(\text{leg } 2) \quad \text{or} \quad A = \frac{1}{2}l_1 l_2 \quad \text{or} \quad A = \frac{l_1 l_2}{2}$$

Continued on following page.

31.   Continued

Using the formula, find the area of each right triangle. Round to tenths.

a)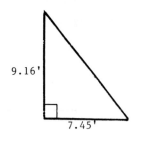

b)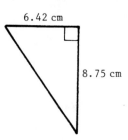

$$A = \frac{(\quad)(\quad)}{2} = \underline{\hspace{1cm}}$$

$$A = \frac{(\quad)(\quad)}{2} = \underline{\hspace{1cm}}$$

---

a) A = 34.1 ft^2

from: $\dfrac{(9.16')(7.45')}{2}$

b) A = 28.1 cm^2

from: $\dfrac{(6.42 \text{ cm})(8.75 \text{ cm})}{2}$

6-4   TRAPEZOIDS AND COMPOSITE FIGURES

In this section, we will discuss the areas of trapezoids and composite figures. Composite figures are those that can be divided into more basic geometric figures.

32.   A "trapezoid" is a four-sided figure with only one pair of parallel sides. To find the area of a trapezoid, we can divide it into two triangles as we have done with the dotted line at the right. The area of the trapezoid is equal to the sum of the areas of triangles I and II.

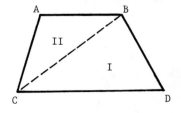

The height "h" of each triangle is equal to the height of the trapezoid, which is the distance between the two parallel sides.

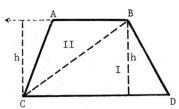

In triangle I, the base is CD. Therefore, the formula for its area is:

$$A = \frac{1}{2}(CD)(h)$$

In triangle II, the base is AB. Therefore, the formula for its area is:

$$A = \frac{1}{2}(\quad)(\quad)$$

$$A = \frac{1}{2}(AB)(h)$$

33. The trapezoid at the right is divided
     into two triangles. The height of
     each triangle equals the height of the
     trapezoid, which is 4". Let's find
     the area of the trapezoid by finding
     the areas of the two triangles and
     then adding.

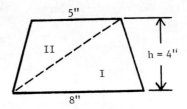

   a) The area of triangle I is $\frac{1}{2}(4")(8") =$ _____

   b) The area of triangle II is $\frac{1}{2}(4")(5") =$ _____

   c) The area of the trapezoid is _____

34. There is a single formula that can be used to find the area of a
     trapezoid. The formula is:

$$A = \frac{1}{2}h(b_1 + b_2) \qquad \text{or} \qquad A = \frac{h(b_1 + b_2)}{2}$$

      where: b_1 and b_2 are the parallel sides, and

              "h" is the height of the trapezoid.

We can see that the above formula makes sense by analyzing the trapezoid
below.

   The area of triangle I is: $\frac{1}{2}hb_1$.

   The area of triangle II is: $\frac{1}{2}hb_2$.

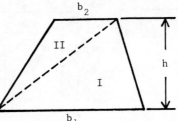

Therefore, the total area of the trapezoid is:

$$A = \frac{1}{2}hb_1 + \frac{1}{2}hb_2 = \frac{1}{2}h(b_1 + b_2) \qquad \text{or} \qquad \frac{h(b_1 + b_2)}{2}$$

Using the formula, find the area of each trapezoid below. Round to tenths
in (a) and to the nearest whole number in (b).

a)

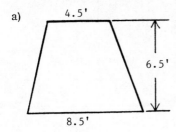

b)

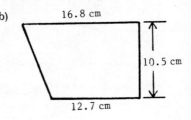

A = _____   A = _____

35.   The figures below are called "composite" figures.   The dotted lines divide
      them into more basic figures.

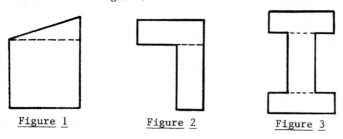

Figure 1   Figure 2   Figure 3

a) Figure 1 is divided into a _____ and a _____ .

b) Figure 2 is divided into two _____ .

c) Figure 3 is divided into three _____ .

a) triangle and a square

b) rectangles

c) rectangles

36.   To find the area of the composite figure below, we divide it into two
      rectangles, find the area of each rectangle, and then add the two areas.
      Two possible ways of dividing the same figure into rectangles are
      shown.

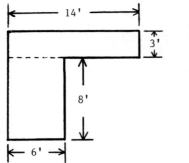
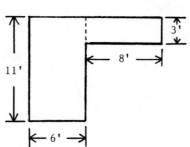

If we compute the total area of the figure at the left, we get:

(3' x 14') + (6' x 8') = 42 ft^2 + 48 ft^2 = 90 ft^2

a) If we compute the total area of the figure at the right, we get:

(3' x 8') + (6' x 11') = _____ + _____ = _____

b) Did we get the same total area both ways? _____

a) 24 ft^2 + 66 ft^2 = 90 ft^2

b) Yes

37. The I-shaped figure at the right
     can be divided into three
     rectangles. Do so and compute
     the area of the total figure.

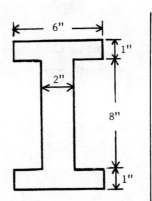

A = _____

38. The composite figure on the
     right is divided into a
     triangle and a rectangle.

     a) The area of the triangle
         is _____ .

     b) The area of the rectangle
         is _____ .

     c) The total area is
         _____ .

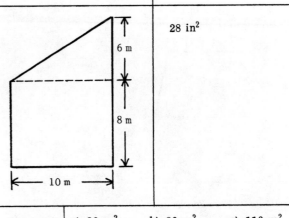

28 in²

a) 30 m²   b) 80 m²   c) 110 m²

SELF-TEST 17 (pp. 226-240)

1. Find the <u>area</u> of the rectangle below.
    Round to tenths.

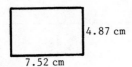

7.52 cm

A = _____

2. Find the <u>perimeter</u> of the square below.

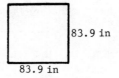

83.9 in

83.9 in

P = _____

3. Find the side of a square whose area is 184 ft². Round to tenths.   s = _____

Do the following conversions involving units of area.

4. 5 ft² = _____ in²

5. 20,000 cm² = _____ m²

6. 40 in² = _____ cm²

7. 250 acres = _____ hectares
    (Round to the nearest whole number.)

Continued on following page.

SELF-TEST 17 - Continued

Find the area of the parallelogram and the two triangles below.

8. Round to hundredths.

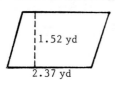

A = _____

9. Round to hundreds.

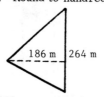

186 m   264 m

A = _____

10. Round to tens.

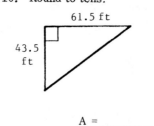

61.5 ft

43.5 ft

A = _____

11. Find the area of the trapezoid below. Round to the nearest whole number.

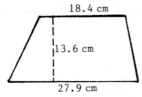

18.4 cm

13.6 cm

27.9 cm

A = _____

12. Find the area of the composite figure below. Round to the nearest whole number.

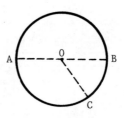

3.4"

5.2"

7.9"

4.8"

8.6"

A = _____

ANSWERS:

1. 36.6 cm^2
2. 335.6 in
3. 13.6 ft
4. 720 in^2

5. 2 m^2
6. 258 cm^2
7. 101 hectares
8. 3.60 yd^2

9. $24,600 \text{ m}^2$
10. $1,340 \text{ ft}^2$
11. 315 cm^2
12. 52 in^2

6-5   CIRCLES

In this section, we will discuss the formulas for the circumference and area of circles.

---

39.   A circle is a closed curved figure. In the circle below:

1) Point 0 is the <u>center</u>.

2) Since a <u>radius</u> (r) is a line connecting the center with any point on the circle, OA, OB, and OC are <u>radii</u> (pronounced "ray-dee-eye"). All radii of the same circle are equal.

3) Since a <u>diameter</u> (d) is a line passing through the center and touching both sides of the circle, AOB is a diameter. All diameters of the same circle are equal.

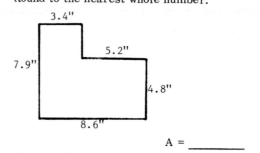

As you can see, a diameter is equal to two radii. That is:

$$d = 2r$$ or $$r = \frac{d}{2}$$

Continued on following page.

39.   Continued

Using the formulas, complete these:

a) Find the diameter of a circle   b) Find the radius of a circle if
    if its radius is 10 cm.   its diameter is 12 ft.

                d = _____   r = _____

---

40.   The distance around a circle is called its "circumference (C)". By break-
ing the circle below at A and bending its circumference into a straight
line, we can see that its circumference is slightly more than 3 times as
long as its diameter.

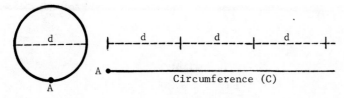

To find the ratio of the circumference of a circle to its diameter, we
divide C by d. For any circle, the ratio of C to d is slightly more than
3. The exact value is the unending number 3.141592... The Greek
letter "π" (pronounced "pie") is used for that number. Therefore:

$$\frac{C}{d} = \pi \qquad \text{or} \qquad \frac{C}{d} = 3.141592...$$

To find the value of π on a calculator, we simply press the $\boxed{\pi}$ key. Do so.

a) d = 20 cm

b) r = 6 ft

---

41.   Rearranging $\frac{C}{d} = \pi$ to solve for C, we get the following "circumference"
formula.

$$\boxed{C = \pi d}$$

Using that formula, we can find the circumference of a circle when its
diameter is 100 cm. To do so on a calculator, the following steps are
used.

Enter	Press	Display
π	$\boxed{\text{x}}$	3.1415927
100	$\boxed{=}$	314.15927

Rounding to the nearest whole number, C = _____ cm.

---

42.   Using a calculator and C = πd, find the circumference of the circles
with the following diameters. Round to tenths.

a) If d = 4.28 ft, C = _____ ft

b) If d = 11.7 m, C = _____ m

314 cm

43. Since d = 2r, we can also get a circumference formula in terms of the radius. That is:

$$C = \pi d$$

$$C = \pi(2r)$$

$$\boxed{C = 2\pi r}$$

Using a calculator and the bottom formula, complete these. Round to tenths.

a) If r = 3.64 in, C = _____ in

b) If r = 4.28 m, C = _____ m

a) 13.4 ft

b) 36.8 m

---

44. The two basic formulas for the circumference of a circle are:

$$C = \pi d \qquad\qquad C = 2\pi r$$

By rearranging them, we can solve for "d" and "r". We get:

$$d = \frac{C}{\pi} \qquad\qquad r = \frac{C}{2\pi}$$

Using the bottom formulas, we can find either the diameter or radius of a circle if its circumference is known. Complete these. Round to hundredths.

a) If C = 14.9 yd, d = _____ yd

b) If C = 23.7 cm, r = _____ cm

a) 22.9 in

b) 26.9 m

---

45. To find the area of a circle, we can use the following formula which states the area in terms of the radius.

$$\boxed{A = \pi r^2}$$

Using the above formula, find the area of the circles with the following radii.

a) If r = 5.75 in, A = _____ (Round to the nearest whole number.)

b) If r = 2.48 cm, A = _____ (Round to tenths.)

a) 4.74 yd

b) 3.77 cm

---

46. Since $r = \dfrac{d}{2}$, we can also get an area formula in terms of the diameter. That is:

$$A = \pi r^2$$

$$A = \pi\left(\frac{d}{2}\right)^2$$

$$A = \pi\left(\frac{d^2}{4}\right) = \frac{\pi d^2}{4} = \left(\frac{\pi}{4}\right)(d^2)$$

$$\boxed{A = 0.7854d^2}, \quad \text{since } \frac{\pi}{4} = 0.7854$$

a) 104 in²

b) 19.3 cm²

Continued on following page.

46.   Continued

Using the bottom formula, find the area of the circles with the following diameters.

a) If d = 16.5 ft, A = _____ (Round to the nearest whole number.)

b) If d = 6.89 cm, A = _____ (Round to tenths.)

---

47.   The two basic formulas for the area of a circle are shown below. The one on the left is used when the radius is given. The one on the right is used when the diameter is given.

$$A = \pi r^2 \qquad\qquad A = 0.7854d^2$$

Find the area. Round to hundredths.

a)

A = _____

b)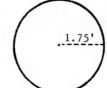

A = _____

a) 214 ft^2

b) 37.3 cm^2

---

48.   The figure at the right involves both a <u>square</u> and a <u>circle</u>. To find the area of the shaded part, <u>we must subtract the area of the circle from the area of the square</u>. Find the shaded area by answering the questions below. Round to tenths.

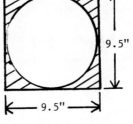

a) For the square, A = _____

b) The diameter of the circle is 9.5" (the same length as the side of the square). Therefore, for the circle, A = _____

c) For the shaded part, A = _____ - _____ = _____

a) 3.66 m^2

b) 9.62 ft^2

---

49.   To find the shaded area at the right, we must subtract the area of the circle from the area of the rectangle. (Round to tenths.)

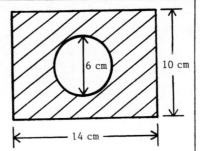

a) The area of the rectangle is _____.

b) The area of the circle is _____.

c) The area of the shaded part is _____.

a) 90.3 in^2

b) 70.9 in^2

c) 19.4 in^2, from: 90.3 in^2 - 70.9 in^2

50. At the right, a semi-circular
     arch with a 10-foot diameter
     has been cut out of a rectangle.
     The area of the semi-circle is
     half the area of a circle with
     the same diameter. Let's
     find the area of the shaded part.
     (Round to tenths.)

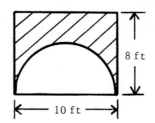
8 ft
10 ft

a) 140 cm²

b) 28.3 cm²

c) 111.7 cm²

   a) The area of the rectangle is _____.

   b) The area of the semi-circle is _____.

   c) The area of the shaded part is _____.

---

51. The shaded figure on the right is
     a cross-section of a large pipe
     Two circles are involved. The
     outer diameter is 17"; the inner
     diameter is 16".

     To find the area of the cross-
     section, we subtract the area
     of the smaller circle from the
     area of the larger circle.
     Round to the nearest whole
     number.

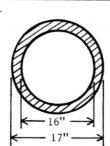
16"
17"

a) 80 ft²

b) 39.3 ft²

c) 40.7 ft²

   a) The area of the outer circle is _____.

   b) The area of the inner circle is _____.

   c) The area of the cross-section is _____

---

52. The composite figure at the right
     is divided into a rectangle and a
     semi-circle. Find its area.
     Round to tenths.

   a) The area of the rectangle is _____.

   b) The area of the semi-circle is _____.

   c) The total area is _____.

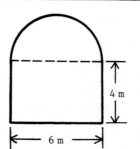
4 m
6 m

a) 227 in²

b) 201 in²

c) 26 in²

---

53. The two basic formulas for the area of a circle are:

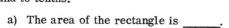

$$A = \pi r^2 \qquad\qquad A = 0.7854d^2$$

     By rearranging them, we can solve for "r" and "d". We get:

$$r = \sqrt{\frac{A}{\pi}} \qquad\qquad d = \sqrt{\frac{A}{0.7854}}$$

a) 24 m²

b) 14.1 m²

c) 38.1 m²

Continued on following page.

53.   Continued

Using the bottom formulas, we can find either the radius or diameter of a circle if its area is known.   Complete these.

a) If A = 1,650 in^2, r = _____ (Round to tenths.)

b) If A = 7.25 m^2, d = _____ (Round to hundredths.)

---

a) 22.9 in   b) 3.04 m

6-6   NON-GEOMETRIC FORMULAS CONTAINING "π"

In this section, we will show how a calculator can be used for evaluations with non-geometric formulas containing "π".

---

54.   The same calculator procedures are used when a formula contains the number "π".   However, we use the $\boxed{\pi}$ key to enter "π".   Do the evaluation below.   Round to tenths.

In $\boxed{X_L = 2\pi f L}$,   when f = 1,250 and L = 0.0065, X_L = _____

---

55.   To perform the evaluation below, we can find the value of the denominator and then press the reciprocal key $\boxed{1/x}$.   The steps are shown.

$X_L = 51.1$

In $\boxed{X_c = \dfrac{1}{2\pi f C}}$,   find X_c when f = 2,500 and C = 3.75 x 10^{-6}

Enter	Press	Display
2	$\boxed{x}$	2.
π	$\boxed{x}$	6.2831853
2,500	$\boxed{x}$	15707.963
3.75	$\boxed{EE}$	3.75 00
6	$\boxed{+/-}$ $\boxed{=}$ $\boxed{1/x}$	1.6977 01

The calculator answer is $X_c = 1.6977$ x 10^1.   As an ordinary number rounded to tenths, X_c = _____.

$X_c = 17.0$

56.   The steps for the evaluation below are shown. Notice that we evaluated $\sqrt{LC}$ first when finding the value of the denominator.

In $\boxed{f = \dfrac{1}{2\pi\sqrt{LC}}}$, find "f" when $L = 1.5 \times 10^{-3}$ and $C = 7.5 \times 10^{-6}$.

Enter	Press	Display	
1.5	$\boxed{EE}$	1.5	00
3	$\boxed{+/-}$ $\boxed{x}$	1.5	-03
7.5	$\boxed{EE}$	7.5	00
6	$\boxed{+/-}$ $\boxed{=}$ $\boxed{\sqrt{x}}$ $\boxed{x}$	1.0607	-04
2	$\boxed{x}$	2.1213	-04
π	$\boxed{=}$ $\boxed{1/x}$	1.5005	03

The calculator answer is $f = 1.5005 \times 10^3$. As an ordinary number rounded to the nearest whole number, f = _____ .

---

57.   Do this evaluation. Report the answer in scientific notation with the first factor rounded to hundredths.

In $\boxed{L = \dfrac{1}{4\pi^2 Cf}}$, when $C = 3.4 \times 10^{-6}$ and $f = 4.8 \times 10^6$,

L = _____ .

f = 1,501

---

58.   Do this evaluation.

In $\boxed{L = \dfrac{\sqrt{Z^2 - R^2}}{2\pi f}}$, when $Z = 750$, $R = 580$, and $f = 1,600$,

L = _____   (Round to thousandths.)

$L = 1.55 \times 10^{-3}$

---

59.   Do these. Round to tenths in (a) and to the nearest whole number in (b).

a) In $\boxed{H = \dfrac{\pi dR(F_1 - F_2)}{33,000}}$, when $d = 48$, $R = 150$, $F_1 = 75$, and

$F_2 = 25$, H = _____ .

b) In $\boxed{R = \dfrac{33,000H}{\pi d(F_1 - F_2)}}$, when $H = 80$, $d = 65$, $F_1 = 90$, and

$F_2 = 30$, R = _____ .

L = 0.047

a) H = 34.3   b) R = 215

6-7   RECTANGULAR PRISMS AND CUBES

In this section, we will review the formulas for the volume of a rectangular prism and a cube.

60. The volume of a solid figure is the number of "unit cubes" needed to fill it. Some "unit cubes" are shown below	

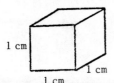 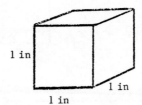

Notice that we use the exponent "3" in the abbreviation for "unit cubes". Therefore:

The abbreviation for "a cubic foot" is "ft³".

The abbreviation for "a cubic meter" is _____.

---

61.   The word names for two volume measures are given below.

      125 cm³   is called   "125 cubic centimeters"

      64 yd³   is called   "64 cubic yards"

Write the abbreviations for each word name.

     a)   9 cubic inches = _____   b)   1,000 cubic meters = _____

Right column answer: m³

---

62.   The figure at the right is called a "rectangular prism" because each side is a rectangle. It takes 24 cubic centimeters to fill the solid figure. Therefore, its volume is 24 cm³. Notice that we can find its volume by multiplying its length, width, and height. That is:

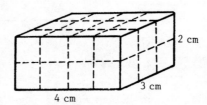

Right column answers:
a) 9 in³
b) 1,000 m³

$$\boxed{V = LWH}$$

When using the above formula, we treat the units as if they were letters. That is, for the rectangular prism above:

$$V = (4 \text{ cm})(3 \text{ cm})(2 \text{ cm})$$

$$= (4)(3)(2)(\text{cm})(\text{cm})(\text{cm})$$

$$= \underline{\hspace{1cm}} \text{ cm}^3$$

Right column answer: 24 cm³

63. Use the formula to find the volume of each rectangular prism below. Round to the nearest whole number.

a)
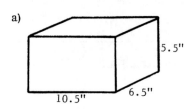
5.5"
10.5"   6.5"

V = _____

b)
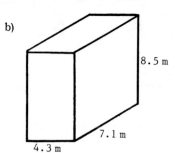
8.5 m
7.1 m
4.3 m

V = _____

---

64. A "cube" is a special type of rectangular prism in which all sides are squares. Therefore, the volume of a cube equals (s)(s)(s). That is:

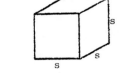

$$V = s^3$$

Using the formula and the $\boxed{y^x}$ key, find the volume of each cube below. Round to the nearest whole number in (a) and to tenths in (b).

a)
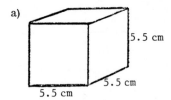
5.5 cm
5.5 cm
5.5 cm

V = _____

b)
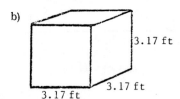
3.17 ft
3.17 ft
3.17 ft

V = _____

a) 375 in³

b) 260 m³

---

65. We can rearrange $V = s^3$ to solve for "s". We get:

$$s = \sqrt[3]{V}$$

Using the above formula, we can find the side of a cube if its volume is known. For example:

If $V = 1$ cm³, $s = \sqrt[3]{1 \text{ cm}^3} = 1$ cm

If $V = 8$ ft³, $s = \sqrt[3]{8 \text{ ft}^3}$ = 2 ft

If $V = 27$ m³, $s = \sqrt[3]{27 \text{ m}^3}$ = _____

a) 166 cm³

b) 31.9 ft³

---

66. Using $s = \sqrt[3]{V}$ and either $\boxed{\sqrt[x]{y}}$ or $\boxed{\text{INV}}\ \boxed{y^x}$, find the side of the cubes with the following volumes.

a) If $V = 75.4$ in³,   s = _____ (Round to hundredths.)

b) If $V = 850$ yd³,   s = _____ (Round to hundredths.)

c) If $V = 5,725$ cm³, s = _____ (Round to tenths.)

3m

---

a) 4.22 in   b) 9.47 yd   c) 17.9 cm

6-8   UNITS OF VOLUME

In this section, we will discuss the conversion facts for the basic units of volume in both the English System and the Metric System.

---

67.   Since 12 in = 1 ft, the figure at the right is 1 cubic foot. We converted 1 ft^3 to cubic inches below by substituting "12 in" for "1 ft".

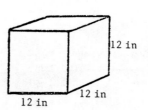

$$1 \text{ ft}^3 = (1 \text{ ft})(1 \text{ ft})(1 \text{ ft})$$
$$= (12 \text{ in})(12 \text{ in})(12 \text{ in})$$
$$= 1,728 \text{ in}^3$$

Since 3 ft = 1 yd, the figure at the right is 1 cubic yard. We converted 1 yd^3 to cubic feet below by substituting "3 ft" for "1 yd".

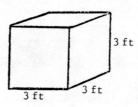

$$1 \text{ yd}^3 = (1 \text{ yd})(1 \text{ yd})(1 \text{ yd})$$
$$= (3 \text{ ft})(3 \text{ ft})(3 \text{ ft})$$
$$= \underline{\hspace{1cm}} \text{ ft}^3$$

---

68.   The basic conversion facts for units of volume in the English system are:

1 cubic foot (ft^3) = 1,728 cubic inches (in^3)
1 cubic yard (yd^3) = 27 cubic feet (ft^3)

Using the facts above, complete these conversions.

a) 10 ft^3 = \underline{\hspace{1.5cm}} in^3 b) 100 yd^3 = \underline{\hspace{1.5cm}} ft^3

Answer (from box): 27 ft^3

---

69.   Using the same conversion facts, complete these. Round to hundredths.

a) 5,000 in^3 = \underline{\hspace{1.5cm}} ft^3 b) 100 ft^3 = \underline{\hspace{1.5cm}} yd^3

Answers (from box):
a) 17,280 in^3
b) 2,700 ft^3

Answers (from box):
a) 2.89 ft^3
b) 3.70 yd^3

70.   Since 10 dm = 1 m, the figure at the right
       is 1 cubic meter. We converted 1 m^3 to
       cubic decimeters below by substituting
       "10 dm" for "1 m".

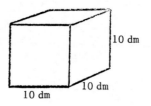

$$1 \ m^3 = (1 \ m)(1 \ m)(1 \ m)$$

$$= (10 \ dm)(10 \ dm)(10 \ dm)$$

$$= 1,000 \ dm^3$$

Since 100 cm = 1 m, the figure at the right
is also 1 cubic meter. We converted 1 m^3
to cubic centimeters below by substituting
"100 cm" for "1 m".

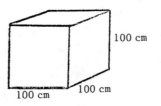

$$1 \ m^3 = (1 \ m)(1 \ m)(1 \ m)$$

$$= (100 \ cm)(100 \ cm)(100 \ cm)$$

$$= \underline{\hspace{3cm}} \ cm^3$$

---

71.   Some basic conversion facts for units of volume in the metric system are: | $1,000,000 \ cm^3$

> 1 cubic meter (m^3) = 1,000 cubic decimeters (dm^3)
>
> 1 cubic meter (m^3) = 1,000,000 cubic centimeters (cm^3)

Using the facts above, complete these:

a) 5 m^3 = _____ dm^3 b) 3 m^3 = _____ cm^3

---

72.   Using the same facts, complete these: | a) 5,000 dm^3

a) 2,500 dm^3 = _____ m^3 b) 7,200,000 cm^3 = _____ m^3 | b) 3,000,000 cm^3

---

73.   The basic conversion facts relating volumes in the English and metric | a) 2.5 m^3
       systems are: | b) 7.2 m^3

> 1 in^3 = 16.39 cm^3
>
> 1 yd^3 = 0.7646 m^3

Using the facts, complete these:

a) 10 in^3 = _____ cm^3 b) 100 yd^3 = _____ m^3

74.   Using the same facts, complete these.   Round to hundredths.

      a) 100 cm³ = _____ in³   b) 5 m³ = _____ yd³

    a) 163.9 cm³

    b) 76.46 m³

    a) 6.10 in³   b) 6.54 yd³

<u>SELF-TEST 18</u> (pp. 240-252)

1. Find the circumference. Round to tenths.

    27.8'

                  C = _____

2. Find the area. Round to one decimal place.

    9.36 cm

                  A = _____

3. Find the area of the shaded figure. Round to two decimal places.

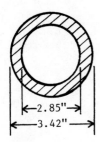

    |←2.85"→|

    |←—3.42"—→|

                  A = _____

4. Find the radius of a circle whose area is 7.28 m². Round to hundredths.

    r

                  r = _____

5. Find "f". Report your answer in scientific notation with the first factor rounded to two decimal places.

$$f = \frac{1}{2\pi\sqrt{LC}}$$

    $L = 475 \times 10^{-6}$

    $C = 34.7 \times 10^{-12}$

                  f = _____

6. Find L. Round to thousandths.

$$L = \frac{\sqrt{Z^2 - R^2}}{2\pi f}$$

    $Z = 538$

    $R = 214$

    $f = 750$

                  L = _____

Continued on following page.

SELF-TEST 18 - Continued

7. Find the volume.   Round to thousands.

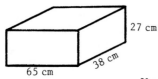

27 cm

38 cm

65 cm

V = _____

8. Find the side of a cube whose volume is 2,400 ft³.
    Round to tenths.

s

s   s

s = _____

9. Round to hundredths.

    4,800 in³ = _____ ft³

10.   5,000 dm³ = _____ m³

11. Round to the nearest whole
     number.

     170 yd³ = _____ m³

ANSWERS: 1. C = 174.7 ft
          2. A = 68.8 cm²
          3. A = 2.81 in²
          4. r = 1.52 m

          5. f = 1.24 x 10⁶
          6. L = 0.105
          7. V = 67,000 cm³
          8. s = 13.4 ft

          9. 2.78 ft³
          10. 5 m³
          11. 130 m³

6-9   RIGHT PRISMS

In an earlier section, we discussed the specific formulas for the volume of a rectangular prism and a cube.
In this section, we will discuss the general formula for the volume of any right prism.

---

75.   Each solid figure below is a right prism.   Notice these points:

       1) The two bases are parallel.   They have the same size and shape.

       2) The height is the distance between the two bases.

When a right prism is vertical, the bases are at the top and the bottom and
the height is a vertical distance.   For example:

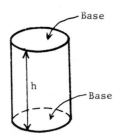

Base

h

Base

Base

h

Base

Base

h

Base

When a right prism is horizontal, the bases are on the sides and the height
is a horizontal distance.

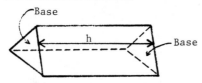

Base

h

Base

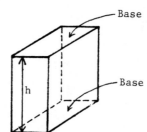

Base

h

Base

76. The bases of a prism can also be composite figures. For example, the
     bases of the prism on the left below are L-shaped. The bases of the
     prism on the right are I-shaped.

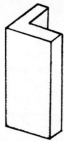 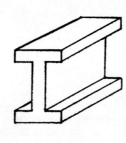

a) Is the L-shaped prism horizontal or vertical? _____

b) Is the I-shaped prism horizontal or vertical? _____

---

77. The general formula for the volume of any right prism is:

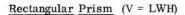

| Volume = Area of Base x Height |

To show that the general formula makes sense, we can use it to derive
the specific formulas for the volume of a rectangular prism and a cube.

Rectangular Prism (V = LWH)

   Area of the rectangular base = LW
   Height = H
   Volume = (LW)(H) or LWH

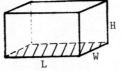

Cube (V = s^3)

   Area of the square base = (s)(s) or s^2
   Height = s
   Volume = $(s^2)(s)$ = _____

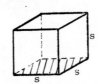

a) vertical

b) horizontal

---

78. The figure at the right is a right
     circular prism or "cylinder".
     Let's use the general formula to
     find its volume. Round to the
     nearest whole number.

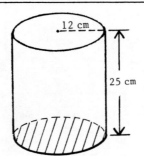

a) Since the radius of each base
    is 12 cm, the area of a base
    is _____.

b) The height of the cylinder is _____.

c) The volume of the cylinder is
       (   )(   ) = _____.

s^3

a) 452 cm^2

b) 25 cm

c) 11,300 cm^3, from:
    (452 cm^2)(25 cm)

79. The figure at the right is a
     right triangular prism. Let's
     use the general formula to
     find its volume.

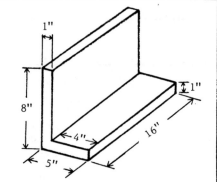

ε.66'

10'   20'

   a) The front triangle is a base.
       The area of that triangle is _____.

   b) The height of the prism is _____.

   c) The volume of the prism is _____.

---

80. The L-shaped figure at the
     right is a right prism. Let's
     find its volume.

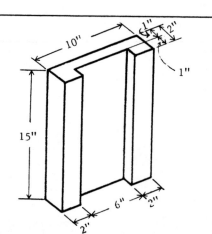

1"

8"

1"

4"

16"

5"

   a) The area of the L-shaped
       base is _____.

   b) The height of the prism
       is _____.

   c) The volume of the prism
       is _____.

a) 43.3 ft^2

b) 20 ft

c) 866 ft^3

---

81. The figure at the right is a
     rectangular piece of steel
     with a slot cut out. It is
     a right prism with a com-
     posite figure as its base.

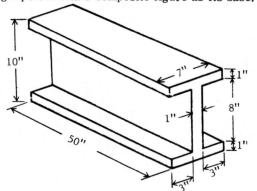

10"   1" 2"

1"

15"

6"   2"

2"

   a) The area of the base
       is _____.

   b) The height is _____.

   c) The volume is _____.

a) 12 in^2

b) 16 in

c) 192 in^3

---

82. The I-beam below is a right prism with a composite figure as its base.

   a) The area of the
       I-shaped base
       is _____.

   b) The volume is
       _____.

10"

7"   1"

1"   8"

50"

1"

3"   3"

a) 14 in^2

b) 15 in

c) 210 in^3

83.   The steel pipe on the right is a
      right prism. Its inside diameter
      is 8.5 cm; its outside diameter
      is 10 cm.

      To find the area of its base, we
      must subtract the area of the
      inner circle from the area of
      the outer circle.

      a) The area of the base to the
          nearest tenth is _____.

      b) The volume of the pipe to the
          nearest whole number is _____.

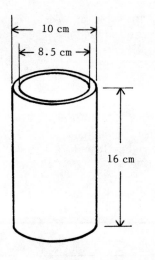

10 cm

8.5 cm

16 cm

a) 22 in²

b) 1,100 in³

a) 21.8 cm²   b) 349 cm³

6-10   CONES AND SPHERES

In this section, we will discuss the formulas for the volume of a cone, a frustum of a cone, and a sphere.
None of the figures are right prisms.

---

84.   The formula for the volume of a <u>right</u> <u>circular</u> <u>cone</u> is:

$$V = \frac{1}{3}Bh \qquad \text{or} \qquad \boxed{V = \frac{Bh}{3}}$$

      where: "B" is the area of the circular base.
               "h" is the height of the cone.

      Using the boxed form of the formula,
      let's find the volume of this cone.

          a) The area of the circular base to
              the nearest whole number is _____.

          b) The height is _____.

          c) The volume rounded to tens
              is _____.

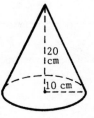

20
cm

10 cm

a) 314 cm²

b) 20 cm

c) 2,090 cm³

85.   The figure below is a "<u>frustum</u>" of a cone. It is the bottom part of a cone
with a smaller cone cut off at the top. The formula for the volume of a
frustum is:

$$V = \frac{\pi h}{3}(R^2 + Rr + r^2)$$ where "h" is the height of the frustum.
"R" is the radius of the large base.
"r" is the radius of the small base.

Using the formula, find the volume
of the frustum on the right. Round
to the nearest whole number.

r = 4"

h = 9"

R = 6"

V = _____

---

86.   The formula for the volume of a <u>sphere</u> (or ball) is:

$$V = \frac{4}{3}\pi r^3 \quad \text{or} \quad \boxed{V = \frac{4\pi r^3}{3}}$$ where "r" is a radius.

Using the boxed form of the formula,
we calculated the volume of the sphere
shown at the right. The calculator
steps are below. Notice that we have
to use the $\boxed{y^x}$ key and the $\boxed{=}$ key
to evaluate r^3 before multiplying by
4 and π.

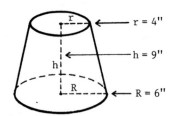

r = 5 cm

Enter	Press	Display
5	$\boxed{y^x}$	5.
3	$\boxed{=}\ \boxed{x}$	125.
4	$\boxed{x}$	500.
π	$\boxed{\div}$	1570. 7963
3	$\boxed{=}$	523. 59877

Rounding to the nearest whole number, V = _____.

---

87.   When using the $\boxed{y^x}$ key and then the $\boxed{=}$ key to evaluate r^3 in the
formula for the volume of a sphere, some calculators are a little slow in
calculating r^3. <u>Be sure to wait until that value appears on the display
before multiplying by 4 and π</u>.

In $\boxed{V = \frac{4\pi r^3}{3}}$: a) when r = 10 in, V = _____.
(Round to the nearest ten.)

b) when r = 2.5 ft, V = _____.
(Round to the nearest tenth.)

V = 716 in³

524 cm³

a) 4,190 in³   b) 65.4 ft³

6-11   SURFACE AREAS

In this section, we will discuss the procedure for finding the total surface area of some right prisms and some non-prisms.

---

88. The total surface area of the rectangular prism at the right is the sum of the areas of the two bases and the four sides.

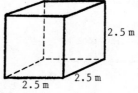

a) The dimensions of each base are 4" by 3". Therefore, the combined area of the two bases is _____.

b) The dimensions of the front and back sides are 4" by 2". Therefore, the combined area of those two sides is _____.

c) The dimensions of the left and right sides are 3" by 2". Therefore, the combined area of those two sides is _____.

d) The total surface area of the prism is _____.

---

89. The total surface area of the cube at the right is the sum of the areas of the two bases and the four sides. Since the dimensions of each side and base are 2.5 m, we can find the total surface area by multiplying the area of any one side or base by 6.

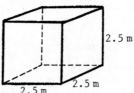

a) The area of each base and side is _____.

b) The total surface area of the cube is _____.

a) 24 in²   b) 16 in²

c) 12 in²   d) 52 in²

---

90. The total surface area of the cylinder below is the sum of the areas of the two circular bases and the area of the circular side. To show what is meant by the circular side, we cut the cylinder vertically and laid the circular side flat as a rectangle. The dimensions of the rectangle are 12" and the circumference of the cylinder which is πd.

a) 6.25 m²

b) 37.5 m²

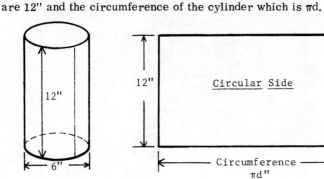

Continued on following page.

90.   Continued

    a) Rounding to tenths, the combined area of the two bases
       is $(2)(.7854)(d^2)$ = _____.

    b) Rounding to tenths, the area of the circular side is _____.

       Note: Since the dimensions of the circular side are
           12" by πd", its area is 12πd in^2.

    c) Rounding to the nearest whole number, the total surface area of
       the cylinder is _____.

---

91.   The <u>total</u> <u>surface</u> <u>area</u> of a cone is <u>the</u> <u>sum</u> <u>of</u> <u>the</u> <u>area</u> <u>of</u> <u>the</u> <u>circular</u>
    <u>base</u> <u>and</u> <u>the</u> <u>area</u> <u>of</u> <u>the</u> <u>curved</u> <u>side</u>.   That is:

$$A = B + \frac{sC}{2}$$

    where: B is the area of the base

             $\frac{sC}{2}$ is the area of the curved side

             "s" is the <u>slant</u> <u>height</u> of the curved side

             "C" is the <u>circumference</u> of the base

Let's use the above formula to
find the total surface area of
this cone.

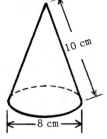

    a) Rounded to tenths, the area of
       the base (B) is _____.

    b) Rounded to tenths, the area of the
       curved side $\left(\frac{sC}{2}\right)$ is _____.

       Note: Since the circumference (C) equals πd,
            $\frac{sC}{2} = \frac{s(\pi d)}{2}$

    c) Rounded to the nearest whole number, the total
       surface area of the cone is _____.

a) 56.5 in^2

b) 226.2 in^2

c) 283 in^2

---

92.   The formula for the <u>total</u> <u>surface</u> <u>area</u>
    of a sphere is:

$$A = 4\pi r^2$$

Using that formula, find the total surface
area of the sphere at the right. Round
to the nearest whole number.

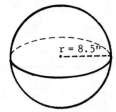

              A = _____

a) 50.3 cm^2

b) 125.7 cm^2

c) 176 cm^2

---

908 in^2

6-12   DENSITY AND WEIGHT

In this section, we will discuss the "density" or "weight per unit volume" of materials and liquids. We will find the weights of various objects or amounts of liquid by using their volumes and the density of the material or liquid.

---

93.   The density of a material is its weight per unit volume.
 	
       In the English system, density is frequently stated in "pounds per cubic inch" or "lb/in^3". For example:
 	
           The density of steel is 0.283 lb/in^3.
           The density of aluminum is 0.0924 lb/in^3.
           The density of copper is 0.322 lb/in^3.
 	
       In the metric system, density is frequently stated in "grams per cubic centimeter" or "g/cm^3". For example:
 	
           The density of steel is 7.83 g/cm^3.
           The density of aluminum is 2.56 g/cm^3.
           The density of copper is 8.91 g/cm^3.
 	
Which of the three materials above has the greatest density? _____	

94.   If the density of steel is 0.283 lb/in^3:
 	
  a) How much does 1 in^3 of steel weigh? _____
 	
  b) How much does 10 in^3 of steel weigh? _____
 	
  c) How much does 33.4 in^3 of steel weigh to the nearest hundredth of a pound? _____
	copper

95.   If the density of aluminum is 2.56 g/cm^3:
 	
  a) What is the weight of 10 cm^3 of aluminum? _____
 	
  b) What is the weight of 1,750 cm^3 of aluminum? _____
	a) 0.283 lb
	
	b) 2.83 lb
	
	c) 9.45 lb

96.   The end of the steel bar below is a square with 2.5" sides. Its length is 40".
 	
  a) What is the volume of the bar? _____
 	
  b) If the density of steel is 0.283 lb/in^3, how much does the steel bar weigh? _____
	a) 25.6 g
	
	b) 4,480 g

97.   The length and width of
      the rectangular aluminum
      sheet at the right are
      given. Its thickness (or
      height) is 0.38 cm.

      100 cm

      230 cm

      a) What is the volume
          of the sheet? _____

      b) If the density of aluminum is 2.56 g/cm^3, what is the weight of
          the sheet to the nearest hundred grams? _____

a) 250 in^3

b) 70.75 lb

---

98.   A piece of circular wire is a cylinder. The diameter of 12-gage copper
      wire is 0.081 in. The density of copper is 0.322 lb/in^3.

      Rounding to the nearest tenth:

      a) Find the volume in cubic inches of 1,000 feet of 12-gage copper
          wire. (Note: 1,000 ft = 12,000 in) _____

      b) Find the weight of 1,000 feet of 12-gage copper wire. _____

a) 8,740 cm^3

b) 22,400 g

---

99.   The density of a liquid is also its weight per unit volume.

      In the English system, density is frequently stated in "pounds
      per cubic foot" or "lb/ft^3". For example:

          The density of water is 62.4 lb/ft^3.

          The density of gasoline is 41.5 lb/ft^3.

          The density of mercury is 849 lb/ft^3.

      In the metric system, density is frequently stated in "grams per
      cubic centimeter" or "g/cm^3". For example:

          The density of water is 1 g/cm^3.

          The density of gasoline is 0.675 g/cm^3.

          The density of mercury is 13.6 g/cm^3.

      Which of the three liquids above has the greatest density? _____

a) 61.8 in^3

b) 19.9 lb

---

100. If the density of water is 62.4 lb/ft^3:

      a) How much does 1 ft^3 of water weigh? _____

      b) How much does 1 yd^3 (or 27 ft^3) of water weigh to the nearest
          pound? _____

mercury

---

a) 62.4 lb

b) 1,685 lb

101.   If the density of gasoline is 0.675 g/cm^3:

    a) How much does 10 cm^3 of gasoline weigh? _____

    b) How much does 1,000 cm^3 of gasoline weigh? _____

---

102.   The following facts relate liquid measures and volume in the metric system.

> 1 liter (ℓ) = 1,000 cubic centimeters (cm^3)
>
> 1 milliliter (mℓ) =   1   cubic centimeter (cm^3)

Therefore: a) 5 ℓ = _____ cm^3 b) 75 mℓ = _____ cm^3

a) 6.75 g

b) 675 g

---

103.   Since the density of water is 1 g/cm^3:

    a) How much does 1 milliliter of water weigh? _____

    b) How much does 1 liter of water weigh? _____

a) 5,000 cm^3

b) 75 cm^3

---

104.   In many countries, gasoline is sold by the liter. Since the density of gasoline is 0.675 g/cm^3:

    a) How much does 1 liter of gasoline weigh in grams? _____

    b) How much does 50 liters of gasoline weigh in kilograms? _____

a) 1 g

b) 1,000 g

---

a) 675 g   b) 33.75 kg, from: 33,750 g

SELF-TEST 19 (pp. 252-262)

1. Find the volume of the right triangular prism below. Round to hundredths.

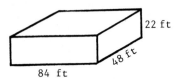

2. Find the volume of the right circular cone below. Round to hundreds.

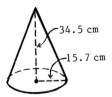

Find the volume of each solid below. Round as directed.

3. Round to thousands.

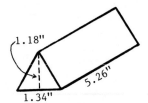

4. Round to tenths.

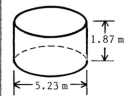

5. Round to hundreds.

6. Find the surface area of the rectangular prism shown in problem 3. Round to thousands.

7. Find the surface area of the the right circular cylinder shown in problem 4. Round to tenths.

8. Find the surface area of the sphere shown in problem 5. Round to the nearest whole number.

_____

9. Find the weight (in pounds) of a 60" by 80" steel plate whose thickness is 0.50". Round to tens.

10. Find the weight (in grams) of a 450 cm length of round aluminum wire whose diameter is 0.165 cm. Round to tenths.

_____

11. Find the weight (in kilograms) of 64 liters of gasoline.

_____

ANSWERS:

1. 4.16 in³	4. 40.2 m³	7. 73.7 m²	10. 24.6 g
2. 8,900 cm³	5. 2,300 in³	8. 845 in²	11. 43.2 kg
3. 89,000 ft³	6. 14,000 ft²	9. 680 lb	

SUPPLEMENTARY PROBLEMS - CHAPTER 6

Assignment 17

Find the <u>area</u> and <u>perimeter</u> of each rectangle and square.

1. Round A to tens.

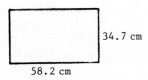

34.7 cm

58.2 cm

2. Round A to hundreds.

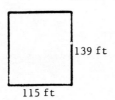

139 ft

115 ft

3. Round A to tenths.

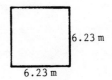

6.23 m

6.23 m

Do these conversions of area units.

4. $648 \text{ in}^2 =$ _____ ft^2

5. $2.93 \text{ m}^2 =$ _____ cm^2

6. $129 \text{ cm}^2 =$ _____ in^2

7. Round to hundredths.

$6 \text{ m}^2 =$ _____ yd^2

8. Round to tenths.

$180 \text{ acres} =$ _____ hectares

9. Round to the nearest whole number.

$28.5 \text{ in}^2 =$ _____ cm^2

Find the <u>area</u> of each parallelogram and triangle.

10. Round to the nearest whole number.

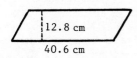

12.8 cm

40.6 cm

11. Round to hundredths.

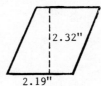

2.32"

2.19"

12. Round to thousands.

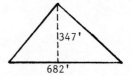

347'

682'

Find the <u>area</u> of each triangle.

13. Round to tenths.

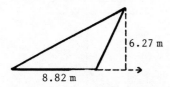

6.27 m

8.82 m

14. Round to the nearest whole number.

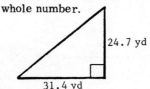

24.7 yd

31.4 yd

15. Round to thousands.

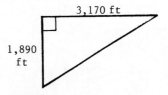

3,170 ft

1,890 ft

Find the <u>area</u> of each trapezoid and composite figure.

16. Round to hundreds.

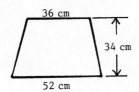

36 cm

34 cm

52 cm

17.

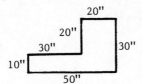

20"

20"

30"

30"

10"

50"

18. Round to tenths.

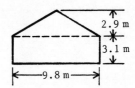

2.9 m

3.1 m

9.8 m

SUPPLEMENTARY PROBLEMS – CHAPTER 6 (Continued)

Assignment 18

Find the <u>circumference</u> of each circle.   Find the <u>area</u> of each circle.

1. Round to tenths.   2. Round to hundreds.   3. Round to hundredths.   4. Round to thousands.

Find the <u>area</u> of each shaded figure.

5. Round to tens.   6. Round to tenths.   7. Round to hundredths.

   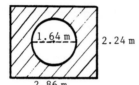   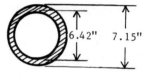

8. Find the <u>radius</u> of a circle whose area is 3.92 cm². Round to hundredths.

9. Find the <u>diameter</u> of a circle whose area is 250 ft². Round to tenths.

10. Find T. Round to   11. Find M. Round to tens.   12. Find "f". Round to the nearest
    hundredths.   whole number.

$$T = 2\pi\sqrt{\dfrac{L}{g}}$$ $L = 25$ $M = \dfrac{2\pi r}{d}$ $r = 23.5$ $f = \dfrac{1}{2\pi\sqrt{LC}}$ $L = 36.2 \times 10^{-3}$
$g = 980$ $d = 0.142$ $C = 2.85 \times 10^{-6}$

Find the <u>volume</u> of each figure.

13. Round to thousands.   14. Round to tens.   15. Round to tenths.

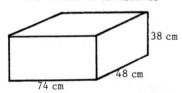   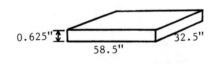   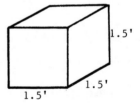

16. If the volume of a cube is 43.7 cm³, find the <u>length</u> of a side. Round to hundredths.

17. Round to hundreds.   18. Round to the nearest   19. Round to tenths.
                                  whole number.

    6.83 ft³ = _____ in³   58 in³ = _____ cm³   20 m³ = _____ yd³

SUPPLEMENTARY PROBLEMS – CHAPTER 6 (Continued)

Assignment 19

Find the volume of each figure.

1. Round to hundreds.

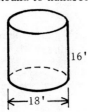

16'

18'

2. Round to tens.

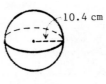

10.4 cm

3. Round to tenths.

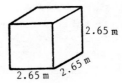

2.65 m

2.65 m   2.65 m

Find the surface area of the figure:

4. In Problem 1.
   Round to hundreds.

5. In Problem 2.
   Round to tens.

6. In Problem 3.
   Round to tenths.

Find the volume of each figure.

7. Round to the nearest
   whole number.

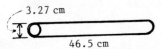

3.27 cm

46.5 cm

8. Round to hundreds.

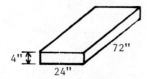

72"

4"

24"

9. Round to the nearest
   whole number.

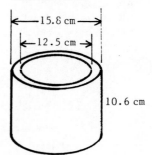

15.8 cm

12.5 cm

10.6 cm

10. Round to hundredths.

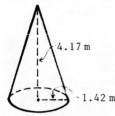

4.17 m

1.42 m

11.

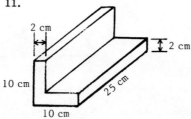

2 cm

2 cm

10 cm

25 cm

10 cm

12. Round to the nearest
    whole number.

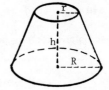

h = 5.4"
R = 3.5"
r = 2.2"

13. Find the weight (in pounds) of a rectangular sheet of aluminum whose length is 36", whose width is 32", and whose thickness is 0.138". Round to tenths.

14. Find the weight (in grams) of a copper wire whose diameter is 0.215 cm and whose length is 2,100 cm. Round to the nearest whole number.

15. Find the weight (in kilograms) of 25 liters of water.

16. Find the weight (in pounds) of 10 ft^3 of gasoline.

Chapter 7

RIGHT TRIANGLE TRIGONOMETRY

In this chapter, we will show how the angle-sum principle, the Pythagorean Theorem, and the three basic trigonometric ratios (sine, cosine, and tangent) can be used to find unknown angles and sides in right triangles.

7-1   THE ANGLE-SUM PRINCIPLE FOR TRIANGLES

In any triangle, the sum of the three angles is 180°. We will discuss that principle in this section.

---

1.   There are three angles in any triangle.   The angles are usually labeled with capital letters.

   Any angle between 0° and 90° is called an <u>acute</u> angle.
   Any angle with exactly 90° is called a <u>right</u> angle.
   Any angle between 90° and 180° is called an <u>obtuse</u> angle.

   In triangle ABC:

   Angle A is an <u>acute angle</u>.

   a) Angle B is an _____ angle.

   b) Angle C is an _____ angle.

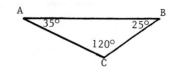

2.   In triangle MPR:

   a) There are two acute angles, angle _____ and angle _____ .

   b) The obtuse angle is angle _____ .

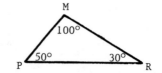

a) acute

b) obtuse

3.   The <u>angle-sum</u> <u>principle</u> for triangles is this:

   | THE SUM OF THE THREE ANGLES OF ANY TRIANGLE IS 180° |

   By subtracting the total number of degrees of the two known angles from 180°, find the number of degrees of the unknown angles in each triangle below.

   a)

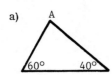

   A = _____

   b)

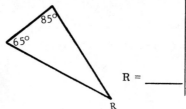

   R = _____

a) Angles P and R

b) Angle M

266

4. The angle–sum principle also applies to right triangles (those that contain a right angle).   Therefore, in right triangle CDE:

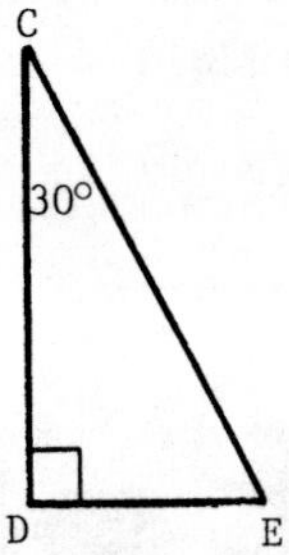

    a) The sum of the angles is _____.

    b) Angle D = _____.

    c) Angle E = _____.

a) A = 80°

b) R = 30°

---

5. In a right triangle, the right angle contains 90°. Since the _sum_ of the other two angles must be 90°, _the other two angles must be acute angles._

In right triangle TBC:

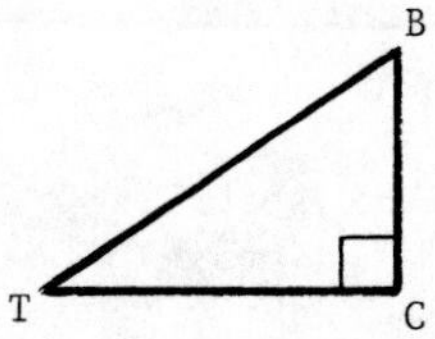

    a) Angles T and B are both _____ angles.

    b) The sum of angles T and B must be _____°.

a) 180°

b) 90°

c) 60°

---

6. Since _the sum of the two acute angles in a right triangle is 90°_, it is easy to find the size of one acute angle if we know the size of the other. To do so, we simply subtract the known angle from 90°.

In right triangle ABC:

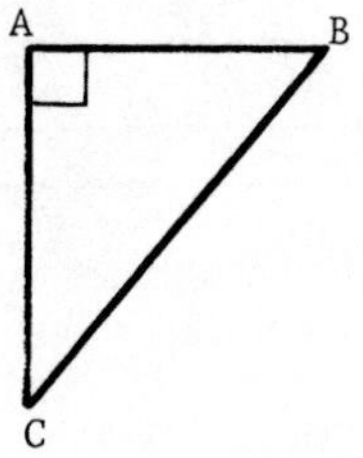

    a) If angle B contains 55°, angle C contains 90° – 55° = _____°.

    b) If angle C contains 37°, angle B contains 90° – 37° = _____°.

a) acute

b) 90°

---

7. By simply subtracting the known acute angle from 90°, find the unknown acute angle in each right triangle.

    a)

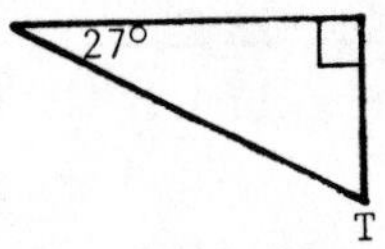

    b)

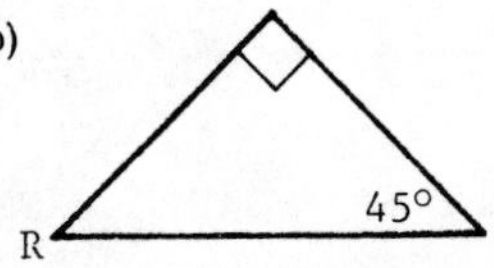

        T = _____   R = _____

a) 35°

b) 53°

a) T = 63°

b) R = 45°

7-2   LABELING ANGLES AND SIDES IN TRIANGLES

In this section, we will discuss the conventional way to label the angles and sides in any triangle.

8.   We labeled the angles and sides
of the triangle at the right in the
conventional way.   Notice these
points:

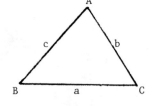

   1) Each <u>angle</u> is labeled with
       a <u>capital</u> <u>letter</u>.

   2) Each <u>side</u> is labeled with the
       <u>small</u> <u>letter</u> corresponding
       to the capital letter of the
       angle opposite it.   That is:

       The side opposite angle A is labeled "a".
       The side opposite angle B is labeled "b".
       The side opposite angle C is labeled "c".

Sometimes we use the two capital letters at each end of the side to represent the side.   For example:

       Instead of "a", we use BC (or CB).

   a) Instead of "b", we use _____ .

   b) Instead of "c", we use _____ .

---

9.   When the sides of a triangle are not labeled with small letters, we must
use two capital letters to represent the sides.   For example, in the
triangle below.

       The side opposite angle M is TV (or VT).

   a) The side opposite angle T is _____ .

   b) The side opposite angle V is _____ .

a) AC (or CA)

b) AB (or BA)

---

10.   In any triangle:

       THE <u>LONGEST</u> <u>SIDE</u> IS OPPOSITE THE <u>LARGEST</u> <u>ANGLE</u>.

       THE <u>SHORTEST</u> <u>SIDE</u> IS OPPOSITE THE <u>SMALLEST</u> ANGLE.

   In triangle DFH:

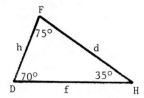

   a) Angle F is the <u>largest</u>
       angle. Therefore,
       _____ is the <u>longest</u> side.

   b) Angle H is the <u>smallest</u>
       angle. Therefore,
       _____ is the <u>shortest</u> side.

a) MV (or VM)

b) MT (or TM)

---

a) f   b) h

11. In the triangle at the right:

   a) The longest side is _____.

   b) The shortest side _____.

---

12. In a right triangle, the side opposite the right angle is called the "hypotenuse". The sides opposite the two acute angles are simply called "legs". For example:

   In right triangle DFH:

   "f" is the hypotenuse, and "d" and "h" are the legs.

   a) The hypotenuse "f" can be labeled in two other ways: _____ or _____.

   b) Leg "d" can be labeled in two other ways: _____ or _____

a) m   b) q

---

13. Ordinarily we use small letters to represent the hypotenuse and legs of a right triangle. Use small letters to complete the questions below.

   In right triangle FET:

   a) The hypotenuse is _____.

   b) The legs are _____ and _____.

a) DH or HD

b) FH or HF

---

14. Since the right angle is the largest angle in any right triangle, the hypotenuse is the longest side of any right triangle.

   In right triangle DMF:

   a) The longest side is _____.

   b) The shortest side is _____.

a) e

b) t and f

---

a) m (the hypotenuse)   b) f

7-3   THE PYTHAGOREAN THEOREM

In this section, we will discuss the Pythagorean Theorem and show how it can be used to find the length of an unknown side in a right triangle.

15.   The Pythagorean Theorem states the following relationship among the
       three sides of a right triangle:

> IN ANY RIGHT TRIANGLE, THE SQUARE OF THE LENGTH
> OF THE HYPOTENUSE IS EQUAL TO THE SUM OF THE
> SQUARES OF THE LENGTHS OF THE TWO LEGS.

For right triangle ABC, the
Pythagorean Theorem says:

$$c^2 = a^2 + b^2$$

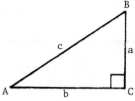

Using small letters, state the Pythagorean Theorem for each right
triangle below.

a)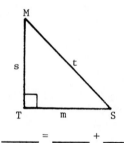

b)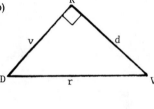

____ = ____ + ____

____ = ____ + ____

---

16.   The lengths of the two legs of the right triangle below are given. We can
       use the Pythagorean Theorem to find the length of the hypotenuse.   The
       steps are:

   1) Write the Pythagorean Theorem.

      $$m^2 = d^2 + p^2$$

   2) Substitute the known values
      and simplify.

      $$m^2 = (4'')^2 + (3'')^2$$
      $$= 16 \text{ in}^2 + 9 \text{ in}^2$$
      $$= 25 \text{ in}^2$$

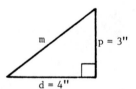

   3) Find "m" by taking the square root of the right side.

      $$m = \sqrt{25 \text{ in}^2} = \underline{\hspace{1cm}}$$

a) $t^2 = s^2 + m^2$

b) $r^2 = v^2 + d^2$

---

17.   We can use the same steps to find the length of the hypotenuse below.

      $$h^2 = t^2 + b^2$$
      $$= (7m)^2 + (4 \text{ m})^2$$
      $$= 49 \text{ m}^2 + 16 \text{ m}^2$$
      $$= 65 \text{ m}^2$$

Use a calculator to find "h" by taking the
square root of the right side.   Round to tenths.

      $$h = \sqrt{65 \text{ m}^2} = \underline{\hspace{1cm}}$$

m = 5 in

18. When using the Pythagorean Theorem to find the hypotenuse of a right triangle, all of the calculations can be done in one process on a calculator. To prepare to do so, we solve for the hypotenuse by taking the square root of the other side before substituting. That is:

$$\text{If } c^2 = a^2 + b^2, \quad c = \sqrt{a^2 + b^2}$$

Using the method above, we set up the solution below so that "t" can be found in one calculator process.

$$t = k^2 + d^2$$
$$t = \sqrt{k^2 + d^2}$$
$$t = \sqrt{(5.68')^2 + (9.75')^2}$$

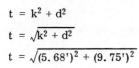

h = 8.1 m

The calculator steps for the solution are shown. Notice that we added the squares of 5.68 and 9.75, pressed $\boxed{=}$ to complete that addition, and then pressed $\boxed{\sqrt{x}}$.

Enter	Press	Display
5.68	$\boxed{x^2}$ $\boxed{+}$	32.2624
9.75	$\boxed{x^2}$ $\boxed{=}$ $\boxed{\sqrt{x}}$	11.283834

Rounding to tenths, we get: t = _____

19. In right triangle TMB, the length of the two legs and the size of two angles are given.

a) Using the angle-sum principle, find the size of angle B. _____

b) Using the Pythagorean Theorem, find the length of the hypotenuse "m". Round to tenths.

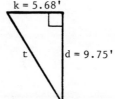

m = _____

t = 11.3 ft

20. The distance between the opposite corners of a rectangle or square is called the "diagonal" of the rectangle or square. The diagonal of each figure is the hypotenuse of a right triangle. Use the Pythagorean Theorem to find the diagonal of the rectangle and square below. Round to tenths.

a)

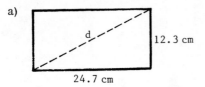

d = _____

b)

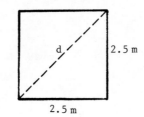

d = _____

a) Angle B = 30°

b) m = 22.9 in

21.   The lengths of the hypotenuse and one leg are given for the triangle
       below. We can use the Pythagorean Theorem to find the length of the
       unknown leg "t".   The steps are:

       1) Write the Pythagorean Theorem.

           $d^2 = t^2 + k^2$

       2) Solve for "t^2".

           $t^2 = d^2 - k^2$

       3) Substitute and simplify.

           $t^2 = (10')^2 - (6')^2$

               $= 100 \text{ ft}^2 - 36 \text{ ft}^2$

               $= 64 \text{ ft}^2$

       4) Find "t" by taking the square root of the right side.

           $t = \sqrt{64 \text{ ft}^2} = $ _____

a) d = 27.6 cm

b) d = 3.5 m

---

22.   We can use the same steps to find the length of leg "v" below.

           $h^2 = v^2 + b^2$

           $v^2 = h^2 - b^2$

           $v^2 = (9 \text{ cm})^2 - (7 \text{ cm})^2$

           $v^2 = 81 \text{ cm}^2 - 49 \text{ cm}^2$

           $v^2 = 32 \text{ cm}^2$

       Find "v" by taking the square root of the right side.   Round to tenths.

           $v = \sqrt{32 \text{ cm}^2} = $ _____

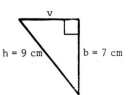

8 ft

---

23.   When using the Pythagorean Theorem to find an unknown leg in a right
       triangle, we can also do all of the calculations in one process on a
       calculator.   To prepare to do so, we solve for the leg before substituting.
       As an example, let's solve for leg "y" below.

           $z^2 = y^2 + x^2$

           $y^2 = z^2 - x^2$

           $y = \sqrt{z^2 - x^2}$

           $y = \sqrt{(18 \text{ m})^2 - (12 \text{ m})^2}$

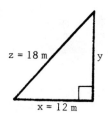

       The calculator steps for the solution are shown.   Notice again that we
       pressed ⎣ = ⎦ before pressing ⎣√x⎦ .

Enter	Press	Display
18	x^2 –	324.
12	x^2 = $\sqrt{x}$	13.416408

       Rounding to tenths, we get: y = _____

v = 5.7 cm

24. We set up the calculator solution for leg "a" below. Complete the
    solution. Round to tenths.

    $$r^2 = m^2 + a^2$$
    $$a^2 = r^2 - m^2$$
    $$a = \sqrt{r^2 - m^2}$$
    $$a = \sqrt{(27.5')^2 - (21.9')^2}$$
    $$a = \underline{\hspace{2cm}}$$

    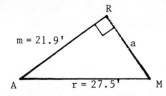

    y = 13.4 m

25. In right triangle RQS, the length of the hypotenuse and one leg and the
    size of two angles are given.

    a) Using the angle-sum principle,
       find the size of angle S. _____

    b) Using the Pythagorean Theorem,
       find the length of leg "s". Round
       to tenths.

    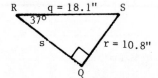

    $$s = \underline{\hspace{2cm}}$$

    a = 16.6 ft

26. Use the Pythagorean Theorem to find
    the width of the rectangle at the right.
    Round to tenths.

    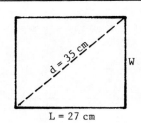

    $$W = \underline{\hspace{2cm}}$$

    a) Angle S = 53°

    b) s = 14.5 in

27. There are two holes in the
    metal plate at the right.
    Use the Pythagorean Theorem
    to find the distance between
    the holes, measured center-
    to-center. Round to tenths.

    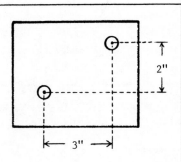

    W = 22.3 cm

    3.6 in

28.   In the rectangular prism below, BD is the diagonal of the prism and BC is
       the diagonal of the base of the prism. To find BD, we must find BC first.
       The Pythagorean Theorem can be used to find both diagonals.

   a) ABC is a right triangle.
      Find BC, the diagonal of
      the base.

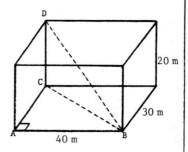

         BC = _____

   b) BCD is a right triangle.
      Find BD, the diagonal of
      the prism. Round to tenths.

         BD = _____

a) BC = 50 m

b) BD = 53.9 m

---

29.   We can find the <u>area</u> of this right
       triangle in two steps.

   a) First find the length of leg "t".
      Round to tenths.

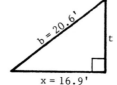

         t = _____

   b) Then use the area formula below,
      rounding to tenths.

      $A = \dfrac{(\text{leg 1})(\text{leg 2})}{2}$

a) t = **11.8** ft   b) A = **99.7** ft^2

7-4   ISOSCELES AND EQUILATERAL TRIANGLES

In this section, we will discuss <u>isosceles</u> and <u>equilateral</u> triangles and solve problems involving triangles
of those types.

---

30.   Triangles with two equal sides are called <u>isosceles</u> triangles. In an
       isosceles triangle, <u>the angles opposite the equal sides are equal</u>.

       The triangle at the right is an
       isosceles triangle because sides
       "c" and "d" are equal.

   a) Since the angles opposite "c"
      and "d" are equal, angle #1 = _____ .

   b) Therefore, angle #2 = _____ .

31. In the isosceles triangle on the right, sides "t" and "p" are equal.

Angles #1 and #2 must be equal because they are opposite the equal sides. How many degrees are there in each of these two equal angles? _____

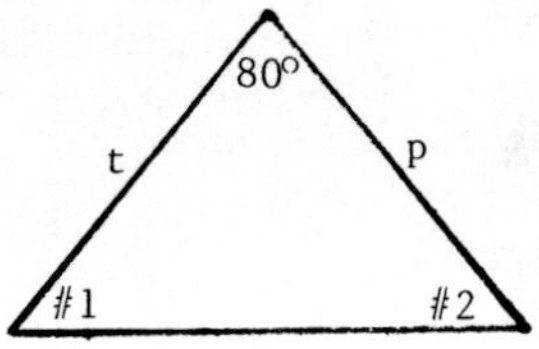

a) 30°   b) 120°

---

32. The triangle at the right is an isosceles **right** **triangle** since it contains a right angle, and sides "c" and "k" are equal.

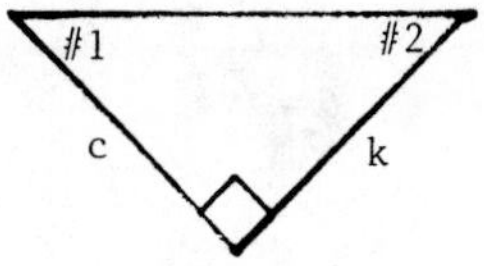

a) Angle #1 = _____ °

b) Angle #2 = _____ °

50°

---

33. Triangles with three equal sides are called equilateral triangles. In an equilateral triangle, all three angles are equal.

The triangle at the right is an equilateral triangle because all three sides are equal. How many degrees are there in each of the three equal angles? _____

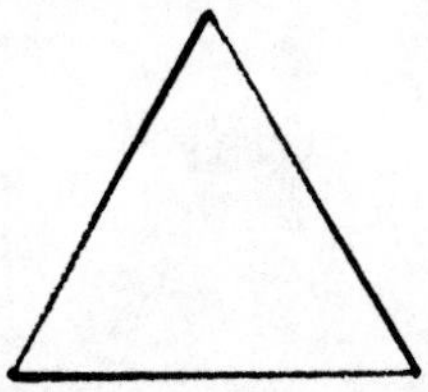

a) 45°

b) 45°

---

34. When a height is drawn to the unequal side in an isosceles triangle, it bisects that side. That is, it cuts that side into two equal parts. We can use that fact to find the height and the area of the triangle when all three sides are known.

Triangle ABC is an isosceles triangle. AD is a height drawn to the unequal side BC.

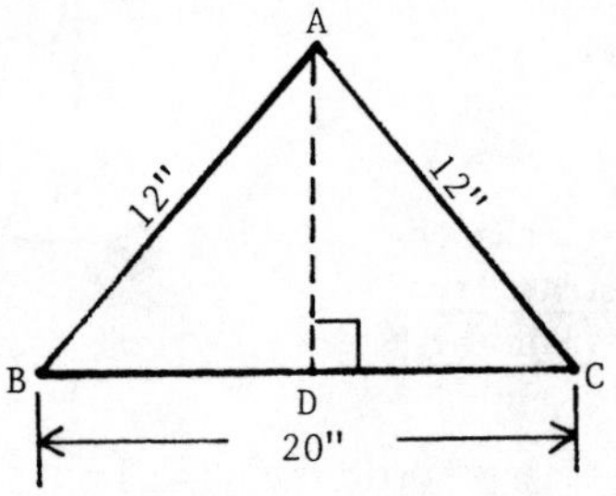

a) Since AD bisects BC, how long are BD and DC? _____

b) Using the Pythagorean Theorem, find the length of AD to the nearest tenth.

AD = _____

c) The area of the triangle is $\dfrac{(BC)(AD)}{2}$ = _____

60°

---

a) 10 in

b) 6.6 in

c) 66 in^2

35.   When a height is drawn to any side in an <u>equilateral</u> triangle, it bisects
that side. We can use that fact to find the height and the area of the
triangle when the length of the sides is known.

Triangle DEF is an equilateral
triangle. FG is a height drawn
to DE.

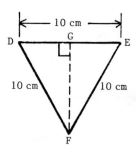

a) Since FG bisects DE, how
long are DG and GE? _____

b) Using the Pythagorean Theorem,
find the length of FG to the nearest
tenth.

FG = _____

c) The area of the triangle is $\dfrac{(DE)(FG)}{2}$ = _____

---

36.   In the rectangular figure at the
right, the shaded part is an
isosceles triangle. Find the
area of the triangle.

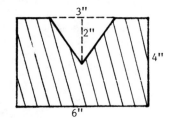

A = _____

a) 5 cm

b) 8.7 cm

c) 43.5 cm^2

---

37.   In the figure at the right, a
V-shaped cut has been made
in the top of a rectangle.
The cut is an isosceles triangle whose base is 3" and
whose height is 2".

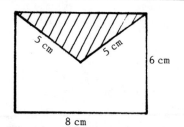

To find the area of the shaded
figure, <u>we</u> <u>must</u> <u>subtract</u> <u>the</u>
<u>area</u> <u>of</u> <u>the</u> <u>triangle</u> <u>from</u> <u>the</u>
<u>area</u> <u>of</u> <u>the</u> <u>rectangle</u>.

a) The area of the rectangle is _____.

b) The area of the triangle is _____.

c) The area of the shaded figure is _____.

A = 12 cm^2

---

a) 24 in^2

b) 3 in^2

c) 21 in^2

SELF-TEST 20 (pp. 266-277)

1. Angle P = _____

2. Which side is <u>shortest</u>? _____

3. Angle F = _____

4. Which side is <u>longest</u>? _____

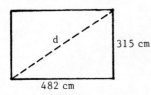

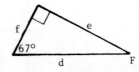

5. Find side "c". Round to tenths.

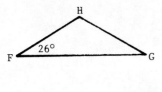

c = _____

6. Find side "p". Round to hundredths.

7. Find the area. Round to hundredths.

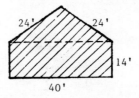

p = _____

A = _____

8. FHG is an <u>isosceles</u> triangle. Find angle G and angle H.

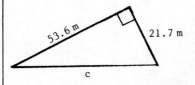

G = _____

H = _____

9. ABC is an <u>isosceles right angle</u>. Find angle A and angle B.

A = _____

B = _____

10. PQR is an <u>equilateral triangle</u>. Find angle P, angle Q, and angle R.

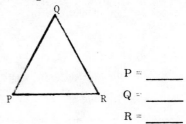

P = _____

Q = _____

R = _____

11. Find "d", the diagonal of the rectangle. Round to the nearest whole number.

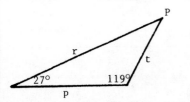

d = _____

12. Find the area of the shaded figure. Round to the nearest whole number.

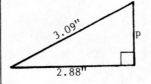

A = _____

7-5   THE TANGENT RATIO

When the Pythagorean Theorem or angle-sum principle cannot be used to find an unknown side or unknown angle in a right triangle, we can usually use one of the three basic trigonometric ratios to do so. In this section, we will define the <u>tangent</u> <u>ratio</u> and show how it can be used to find an unknown side or angle in a right triangle.

38.   In right triangle ABC, "c" is the hypotenuse, and "a" and "b" are the two legs.

"a" is the <u>side opposite</u> angle A.

"b" is the <u>side adjacent</u> to angle A.

Note: The word "adjacent" means "next to". Though both "c" and "b" are "next to" angle A, only "b" is the "side adjacent" because "c" is the hypotenuse.

In the same triangle: a) the <u>side opposite</u> angle B is _____.

b) the <u>side adjacent</u> to angle B is _____.

39.   In right triangle PRT:

a) The hypotenuse is _____.

b) The side opposite angle T is _____.

c) The side adjacent to angle T is _____.

d) The side opposite angle R is _____.

e) The side adjacent to angle R is _____.

a) b   b) a

40.   In right triangle HSV:

a) The hypotenuse is _____.

b) The side opposite angle V is _____.

c) The side adjacent to angle V is _____.

d) The side opposite angle H is _____.

e) The side adjacent to angle H is _____.

a) p   d) r

b) t   e) t

c) r

41.   The <u>tangent</u> <u>ratio</u> is one of the three basic trigonometric ratios. In a right triangle, the "<u>tangent of an acute angle</u>" is a comparison of the "<u>length of the side opposite</u>" to the "<u>length of the side adjacent</u>" to the angle. That is:

$$\text{THE TANGENT OF AN ANGLE} = \frac{\text{SIDE OPPOSITE}}{\text{SIDE ADJACENT}}$$

a) s   d) h

b) v   e) v

c) h

Continued on following page.

41.   Continued

In right triangle ABC:

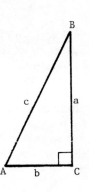

The side opposite angle A is "a".

The side adjacent to angle A is "b".

Therefore, the <u>tangent</u> <u>of</u> <u>angle</u> <u>A</u> is $\frac{a}{b}$.

The side opposite angle B is "b".

The side adjacent to angle B is "a".

Therefore, the tangent of angle B is _____ .

---

42.   In right triangle MPT:

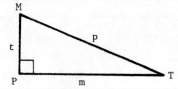

a) The tangent of angle T is _____

b) The tangent of angle M is _____ .

$\frac{b}{a}$

---

43.   Using capital letters to label the sides, complete these for right triangle CDF.

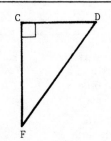

a) The tangent of angle D is _____ .

b) The tangent of angle F is _____ .

a) $\frac{t}{m}$ b) $\frac{m}{t}$

---

44.   In right triangle ADV:

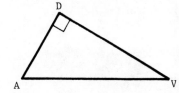

a) The tangent of angle A is _____ .

b) The tangent of angle V is _____ .

a) $\frac{CF}{CD}$ b) $\frac{CD}{CF}$

---

45.   The tangent of any specific angle has the same numerical value in right triangles of any size. For example, in the diagram below, there are three right triangles: ABG, ACF, and ADE. Each right triangle contains the common angle A which is a 37° angle.

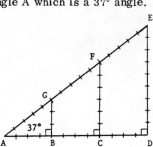

a) $\frac{DV}{AD}$ b) $\frac{AD}{DV}$

Continued on following page.

45.   Continued

To show that the tangent of a 37° angle is the same in right triangles of any size, we computed the tangent of angle A in each of the three right triangles by counting units to approximate the lengths of the sides. As you can see, the tangent of 37° is approximately $\frac{3}{4}$ in each triangle.

In triangle ABG, the tangent of A is $\dfrac{BG}{AB} = \dfrac{3}{4}$

In triangle ACF, the tangent of A is $\dfrac{CF}{AC} = \dfrac{6}{8}$ or $\dfrac{3}{4}$

In triangle ADE, the tangent of A is $\dfrac{DE}{AD} = \dfrac{9}{12}$ or $\dfrac{3}{4}$

Tangents are usually expressed as decimal numbers. Therefore, instead of saying that the tangent of 37° is approximately $\frac{3}{4}$, we would say that the tangent of 37° is approximately _____.

---

46.   Instead of "the tangent of a 37° angle", the abbreviation "tan 37°" is used. Therefore:

tan 15° means: the tangent of a 15° angle

tan 74° means: _____

0.75

---

47.   A partial table of tangents is shown at the right. Each tangent is rounded to four decimal places. Notice this fact:

As the size of the angle increases from 0° to 89°, the size of the tangent <u>increases</u> from 0.0000 to 57.2900 .

Angle A	tan A
0°	0.0000
10°	0.1763
20°	0.3640
30°	0.5774
40°	0.8391
50°	1.1918
60°	1.7321
70°	2.7475
80°	5.6713
89°	57.2900

Using the table, complete these:

a) The tangent of any 70° angle is _____.

b) 0.1763 is the tangent of any _____ angle.

the tangent of a 74° angle

---

48.   A calculator can be used to find the tangent of any angle. To do so, we simply enter the angle and press [tan] .

Note: Some calculators are designed to give tangents of angles measured in degrees, radians, and even grads. <u>Be sure that your calculator is set to give the tangents of angles measured in degrees.</u>

a) 2.7475

b) 10°

Continued on following page.

48. Continued

Following the steps below, find tan 15°, tan 47°, and tan 86°.

Enter	Press	Display
15	tan	.26794919
47	tan	1.0723687
86	tan	14.300666

Tangents are usually rounded to four decimal places. Therefore:

a) tan 15° = _____   b) tan 47° = _____   c) tan 86° = _____

---

49. Use a calculator for these. Round each tangent to four decimal places.

a) tan 34° = _____   b) tan 78° = _____

a) 0.2679

b) 1.0724

c) 14.3007

---

50. A calculator can also be used to find the size of an angle whose tangent is known. Either [INV] [tan] , [arc] [tan] , or [tan⁻¹] is used. Following the steps below, find the angle whose tangent is 0.3173.

Enter	Press	Display
0.3173	[INV] [tan]	17.604233
	or [arc] [tan]	17.604233
	or [tan⁻¹]	17.604233

The angle is usually rounded to the nearest whole number. Therefore:

The angle whose tangent is 0.3173 is _____.

a) 0.6745

b) 4.7046

---

51. Complete these. Round each angle to the nearest whole number.

a) If tan F = 0.7541,   b) If tan A = 2.3555,

F = _____   A = _____

18°

---

52. We cannot use the Pythagorean Theorem to find side MR in the right triangle below because the length of only one side is known. However, we can use tan P to do so. The steps are:

$$\tan P = \frac{MR}{PR}$$

$$\tan 25° = \frac{MR}{52.5}$$

$$MR = (52.5)(\tan 25°)$$

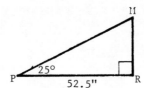

a) 37°   b) 67°

Continued on following page.

52.   Continued

To evaluate $(52.5)(\tan 25°)$ on a calculator, follow the steps below. Wait until $\tan 25°$ appears on the display before pressing $\boxed{=}$.

Enter	Press	Display
52.5	$\boxed{\times}$	52.5
25	$\boxed{\tan}$ $\boxed{=}$	24.481152

Rounding to the nearest tenth of an inch, MR = _____

---

53.   We cannot use the Pythagorean Theorem to find side CE in the right triangle below because only one side is known. However, we can use $\tan C$ to do so. The steps are:

$$\tan C = \frac{DE}{CE}$$

$$\tan 55° = \frac{125}{CE}$$

$$(\tan 55°)(CE) = 125$$

$$CE = \frac{125}{\tan 55°}$$

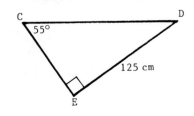

To complete the evaluation on a calculator, follow the steps below. Wait until $\tan 55°$ appears on the display before pressing $\boxed{=}$.

Enter	Press	Display
125	$\boxed{\div}$	125.
55	$\boxed{\tan}$ $\boxed{=}$	87.525942

Rounding to the nearest tenth of a centimeter. CE = _____

24.5 in

---

54.   We cannot use the angle-sum principle to find angle R in the right triangle at the right because angle T is not known. However, we can use $\tan R$ to do so. The steps are:

$$\tan R = \frac{ST}{RS}$$

$$\tan R = \frac{19.8}{34.3}$$

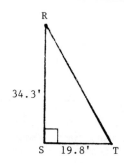

To find R, follow the steps below. Notice that we pressed $\boxed{=}$ to complete the division before pressing $\boxed{INV}$ $\boxed{\tan}$, $\boxed{arc}$ $\boxed{\tan}$, or $\boxed{\tan^{-1}}$.

Enter	Press	Display
19.8	$\boxed{\div}$	19.8
34.3	$\boxed{=}$ $\boxed{INV}$ $\boxed{\tan}$	29.996098
	or $\boxed{arc}$ $\boxed{\tan}$	
	or $\boxed{\tan^{-1}}$	

Rounding to the nearest whole number degree, R = _____

87.5 cm

30°

7-6   THE SINE RATIO

The <u>sine ratio</u> is also one of the three basic trigonometric ratios. In this section, we will define the sine ratio and show how it can be used to find an unknown side or angle in a right triangle.

---

55.   In a right triangle, the "sine <u>of an acute angle</u>" is a comparison of the "<u>length of the side opposite</u>" the angle to the "<u>length of the hypotenuse</u>". That is:

THE SINE OF AN ANGLE $= \dfrac{\text{SIDE OPPOSITE}}{\text{HYPOTENUSE}}$

    <u>Note</u>: The word "sine" is pronounced "sign". It is not pronounced "**sin**".

In right triangle ABC:

    The side opposite angle A is "a".

    The hypotenuse is "c".

    Therefore, the <u>sine of angle A</u> is $\frac{a}{c}$.

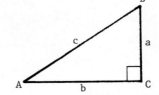

What is the sine of angle B? _____

---

56.   In right triangle XYT:

    a) The sine of angle T is _____.

    b) The sine of angle Y is _____.

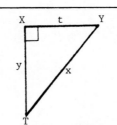

$\dfrac{b}{c}$

---

57.   Using capital letters to label the sides, complete these for right triangle PQS.

    a) The sine of angle P is _____.

    b) The sine of angle S is _____.

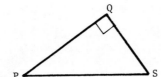

a) $\dfrac{t}{x}$

b) $\dfrac{y}{x}$

---

58.   To show that the sine of a specific angle has the same numerical value in right triangles of any size, we will again use a 37° angle as an example. Let's count units to find the sine of angle A in each of the three triangles at the right.

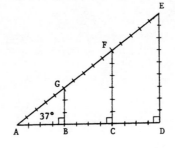

a) $\dfrac{QS}{PS}$

b) $\dfrac{PQ}{PS}$

Continued on following page.

58.   Continued

In triangle ABG, the sine of A is $\dfrac{BG}{AG}$ = $\dfrac{3}{5}$

In triangle ACF, the sine of A is $\dfrac{CF}{AF}$ = $\dfrac{6}{10}$ or $\dfrac{3}{5}$

In triangle ADE, the sine of A is $\dfrac{DE}{AE}$ = $\dfrac{9}{15}$ or $\dfrac{3}{5}$

Sines are usually expressed as decimal numbers. Therefore, instead of saying that the sine of 37° is approximately $\dfrac{3}{5}$, we would say that the sine of 37° is approximately _____ .

---

59.   Instead of "<u>the sine of a 37° angle</u>", the abbreviation "<u>sin 37°</u>" is used. Though the abbreviation is "sin", it is still pronounced "sign".

0.6

sin 15° means: the sine of a 15° angle

sin 48° means: _____

---

60.   A partial table of sines is shown at the right. Each sine is rounded to four decimal places. Notice this fact:

the sine of a 48° angle

As the size of the angle increases from 0° to 90°, the size of the sine <u>increases</u> from 0.0000 to 1.0000.

Angle A	sin A
0°	0.0000
10°	0.1736
20°	0.3420
30°	0.5000
40°	0.6428
50°	0.7660
60°	0.8660
70°	0.9397
80°	0.9848
90°	1.0000

Using the table, complete these:

a) The sine of any 50° angle is _____ .

b) 0.9848 is the sine of any _____ angle.

---

61.   To find <u>the</u> sine of an angle on a calculator, we enter the angle and press ⌑sin⌑. Use a calculator for these. Round to four decimal places.

a) 0.7660

b) 80°

a) sin 7° = _____   b) sin 41° = _____   c) sin 83° = _____

---

62.   To find an angle whose <u>sine is known</u> on a calculator, we enter the sine and then press either ⌑INV⌑ ⌑sin⌑ or ⌑arc⌑ ⌑sin⌑ or ⌑sin⁻¹⌑. Use a calculator for these. Round to the nearest whole-number degree.

a) 0.1219

b) 0.6561

c) 0.9925

a) If sin D = 0.2666,   b) If sin G = 0.8511,

     D = _____   G = _____

---

a) 15°   b) 58°

63. We cannot use the Pythagorean Theorem to find side ST in the right triangle below because only one side is known. However, we can use sin Q to do so. The steps are:

$$\sin Q = \frac{ST}{SQ}$$

$$\sin 35° = \frac{ST}{96}$$

$$ST = (96)(\sin 35°)$$

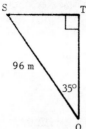

To evaluate (96)(sin 35°) on a calculator, follow the steps below. Wait until sin 35° appears on the display before pressing $\boxed{=}$.

Enter	Press	Display
96	$\boxed{x}$	96.
35	$\boxed{\sin}$ $\boxed{=}$	55. 063338

Rounding to the nearest tenth of a meter, ST = _____.

---

64. We cannot use the Pythagorean Theorem to find hypotenuse CF in the right triangle below because only one side is known. However, we can use sin F to do so. The steps are:

$$\sin F = \frac{CR}{CF}$$

$$\sin 50° = \frac{57.8}{CF}$$

$$(\sin 50°)(CF) = 57.8$$

$$CF = \frac{57.8}{\sin 50°}$$

55.1 m

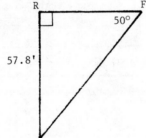

To complete the evaluation on a calculator, follow the steps below. Wait until sin 50° appears on the display before pressing $\boxed{=}$.

Enter	Press	Display
57.8	$\boxed{\div}$	57.8
50	$\boxed{\sin}$ $\boxed{=}$	75. 452541

Rounding to the nearest tenth of a foot, CF = _____.

---

65. We cannot use the angle-sum principle to find angle D below because angle R is not known. However, we can use sin D to do so. The steps are:

$$\sin D = \frac{PR}{DR}$$

$$\sin D = \frac{60.1}{85.8}$$

75.5 ft

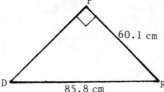

Continued on following page.

65.   Continued

To find D on a calculator, following the steps below. Notice that we pressed ⬚= to complete the division before pressing ⬚INV ⬚sin , ⬚arc ⬚sin , or ⬚sin⁻¹

Enter	Press	Display
60.1	⬚÷	60.1
85.8	⬚= ⬚INV ⬚sin	44.464419
	or ⬚arc ⬚sin	
	or ⬚sin⁻¹	

Rounding to the nearest whole-number degree, D = _____

44°

7-7   THE COSINE RATIO

The <u>cosine ratio</u> is the last of the three basic trigonometric ratios. In this section, we will define the cosine ratio and show how it can be used to find an unknown side or angle in a right triangle.

---

66.   In a right triangle, the "<u>cosine of an acute angle</u>" is a comparison of the "<u>length of the side adjacent</u>" to the angle to the "<u>length of the hypotenuse</u>". That is:

> THE COSINE OF AN ANGLE = $\dfrac{\text{SIDE ADJACENT}}{\text{HYPOTENUSE}}$

<u>Note</u>: The word "cosine" is pronounced "co-sign".

In right triangle CDE:

The side adjacent to angle C is "d".

The hypotenuse is "e".

Therefore, the cosine of angle C is $\dfrac{d}{e}$

What is the cosine of angle D?   _____

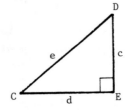

$\dfrac{c}{e}$

---

67.   The abbreviation for "cosine" is "cos". Therefore, in right triangle MFT:

a) cos M = _____

b) cos T = _____

a) $\dfrac{t}{f}$ b) $\dfrac{m}{f}$

68. Using capital letters to label the sides, complete these.

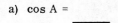

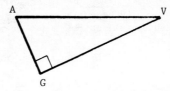

a) cos A = _____

b) cos V = _____

a) $\dfrac{AG}{AV}$ b) $\dfrac{GV}{AV}$

69. To show that the cosine of a specific angle has the same numerical value in right triangles of any size, we will again use a 37° angle as the example. Let's count units to find the cosine of angle A in each of the three triangles at the right.

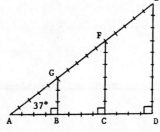

In triangle ABG, cos A $= \dfrac{AB}{AG} = \dfrac{4}{5}$

In triangle ACF, cos A $= \dfrac{AC}{AF} = \dfrac{8}{10}$ or $\dfrac{4}{5}$

In triangle ADE, cos A $= \dfrac{AD}{AE} = \dfrac{12}{15}$ or $\dfrac{4}{5}$

Expressed as a decimal number, cos 37° in each of the three triangles is approximately _____.

70. A partial table of cosines is shown at the right. Each cosine is rounded to four decimal places. Notice this fact:

As the size of the angle increases from 0° to 90°, the size of the cosine <u>decreases</u> from 1.0000 to 0.0000.

Using the table, complete these:

Angle A	cos A
0°	1.0000
10°	0.9848
20°	0.9397
30°	0.8660
40°	0.7660
50°	0.6428
60°	0.5000
70°	0.3420
80°	0.1736
90°	0.0000

0.8

a) cos 50° = _____

b) If cos D = 0.1736, D = _____

71. To find the cosine of an angle on a calculator, we enter the angle and press `cos`. Use a calculator for these. Round to four decimal places.

a) cos 9° = _____   b) cos 27° = _____   c) cos 81° = _____

a) 0.6428

b) 80°

a) 0.9877

b) 0.8910

c) 0.1564

72.   To find an angle whose cosine is known on a calculator, we enter the cosine and then press either $\boxed{\text{INV}}$ $\boxed{\text{cos}}$ or $\boxed{\text{arc}}$ $\boxed{\text{cos}}$ or $\boxed{\text{cos}^{-1}}$. Use a calculator for these. Round to the nearest whole-number degree.

   a) If cos H = 0.7561,   b) If cos V = 0.2099,

   H = _____   V = _____

---

73.   We cannot use the Pythagorean Theorem to find side AC in the right triangle below because only one side is known. However, we can use cos A to do so. The steps are:

   $$\cos A = \frac{AC}{AB}$$

   $$\cos 53° = \frac{AC}{60.7}$$

   $$AC = (60.7)(\cos 53°)$$

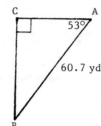

   To evaluate (60.7)(cos 53°) on a calculator, follow the steps below. Wait until cos 53° appears before pressing $\boxed{=}$.

Enter	Press	Display
60.7	$\boxed{\text{x}}$	60.7
53	$\boxed{\text{cos}}$ $\boxed{=}$	36.530172

   Rounding to the nearest tenth of a yard, AC = _____

a) 41°   b) 78°

---

74.   We cannot use the Pythagorean Theorem to find hypotenuse FM in the right triangle below because only one side is known. However, we can use cos F to do so. The steps are:

   $$\cos F = \frac{FG}{FM}$$

   $$\cos 33° = \frac{256}{FM}$$

   $$(\cos 33°)(FM) = 256$$

   $$FM = \frac{256}{\cos 33°}$$

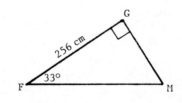

   To complete the evaluation on a calculator, follow the steps below. Wait until cos 33° appears before pressing $\boxed{=}$.

Enter	Press	Display
256	$\boxed{÷}$	256.
33	$\boxed{\text{cos}}$ $\boxed{=}$	305.245

   Rounding to the nearest centimeter, FM = _____

36.5 yd

305 cm

75. We cannot use the angle-sum principle to find angle D below because angle S is not known. However, we can use cos D to do so. The steps are:

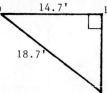

$$\cos D = \frac{DH}{DS}$$

$$\cos D = \frac{14.7}{18.7}$$

Find D on a calculator. Be sure to press ☐= before pressing INV cos or arc cos or cos⁻¹ .

To the nearest whole-number degree, D = _____

38°

7-8   CONTRASTING THE BASIC TRIGONOMETRIC RATIOS

In this section, we will give some exercises contrasting the definitions of the three basic trigonometric ratios.

76. The definitions of the three basic trigonometric ratios for angle A in the right triangle are:

$$\tan A = \frac{\text{side opposite angle A}}{\text{side adjacent to angle A}}$$

$$\sin A = \frac{\text{side opposite angle A}}{\text{hypotenuse}}$$

$$\cos A = \frac{\text{side adjacent to angle A}}{\text{hypotenuse}}$$

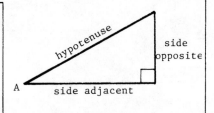

When writing the trig ratios for an angle in a right triangle, it is helpful to locate the hypotenuse first. For example, in triangle BFT:

a) The hypotenuse is _____.

b) sin B = _____

c) cos B = _____

d) tan B = _____

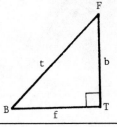

a) t   c) $\frac{f}{t}$

b) $\frac{b}{t}$ d) $\frac{b}{f}$

77. Using capital letters for the sides, complete these:

 a) cos P = ____
 b) tan S = ____
 c) sin P = ____
 d) cos S = ____

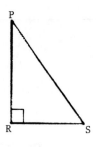

78. Using capital letters for the sides, complete these:

 a) tan H = ____
 b) tan V = ____
 c) sin H = ____
 d) sin V = ____

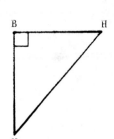

a) $\frac{PR}{PS}$ c) $\frac{RS}{PS}$

b) $\frac{PR}{RS}$ d) $\frac{RS}{PS}$

79. Complete:

 a) tan T = ____
 b) cos F = ____
 c) sin T = ____
 d) tan F = ____

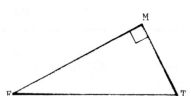

a) $\frac{BV}{BH}$ c) $\frac{BV}{HV}$

b) $\frac{BH}{BV}$ d) $\frac{BH}{HV}$

80. Using small letters for the sides, complete these:

 a) sin D = ____
 b) tan C = ____
 c) sin C = ____
 d) cos D = ____

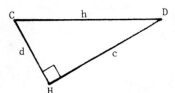

a) $\frac{FM}{MT}$ c) $\frac{FM}{FT}$

b) $\frac{FM}{FT}$ d) $\frac{MT}{FM}$

a) $\frac{d}{h}$ c) $\frac{c}{h}$

b) $\frac{c}{d}$ d) $\frac{c}{h}$

81. If a ratio does not involve the
hypotenuse, it is a "<u>tangent</u>"
ratio. For example, in right
triangle FGH:

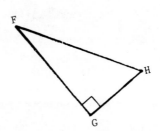

  a) The ratio $\dfrac{FG}{GH}$ is the tangent
     of angle _____ .

  b) The ratio $\dfrac{GH}{FG}$ is the tangent
     of angle _____ .

---

82. If a ratio involves the hypotenuse,
it is the sine of one angle and the
cosine of the other. For example,
in the triangle on the right:

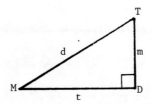

  $\dfrac{m}{d}$ is the <u>sine</u> of angle M and
  the <u>cosine</u> of angle T.

  $\dfrac{t}{d}$ is the <u>sine</u> of angle _____ and
  the <u>cosine</u> of angle ____ .

a) H   b) F

---

83. In this right triangle:

  $\dfrac{MN}{NP}$ is tan P

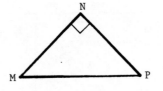

  a) $\dfrac{NP}{MN}$ is _____

  b) $\dfrac{MN}{MP}$ is either _____ or _____

sine of angle T,
cosine of angle M

---

84. In this right triangle:

  a) $\dfrac{a}{b}$ = _____

  b) $\dfrac{b}{a}$ = _____

  c) $\dfrac{a}{c}$ = _____ or _____

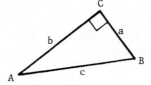

a) tan M

b) sin P or cos M

---

85. In this right triangle:

  a) $\dfrac{m}{t}$ = _____ or _____

  b) $\dfrac{f}{t}$ = _____ or _____

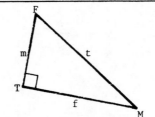

a) tan A

b) tan B

c) sin A or cos B

a) sin M or cos F   b) sin F or cos M

SELF-TEST 21 (pp. 278-292)

Find the numerical value of each trigonometric ratio.   Round to four decimal places.

1. cos 35° = _____

2. tan 23° = _____

3. sin 66° = _____

Find each angle.   Round to the nearest whole-number degree.

4. If sin B = 0.1614,

         B = _____

5. If cos F = 0.2597,

         F = _____

6. If tan R = 1.3318,

         R = _____

Evaluate each of the following.   Round as directed.

7.   Round to tenths.

$$\frac{27.8}{\cos 19°} = \text{_____}$$

8.   Round to hundreds.

(42,900)(tan 61°) = _____

9.   Round to hundredths.

$$\frac{1.53}{\sin 40°} = \text{_____}$$

Find each angle.   Round to the nearest whole-number degree.

10. $\tan A = \dfrac{5,730}{1,980}$

         A = _____

11. $\sin P = \dfrac{2.19}{7.55}$

         P = _____

12. $\cos H = \dfrac{66.7}{78.4}$

         H = _____

13. In right triangle PRT, define the following trigonometric ratios. Use capital letters for labeling the sides.

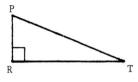

a) sin T = _____

c) tan P = _____

b) tan T = _____

d) cos P = _____

14. In right triangle DEH, define the following trigonometric ratios. Use small letters for labeling the sides.

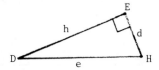

a) cos H = _____

c) sin D = _____

b) sin H = _____

d) tan H = _____

ANSWERS:
1. 0.8192	7. 29.4	13. a) $\frac{PR}{PT}$	14. a) $\frac{d}{e}$
2. 0.4245	8. 77,400		
3. 0.9135	9. 2.38	b) $\frac{PR}{RT}$	b) $\frac{h}{e}$
4. B = 9°	10. A = 71°		
5. F = 75°	11. P = 17°	c) $\frac{RT}{PR}$	c) $\frac{d}{e}$
6. R = 53°	12. H = 32°		
		d) $\frac{PR}{PT}$	d) $\frac{h}{d}$

7-9   FINDING UNKNOWN SIDES IN RIGHT TRIANGLES

Either the Pythagorean Theorem or the trig ratios are used to find unknown sides in right triangles. We will discuss the use of both methods in this section.

86.   When only two sides of a right triangle are known, we use the Pythagorean Theorem to find the third side.

In right triangle ABC, AB and BC are known. Let's use the Pythagorean Theorem to find AC.

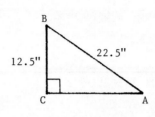

$$(AB)^2 = (BC)^2 + (AC)^2$$

$$(22.5)^2 = (12.5)^2 + (AC)^2$$

$$(AC)^2 = (22.5)^2 - (12.5)^2$$

$$AC = \sqrt{(22.5)^2 - (12.5)^2}$$

Use a calculator to complete the solution. Round to the nearest tenth of an inch.   AC = _____

87.   When only one side and one acute angle of a right triangle are known, we use a trig ratio to find an unknown side.

18.7 in

In right triangle CDF, side CD and angle C are known. We can use the <u>sine ratio</u> to find DF.

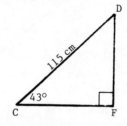

$$\sin C = \frac{DF}{CD}$$

$$\sin 43° = \frac{DF}{115}$$

$$DF = (115)(\sin 43°)$$

Use a calculator to complete the solution. Round to the nearest tenth of a centimeter.   DF = _____

88.   To find PS in this right triangle, we can use the <u>cosine ratio</u>.

78.4 cm

$$\cos P = \frac{PR}{PS}$$

$$\cos 75° = \frac{400}{PS}$$

$$(\cos 75°)(PS) = 400$$

$$PS = \frac{400}{\cos 75°}$$

Use a calculator to complete the solution. Round to the nearest yard.
PS = _____

89.   Before using a trig ratio to find an unknown side in a right triangle, we must decide whether to use the sine, cosine, or tangent of the given angle. The following strategy can be used.

1,545 yd

Find "c" in right triangle CHM.

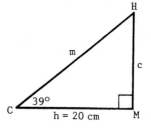

1) Identify the side you want to find ("c") and the known side ("h") in terms of the known angle (C).

"c" is the side opposite angle C.

"h" is the side adjacent to angle C.

2) Identify the ratio that includes both "side opposite" and "side adjacent".

The ratio is the tangent of angle C.

3) Use that ratio to set up an equation.

$$\tan C = \frac{c}{h} \quad \text{or} \quad \tan 39° = \frac{c}{20}$$

Use a calculator to complete the solution. Round to the nearest tenth of a centimeter.   c = _____

90.   Let's use the same steps to solve for DM in this right triangle.

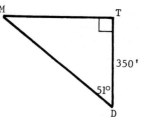

16.2 cm, from:
c = (20)(tan 39°)

1) Identify the side you want to find (DM) and the known side (DT) in terms of the known angle (D).

DM is the hypotenuse.

DT is the side adjacent to angle D.

2) Identify the ratio that includes both "side adjacent" and "hypotenuse".

The ratio is the cosine of angle D.

3) Use that ratio to set up an equation.

$$\cos D = \frac{DT}{DM} \quad \text{or} \quad \cos 51° = \frac{350}{DM}$$

Use a calculator to complete the solution. Round to the nearest foot.
DM = _____

556 ft, from:
$$DM = \frac{350}{\cos 51°}$$

91. We want to find GF in the triangle
     at the right.   The known angle is G.

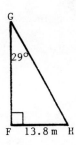

          GF is the <u>side adjacent</u> to angle G.

          FH is the <u>side opposite</u> angle G.

     a) Should we use sin 29°, cos 29°, or
         tan 29° to solve for GF? _____

     b) Complete the solution.   Round to the nearest
         tenth of a meter.

                    GF = _____

---

92. We want to find "d" in the
     triangle at the right.   The
     known angle is B.

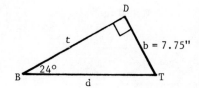

          "d" is the <u>hypotenuse</u>.

          "b" is the <u>side opposite</u> angle B.

     a) Should we use sin 24°, cos 24°, or
         tan 24° to solve for "d" ? _____

     b) Complete the solution.   Round to the
         nearest tenth of an inch.

                    d = _____

a) tan 29°

b) 24.9 m, from:

$$\tan 29° = \frac{13.8}{GF}$$

---

93. We want to find "f" in the
     triangle at the right.   The
     known angle is S.

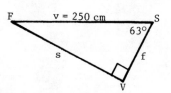

          "f" is the <u>side adjacent</u> to angle S.

          "v" is the <u>hypotenuse</u>.

     a) Should we use sin 63°, cos 63° or
         tan 63° to solve for "f" ? _____

     b) Complete the solution.   Round to the
         nearest centimeter.

                    f = _____

a) sin 24°

b) 19.1 in, from:

$$\sin 24° = \frac{7.75}{d}$$

---

94. Let's solve for "p" in this triangle.

     a) Should we use sin 65°, cos 65°, or
         tan 65° ? _____

     b) Complete the solution.   Round to
         the nearest foot.

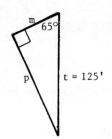

                    p = _____

a) cos 63°

b) 113 cm, from:

$$\cos 63° = \frac{f}{250}$$

95.   Let's solve for TV in this triangle.

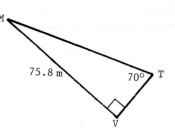

a) Should we use sin 70°, cos 70°, or tan 70°? _____

b) Complete the solution. Round to the nearest tenth of a meter.

TV = _____

a) sin 65°

b) 113 ft, from:

$$\sin 65° = \frac{p}{125}$$

---

96.   Let's solve for AD in this triangle.

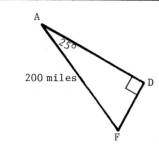

a) Should we use sin 25°, cos 25°, or tan 25°? _____

b) Complete the solution. Round to the nearest mile.

AD = _____

a) tan 70°

b) 27.6 m, from:

$$\tan 70° = \frac{75.8}{TV}$$

---

97.   If only one side and an acute angle of a right triangle are known, we must use one of the trig ratios to find an unknown side.

If only two sides of a right triangle are known, we should use the Pythagorean Theorem to find the third side.

In which triangles below would we use the Pythagorean Theorem in order to find MP? _____

a)

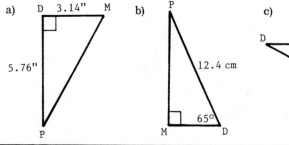

b)

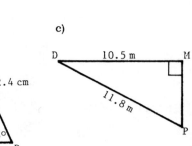

c)

a) cos 25°

b) 181 miles, from:

$$\cos 25° = \frac{AD}{200}$$

---

98.   In which triangles below would we have to use a trigonometric ratio to find CD? _____

a)

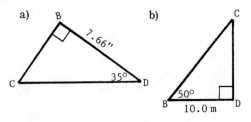

b)

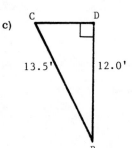

c)

In (a) and (c)

In (a) and (b)

99. If we know <u>two</u> <u>sides</u> <u>and</u> <u>one</u> <u>acute</u> <u>angle</u> <u>of</u> <u>a</u> <u>right</u> <u>triangle</u>, we can use
either of two trig ratios or the Pythagorean Theorem to find the third
side. Here is an example. Round each answer to the nearest tenth of
an inch.

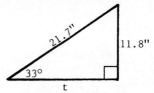

a) Use cos 33° to find "t".

t = _____

b) Use tan 33° to find "t".

t = _____

c) Use the Pythagorean Theorem to find "t .

t = _____

a) t = 18.2 in, from:	b) t = 18.2 in, from:	c) t = 18.2 in, from:
$\cos 33° = \dfrac{t}{21.7}$	$\tan 33° = \dfrac{11.8}{t}$	$t = \sqrt{(21.7)^2 - (11.8)^2}$

7-10   FINDING UNKNOWN ANGLES IN RIGHT TRIANGLES

Either the angle-sum principle or the trig ratios are used to find unknown angles in right triangles. We
will discuss the use of both methods in this section.

100. If one acute angle in a right triangle is known, we can subtract it from 90°
to find the other acute angle. Find angle A in each right triangle below.

a)

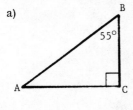

A = _____

b)

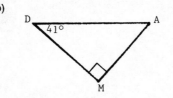

A = _____

a) 35°

b) 49°

101.   When neither acute angle in a right triangle is known, we can use a trig
ratio to find an acute angle <u>if two sides of the triangle are known</u>. An
example is shown.

To find angle A at the right,
we can use the <u>sine ratio</u>.

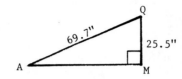

$$\sin A = \frac{MQ}{AQ}$$

$$\sin A = \frac{25.5}{69.7}$$

Use a calculator to complete the solution. Be sure to press $\boxed{=}$ to
complete the division before pressing $\boxed{INV}$ $\boxed{\sin}$ or $\boxed{arc}$ $\boxed{\sin}$ or
$\boxed{\sin^{-1}}$. Round to the nearest degree.   A = _____

---

102.   To find angle P at the right,
we can use the <u>cosine ratio</u>.

$$\cos P = \frac{PT}{PQ}$$

$$\cos P = \frac{6.5}{9.5}$$

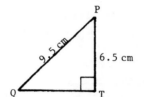

Use a calculator to complete the solution. Press $\boxed{INV}$ $\boxed{\cos}$ or $\boxed{arc}$ $\boxed{\cos}$
or $\boxed{\cos^{-1}}$ after dividing. Round to the nearest degree.   P = _____

21°

---

103.   To decide which trig ratio to use to find an unknown angle in a right
triangle, we identify the known sides in terms of the desired angle.
An example is shown.

To find angle M at the right
we identify PV and MP in
terms of angle M.

PV is the <u>side opposite</u> angle M.

MP is the <u>side adjacent</u> to angle M.

Therefore, we can use tan M to find angle M.

$$\tan M = \frac{PV}{MP}$$

$$\tan M = \frac{13.2}{10.4}$$

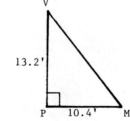

Use $\boxed{INV}$ $\boxed{\tan}$ or $\boxed{arc}$ $\boxed{\tan}$ or $\boxed{\tan^{-1}}$ to complete the solution.
Round to the nearest degree.   M = _____

47°

52°

104. To find angle T at the right, we identify TV and DT in terms of angle T.

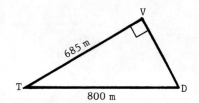

TV is the side adjacent to angle T.

DT is the hypotenuse.

a) Should we use sin T, cos T, or tan T to find angle T? _____

b) To the nearest degree, angle T = _____

---

105. We want to find angle M in this right triangle.

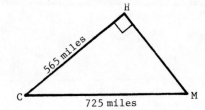

a) Should we use sin M, cos M, or tan M? _____

b) To the nearest degree, angle M = _____

a) cos T

b) 31°, from:

$$\cos T = \frac{685}{800}$$

---

106. We want to find angle C in this right triangle.

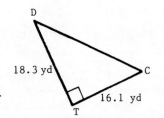

a) Should we use sin C, cos C, or tan C? _____

b) To the nearest degree, angle C = _____

a) sin M

b) 51°, from:

$$\sin M = \frac{565}{725}$$

---

107. After finding one acute angle in a right triangle, we can find the other acute angle by subtracting from 90°.

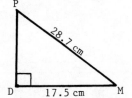

In right triangle PDM:

a) To the nearest degree, angle P = _____

b) To the nearest degree, angle M = _____

a) tan C

b) 49°, from:

$$\tan C = \frac{18.3}{16.1}$$

---

a) 38°, from:

$$\sin P = \frac{17.5}{28.7}$$

b) 52°, from:

90° − 38°

108.   In right triangle FGR:

a) Angle F = _____

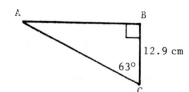

b) To the nearest tenth of a foot,
FG = _____

---

109.   In right triangle ABC:

a) Angle A = _____

b) To the nearest tenth of a centimeter,
AB = _____

a) 41°

b) 64.8 ft

---

a) 27°   b) 25.3 cm

7-11   APPLIED PROBLEMS

In this section, we will discuss some applied problems that involve solving a right triangle.

---

110.   To measure the height of a tall building, a surveyor set his transit (angle-measuring instrument) 100 feet horizontally from the base of the building, measured the angle of elevation to the top of the building, and found it to be 72°.

Find "h", the height of the building.
Round to the nearest foot.

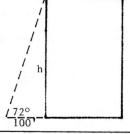

h = _____

---

111.   A right triangle called an "impedance" triangle is used in analyzing alternating current circuits. Angle A, shown in the diagram, is the "phase angle" of the circuit.

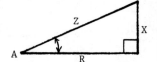

In a particular circuit, X = 2,630 and Z = 6,820.
Find angle A to the nearest degree.

A = _____

h = 308 ft, from:

$$\tan 72° = \frac{h}{100}$$

112. The end-view of the roof of
a building is an isosceles
triangle, as shown at the
right.

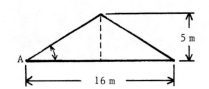

Find angle A, the angle of
slope of the roof to the
nearest degree.

A = _____

A = 23°, from:

$$\sin A = \frac{2,630}{6,820}$$

---

113. In this metal bracket,
find D, the distance
between the two holes.
Round to the nearest
hundredth of an inch.

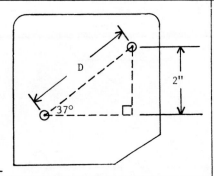

D = _____

A = 32°, from:

$$\tan A = \frac{5}{8}$$

---

114. The metal shape at the right is
called a "template". Figure ABCD
is a rectangle. We want to find
side BE to the nearest tenth of a
centimeter. (<u>Note</u>: AE can be
found by subtracting BC from DE.)

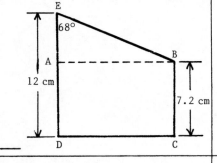

BE = _____

D = 3.32 in, from:

$$\sin 37° = \frac{2}{D}$$

---

115. Here is a cross-sectional view of a metal shaft with a tapered end.

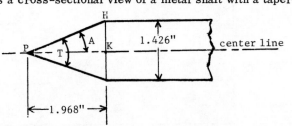

Find T, the taper angle, to the nearest degree.   T = _____

    <u>Note</u>: Angle A is half of angle T. Angle A is an
        acute angle in right triangle PHK.

T = _____

BE = 12.8 cm, from:

$$\cos 68° = \frac{4.8}{BE}$$

T = 40°, since A = 20°, from: $\tan A = \dfrac{0.713}{1.968}$

SELF-TEST 22 (pp. 293-302)

1. Find sides "h" and "v". Round to the nearest tenth of a centimeter.

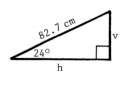

h = _____
v = _____

2. Find angles F and H. Round to the nearest degree.

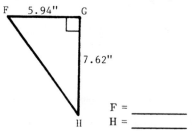

F = _____
H = _____

3. Find angle T. Round to the nearest degree.

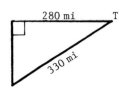

T = _____

4. Find side GP. Round to hundredths.

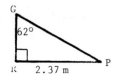

GP = _____

5. Find side "w". Round to tenths.

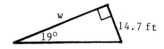

w = _____

6. Find angle A. Round to the nearest degree.

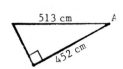

A = _____

7. Find "a" in isosceles triangle DEF. Round to hundredths.

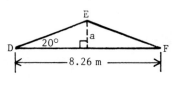

a = _____

8. Find angle P in parallelogram PQRS. Round to the nearest degree.

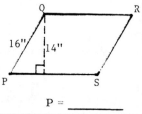

P = _____

ANSWERS:
1. h = 75.6 cm
   v = 33.6 cm
2. F = 52°
   H = 38°
3. T = 32°
4. GP = 2.68 m
5. w = 42.7 ft
6. A = 28°
7. a = 1.50 m
8. P = 61°

SUPPLEMENTARY PROBLEMS - CHAPTER 7

Assignment 20

Use the angle-sum principle to find each unknown angle.

1.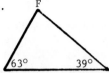

2.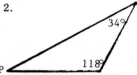

3.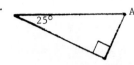

Use the Pythagorean Theorem to find each unknown dimension.   Round as directed.

4. Round to hundreds.

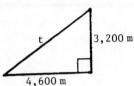

5. Round to tenths.

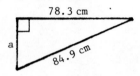

6. Round to hundredths.

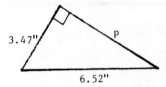

7. Round to tenths.

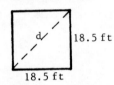

8. Round to the nearest whole number.

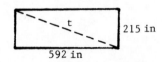

9. Round to thousandths.

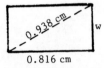

Each triangle is an isosceles triangle.   Find each unknown angle.

10.

11.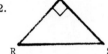

12.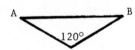

Find the area of each figure.   Round as directed.

13. Round to hundreds.

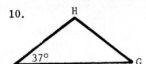

14. Round to the nearest whole number.

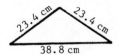

15. Round to tenths.

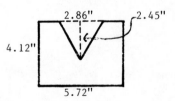

Assignment 21

Find the numerical value of each trig ratio.   Round to four decimal places.

1. tan 12°

2. sin 87°

3. cos 71°

4. cos 0°

5. sin 30°

6. cos 30°

7. tan 65°

8. sin 1°

Continued on following page.

SUPPLEMENTARY PROBLEMS - CHAPTER 7 (Continued)

Assignment 21 (Continued)

Find each angle. Round to the nearest degree.

9. cos A = 0.9715   10. tan P = 3.2168   11. sin H = 0.9999   12. cos Q = 0.3014

13. tan T = 0.1172   14. cos F = 0.1382   15. tan A = 1.0000   16. sin B = 0.2105

Evaluate each of the following. Round to tenths.

17. (31.8)(cos 24°)   18. $\dfrac{136}{\tan 85°}$ 19. $\dfrac{51.9}{\sin 42°}$ 20. (71.3)(tan 30°)

Evaluate each of the following. Round to the nearest whole number.

21. $\dfrac{429}{\tan 51°}$ 22. (1,136)(sin 14°)   23. (250)(cos 64°)   24. $\dfrac{98.4}{\cos 79°}$

Find each angle. Round to the nearest degree.

25. $\cos A = \dfrac{38.4}{41.7}$ 26. $\sin G = \dfrac{7,850}{9,160}$ 27. $\tan R = \dfrac{5.62}{1.93}$ 28. $\sin E = \dfrac{0.138}{0.897}$

Using sides "r", "s" and "t" in right triangle RST, define the following trig ratios.

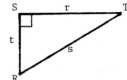

29. sin R   30. cos R   31. tan R

32. sin T   33. cos T   34. tan T

Using sides "a", "b", and "c" in right triangle ABC, define the following trig ratios.

35. cos B   36. sin A   37. tan B

38. tan A   39. cos A   40. sin B

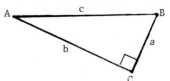

Assignment 22

In problems 1-3, round each answer to hundredths.

1. Find "h".

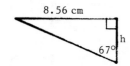

2. Find "w".

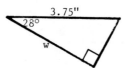

3. Find "p".

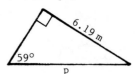

In problems 4-6, round each answer to the nearest whole number.

4. Find "d".

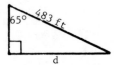

5. Find "t".

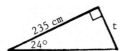

6. Find "b".

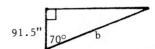

Continued on following page.

SUPPLEMENTARY PROBLEMS - CHAPTER 7 (Continued)

Assignment 22 (Continued)

In problems 7-9, round each angle to the nearest degree.

7. Find angles F and G.

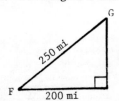

8. Find angles A and B.

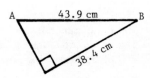

9. Find angles R and S.

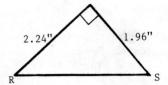

10. Find "h" and "v" in the diagram below. Round to hundredths.

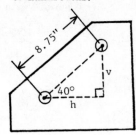

11. The roof diagram below is an isosceles triangle. Find angle A. Round to the nearest degree.

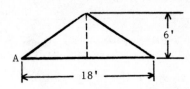

12. In the figure below, the bottom is rectangular. Find "w". Round to tenths.

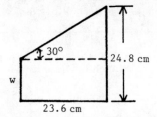

Chapter 8

POWERS, ROOTS, LOGARITHMS

In this chapter, we will discuss powers, roots, common (base "10") logarithms, and natural (base "e") logarithms. The calculator procedures related to those four topics are included. Evaluations are performed with formulas containing powers-of-"e", other powers, "log" expressions, and "ln" expressions.

8-1   POWERS

In this section, we will define "powers" and show how a calculator can be used to convert powers with whole-number exponents to ordinary numbers.

---

1.   In an earlier chapter, we saw that any base-exponent expression in which the base is "10" is called a "power of 10". For example:

10^2, 10^4, and 10^7 are called "powers of 10".

Base-exponent expressions with a base other than 10 are also called "powers" of that base. For example:

4^3, 4^5, and 4^8 are called "powers of 4".

9^2, 9^4, and 9^{10} are called "powers of _____".

---

2.   The word names for some powers are given below.

7^4 is called "7 to the fourth power" or "7 to the fourth".

3^9 is called "3 to the ninth power" or "3 to the ninth".

Write the power corresponding to each word name.

a) 8 to the fifth power = _____   b) 12 to the tenth = _____

9	

---

3.   Any power with a whole-number exponent stands for a multiplication of identical factors. For example:

5^3 = (5)(5)(5)   (Exponent is "3". Multiply three 5's.)

2^4 = (2)(2)(2)(2)   (Exponent is "4". Multiply four 2's.)

Write each power as a multiplication of identical factors.

a) 7^2 = _____   b) 9^5 = _____

a) 8^5

b) 12^{10}

306

4. We can convert a power to an ordinary number by performing the equivalent multiplication. For example:

$$4^3 = (4)(4)(4) = 64$$

Convert these to ordinary numbers by performing the multiplication.

a) $3^4 = (3)(3)(3)(3) = $ _____   b) $2^5 = (2)(2)(2)(2)(2) = $ _____

a) (7)(7)

b) (9)(9)(9)(9)(9)

---

5. We saw earlier that the $\boxed{x^2}$ key and the "power" key $\boxed{y^x}$ can be used to convert any square (second power) or cube (third power) to an ordinary number. The "power" key $\boxed{y^x}$ can also be used to convert larger powers to ordinary numbers.

The calculator steps for $12^4 = 20,736$ are:

Enter	Press	Display
12	$\boxed{y^x}$	12.
4	$\boxed{=}$	20736.

Use $\boxed{y^x}$ to convert each power below to an ordinary number.

a) $15^5 = $ _____   b) $3^9 = $ _____

a) 81   b) 32

---

6. The base of a power can be a decimal number. For example:

$$(1.5)^4 \text{ is a power of } 1.5 .$$

$$(0.8)^5 \text{ is a power of } 0.8 .$$

Powers with decimal bases also stand for multiplications of identical factors. That is:

$$(1.5)^4 = (1.5)(1.5)(1.5)(1.5) \qquad (0.8)^5 = \text{\underline{\hspace{3cm}}}$$

a) 759,375

b) 19,683

---

7. The $\boxed{y^x}$ key can also be used to convert powers with decimal bases to ordinary numbers.

The calculator steps for $(2.2)^4 = 23.4256$ are:

Enter	Press	Display
2.2	$\boxed{y^x}$	2.2
4	$\boxed{=}$	23.4256

Use $\boxed{y^x}$ to convert each power to an ordinary number.

a) $(1.8)^5 = $ _____   b) $(0.9)^7 = $ _____

(0.8)(0.8)(0.8)(0.8)(0.8)

---

a) 18.89568

b) 0.4782969

8.   A power can equal a very large or very small number. In such cases, a
      calculator gives the ordinary number in scientific notation. Do these.
      Round the first factor in scientific notation to two decimal places.

      a) $(247)^8$ = _____   b) $(0.039)^{20}$ = _____

---

a) 1.39×10^{19} b) 6.63×10^{-29}

8-2   ROOTS

In this section, we will define "roots" and show how they can be found on a calculator.

---

9.   In earlier chapters, we discussed the square roots and cube roots of
      numbers.

      To find the <u>square</u> root (or <u>second</u> root) of a number, we must find
      one of <u>two</u> identical factors whose product is that number.

      Since $(7)(7) = 49$, the <u>square</u> root of 49 is 7 .

      To find the <u>cube</u> root (or <u>third</u> root) of a number, we must find
      one of <u>three</u> identical factors whose product is that number.

      Since $(5)(5)(5) = 125$, the <u>cube</u> root of 125 is _____ .

---

10.   Roots beyond the third root do not have special names like "square" root
      or "cube" root. They are simply called <u>fourth</u> root, <u>fifth</u> root, <u>sixth</u> root,
      and so on.

      To find the <u>fourth</u> root of a number, we must find one of <u>four</u>
      identical factors whose product is that number.

      Since $(3)(3)(3)(3) = 81$, the <u>fourth</u> root of 81 is 3 .

      To find the <u>sixth</u> root of a number, we must find one of <u>six</u> identical
      factors whose product is that number.

      Since $(4)(4)(4)(4)(4)(4) = 4,096$, the <u>sixth</u> root of 4,096 is _____ .

5

---

11.   We have seen that radicals are used for square roots and cube roots.
      For example:

                $\sqrt{64}$ means: find the <u>square</u> root of 64 .
                $\sqrt[3]{27}$ means: find the <u>cube</u> root of 27 .

      The small "3" in a cube root is called the <u>index</u>. A radical with an index is
      also used for roots beyond cube roots. That is:

                $\sqrt[4]{81}$ means: find the <u>fourth</u> root of 81 .
                $\sqrt[5]{32}$ means: find the <u>fifth</u> root of 32 .
                $\sqrt[7]{1,089}$ means: find the _____ root of 1,089 .

4

12. Though the index of a square root is actually "2", the "2" is not usually written. That is:

Instead of $\sqrt[2]{824}$, we simply write $\sqrt{824}$.

Write each of these in radical form.

a) the ninth root of 45 _____   b) the square root of 33 _____

---

13. Since $(4)(4)(4) = 64$, $\sqrt[3]{64} = 4$

Since $(6)(6)(6)(6) = 1,296$, $\sqrt[4]{1,296} = 6$

Since $(8)(8)(8)(8)(8) = 32,768$, $\sqrt[5]{32,768} = $ _____

---

14. Roots can be found on a calculator. As an example, let's do $\sqrt[4]{625} = 5$.

If your calculator has a "root" key $\boxed{\sqrt[x]{y}}$, the steps are:

Enter	Press	Display
625	$\boxed{\sqrt[x]{y}}$	625.
4	$\boxed{=}$	5.

If your calculator does not have a $\boxed{\sqrt[x]{y}}$ key, use $\boxed{INV}$ $\boxed{y^x}$. The steps are:

Enter	Press	Display
625	$\boxed{INV}$ $\boxed{y^x}$	625.
4	$\boxed{=}$	5.

Using either $\boxed{\sqrt[x]{y}}$ or $\boxed{INV}$ $\boxed{y^x}$ and the proper index, do these.

a) $\sqrt[5]{59,049} = $ _____   b) $\sqrt[8]{6,561} = $ _____

---

15. Do these on a calculator. Round to tenths.

a) $\sqrt[4]{256,740} = $ _____   b) $\sqrt[7]{12,250,000} = $ _____

---

16. Do these on a calculator. Round to hundredths.

a) $\sqrt[5]{15.89} = $ _____   b) $\sqrt[9]{0.00466} = $ _____

Answer column (right side):

seventh

a) $\sqrt[9]{45}$

b) $\sqrt{33}$ (not $\sqrt[2]{33}$)

8

a) 9   b) 3

a) 22.5   b) 10.3

a) 1.74   b) 0.55

8-3   POWERS WITH DECIMAL EXPONENTS

We will begin this section by showing how any radical can be converted to a power with either a whole number or decimal exponent. Then we will show how a calculator can be used to convert powers with decimal exponents to ordinary numbers.

---

17.   To multiply powers with the same base, we simply add the exponents. For example:

$$(3^4)(3^4) \;=\; 3^{4\,+\,4} \;=\; 3^8 \qquad\qquad (5^2)(5^2)(5^2) \;=\; 5^{2\,+\,2\,+\,2} \;=\; 5^6$$

Do these multiplications by simply adding the exponents.

a) $(7^6)(7^6)$ = _____   b) $(2^5)(2^5)(2^5)(2^5)$ = _____

---

18.   We can also find roots of powers. For example:

$\sqrt{2^6}$ means: find one of <u>two</u> identical factors whose product is 2^6.

Since $(2^3)(2^3) = 2^6$, $\sqrt{2^6} = 2^3$

$\sqrt[3]{7^{12}}$ means: find one of <u>three</u> identical factors whose product is 7^{12}.

Since $(7^4)(7^4)(7^4) = 7^{12}$, $\sqrt[3]{7^{12}} = 7^4$

$\sqrt[4]{5^8}$ means: find one of <u>four</u> identical factors whose product is 5^8.

Since $(5^2)(5^2)(5^2)(5^2) = 5^8$, $\sqrt[4]{5^8}$ = _____

a) 7^{12} b) 2^{20}

---

19.   In the last frame, we saw these facts:

$$\sqrt{2^6} \;=\; 2^3 \qquad\qquad \sqrt[3]{7^{12}} \;=\; 7^4 \qquad\qquad \sqrt[4]{5^8} \;=\; 5^2$$

To find the root of a power, we can simply divide the exponent of the power by the index of the radical. That is:

$$\sqrt{2^6} \;=\; 2^{\frac{6}{2}} \;=\; 2^3 \qquad \sqrt[3]{7^{12}} \;=\; 7^{\frac{12}{3}} \;=\; 7^4 \qquad \sqrt[4]{5^8} \;=\; 5^{\frac{8}{4}} \;=\; 5^2$$

<u>Note</u>: Remember that the index of a square root is "2".

Using the same procedure, complete these.

a) $\sqrt{4^{10}} = 4^{\frac{10}{2}}$ = _____   b) $\sqrt[5]{12^{30}} = 12^{\frac{30}{5}}$ = _____

5^2

---

20.   To find the root of a power, we convert the radical to a power. Convert each of these to a power by dividing the exponent by the index of the radical.

a) $\sqrt[3]{2^{15}}$ = _____   b) $\sqrt[7]{8^{14}}$ = _____

a) 4^5 b) 12^6

---

a) 2^5, from: $2^{\frac{15}{3}}$ b) 8^2, from: $8^{\frac{14}{7}}$

21. When converting a radical to a power by dividing the exponent by the index:

Sometimes we get a power with a whole-number exponent.

$$\sqrt[3]{5^6} = 5^{\frac{6}{3}} = 5^2 \qquad\qquad \sqrt{9^8} = 9^{\frac{8}{2}} = 9^4$$

Sometimes we get a power with a fractional exponent.

$$\sqrt[4]{11^3} = 11^{\frac{3}{4}} \qquad\qquad \sqrt{15^5} = 15^{\frac{5}{2}}$$

When a power has a fractional exponent, we can convert the fractional exponent to a decimal exponent. That is:

$11^{\frac{3}{4}}$ can be converted to $11^{0.75}$

$15^{\frac{5}{2}}$ can be converted to _____

---

22. Convert each radical to a power with a decimal exponent.

a) $\sqrt[5]{7^9}$ = _____   b) $\sqrt[8]{25^3}$ = _____

$15^{2.5}$

---

23. Any ordinary number is equal to a <u>first</u> power. For example:

$$7 = 7^1 \qquad\qquad 23 = 23^1$$

Therefore, any radical containing an ordinary number can be written as a radical containing a <u>first</u> power. For example:

$$\sqrt{7} = \sqrt{7^1} \qquad\qquad \sqrt[4]{23} = \sqrt[4]{23^1}$$

A radical containing a first power can be converted to a power with a decimal exponent. That is:

$$\sqrt{7} = \sqrt{7^1} = 7^{\frac{1}{2}} = 7^{0.5} \qquad \sqrt[4]{23} = \sqrt[4]{23^1} = 23^{\frac{1}{4}} = \text{_____}$$

a) $7^{1.8}$, from: $7^{\frac{9}{5}}$

b) $25^{0.375}$, from: $25^{\frac{3}{8}}$

---

24. Convert each radical to a power with a decimal exponent.

a) $\sqrt[8]{3}$ = _____   b) $\sqrt[5]{47}$ = _____

$23^{0.25}$

---

25. We use the $\boxed{y^X}$ key to convert a power with a decimal exponent to an ordinary number. For example:

To do $36^{0.5} = 6$, follow these steps.

Enter	Press	Display
36	$\boxed{y^X}$	36.
0.5	$\boxed{=}$	6.

Continued on following page.

a) $3^{0.125}$, from: $3^{\frac{1}{8}}$

b) $47^{0.2}$, from: $47^{\frac{1}{5}}$

25.   Continued

Use a calculator for these.

   a) $4^{1.5}$ = _____   b) $81^{0.25}$ = _____

---

26.   Do these on a calculator. Round to hundredths.

   a) $15^{0.4}$ = _____   b) $4.8^{1.3}$ = _____

| a) 8 | b) 3 |

---

27.   Do these on a calculator. Round to the nearest whole number.

   a) $75^{2.4}$ = _____   b) $19.4^{1.8}$ = _____

| a) 2.95 | b) 7.68 |

---

28.   To evaluate $(1 + 0.05)^{12}$, we perform the addition within the parentheses first. The calculator steps are shown. Notice that we pressed $\boxed{=}$ to complete the addition before pressing $\boxed{y^X}$. Rounded to hundredths, the answer is 1.80 .

| a) 31,633 | b) 208 |

Enter	Press	Display
1	$\boxed{+}$	1.
0.05	$\boxed{=}$ $\boxed{y^X}$	1.05
12	$\boxed{=}$	1.7958563

Do these on a calculator. Round to hundredths.

   a) $(1 + 0.06)^{20}$ = _____   b) $(27 + 3.5)^{0.1}$ = _____

---

29.   To evaluate $\left(\dfrac{342}{295}\right)^{1.67}$, we perform the division within the parentheses first. The calculator steps are shown. Notice that we pressed $\boxed{=}$ to complete the division before pressing $\boxed{y^X}$. Rounded to hundredths, the answer is 1.28 .

| a) 3.21 | b) 1.41 |

Enter	Press	Display
342	$\boxed{\div}$	342.
295	$\boxed{=}$ $\boxed{y^X}$	1.159322
1.67	$\boxed{=}$	1.280032

Use a calculator for these. Round to hundredths.

   a) $\left(\dfrac{450}{299}\right)^{1.9}$ = _____   b) $\left(\dfrac{228}{1.22}\right)^{0.2}$ = _____

| a) 2.17 | b) 2.85 |

8-4   POWERS WITH NEGATIVE EXPONENTS

In this section, we will define powers with negative exponents and show how a calculator can be used to convert powers with negative exponents to ordinary numbers.

---

30.   The following definition is used for powers with negative exponents.

$$x^{-n} = \frac{1}{x^n}$$

Therefore: $9^{-2} = \frac{1}{9^2}$ $8^{-1.4} = \frac{1}{8^{1.4}}$ $7^{-0.5} =$ _____

---

31.   Using the definition, we converted each power below to an ordinary number.

$$2^{-1} = \frac{1}{2^1} = \frac{1}{2} = 0.5$$

$$5^{-2} = \frac{1}{5^2} = \frac{1}{25} = 0.04$$

We can use the $\boxed{y^X}$ key and the $\boxed{+/-}$ key to perform the same conversions on a calculator. The steps are:

Enter	Press	Display
2	$\boxed{y^X}$	2.
1	$\boxed{+/-}$ $\boxed{=}$	0.5

Enter	Press	Display
5	$\boxed{y^X}$	5.
2	$\boxed{+/-}$ $\boxed{=}$	0.04

Use a calculator to convert these to ordinary numbers.

a) $4^{-1} =$ _____   b) $20^{-2} =$ _____   c) $5^{-3} =$ _____

[margin answer for 31:] $\dfrac{1}{7^{0.5}}$

---

32.   Use a calculator for these. Round to millionths.

a) $12^{-4} =$ _____   b) $9^{-5} =$ _____

[margin answers for 32:]
a) 0.25
b) 0.0025
c) 0.008

---

33.   The same calculator procedure is used for powers with negative decimal exponents. Do these. Round to four decimal places.

a) $23^{-1.4} =$ _____   b) $85^{-0.8} =$ _____

[margin answers for 33:]
a) 0.000048
b) 0.000017

34.   Use a calculator for these. Round to four decimal places.

   a) $17.5^{-1.2}$ = _____   b) $3.06^{-2.4}$ = _____

a) 0.0124	b) 0.0286
a) 0.0322	b) 0.0683

8-5   EVALUATING FORMULAS CONTAINING POWERS

In this section, we will use a calculator for evaluations with formulas containing powers.

35.   We used $\boxed{y^X}$ for the evaluation below. The steps are shown.

In $\boxed{y = x^a}$, find "y" when x = 5 and a = 4 .

Enter	Press	Display
5	$\boxed{y^X}$	5.
4	$\boxed{=}$	625. (y = 625)

Use a calculator for these. Round to tenths.

   a) In $\boxed{M = T^a}$, when T = 18 and a = 1.3 , M = _____

   b) In $\boxed{P = V^k}$, when V = 15.9 and k = 1.64 , P = _____

36.   Do these. Round to hundredths.

   a) In $\boxed{t = m^{0.4}}$, when m = 27.1 , t = _____

   b) In $\boxed{H = 1.14^d}$, when d = 3.75 , H = _____

a) 42.8
b) 93.4

37.   In the evaluation below, we divide first. Notice that we press $\boxed{=}$ to complete the division before using $\boxed{y^X}$. We rounded the answer to hundredths.

In $\boxed{D = \left(\dfrac{a}{b}\right)^{1.8}}$, find D when a = 72.9 and b = 58.6 .

Enter	Press	Display
72.9	$\boxed{\div}$	72.9
58.6	$\boxed{=}\ \boxed{y^X}$	1.2440273
1.8	$\boxed{=}$	1.4814733 (D = 1.48)

Use the same steps for this one. Round to thousandths.

In $\boxed{F = \left(\dfrac{S}{T}\right)^{0.1}}$, when S = 13.9 and T = 47.6 , F = _____

a) 3.74
b) 1.63

38. In the evaluation below, we add first. Notice that we press $\boxed{=}$ to complete the addition before using $\boxed{y^X}$. We rounded the answer to the nearest whole number.

In $\boxed{V = (x + 1)^n}$, find V when $x = 7.8$ and $n = 2.5$.

Enter	Press	Display	
7.8	$\boxed{+}$	7.8	
1	$\boxed{=}$ $\boxed{y^X}$	8.8	
2.5	$\boxed{=}$	229.72416	(V = 230)

Use the same steps for this one. Round to hundredths.

In $\boxed{F = (c + d)^{0.2}}$, when $c = 11.9$ and $d = 48.3$,. $F =$ _____

0.884

---

39. In the formula below, Q^a is a factor in a multiplication. The steps for the evaluation are shown. We rounded the answer to tenths. (<u>Note</u>: On some calculators, you have to evaluate Q^a first and then multiply by P.)

In $\boxed{R = PQ^a}$, find R when $P = 1.59$, $Q = 2.24$, and $a = 3.18$.

Enter	Press	Display	
1.59	$\boxed{x}$	1.59	
2.24	$\boxed{y^X}$	2.24	
3.18	$\boxed{=}$	20.662641	(R = 20.7)

Use the same steps for these. Round to ten-thousandths in (a) and to the nearest whole number in (b).

a) In $\boxed{I = KP^{1.5}}$, when $K = 0.001$ and $P = 19$, $I =$ _____

b) In $\boxed{b = TV^{0.4}}$, when $T = 400$ and $V = 100$, $b =$ _____

2.27

---

40. Two methods for the evaluation below are shown. We rounded the answer to tenths.

In $\boxed{M = T(a + b)^{0.1}}$, find M when $T = 25$, $a = 37$, and $b = 49$.

1) If a calculator has parentheses symbols, we can use the parentheses to evaluate $(a + b)$ before pressing $\boxed{y^X}$.

Enter	Press	Display	
25	$\boxed{x}$ $\boxed{(}$	25.	
37	$\boxed{+}$	37.	
49	$\boxed{)}$ $\boxed{y^X}$	86.	
0.1	$\boxed{=}$	39.029218	(M = 39.0)

Continued on following page.

a) 0.0828

b) 2,524

40.   Continued

2) If a calculator does not have parentheses symbols, we must evaluate $(a + b)^{0.1}$ first and then multiply by the value of T.

Enter	Press	Display
37	$+$	37.
49	$=$ y^x	86.
0.1	x	1.5611687
25	$=$	39.029218 (M = 39.0)

Use a calculator for these. Round to the nearest whole number.

a) In $\boxed{A = P(1 + i)^n}$, when P = 1,000 , i = 0.085 , and n = 10 ,

A = _____

b) In $\boxed{D = K(t + 1)^n}$, when K = 250 , t = 60 , and n = 0.5 ,

D = _____

a) 2,261   b) 1,953

SELF-TEST 23 (pages 306-317)

Evaluate these powers. Record all digits shown on the display.

1. 2^{14} = _____ | 2. $(3.6)^4$ = _____ | 3. $(0.5)^7$ = _____

Evaluate these powers. Report each answer in scientific notation with the first factor rounded to two decimal places.

4. $(592)^6$ = _____ | 5. $(0.178)^{18}$ = _____

Evaluate these roots. Round as directed.

6. Round to tenths. | 7. Round to hundredths. | 8. Round to thousandths.

$\sqrt[5]{260,000}$ = _____ | $\sqrt[8]{68.5}$ = _____ | $\sqrt[4]{0.0973}$ = _____

Convert each radical to a power with a decimal exponent.		Evaluate these powers. Round as directed.	
9. $\sqrt{8^3}$ = _____	10. $\sqrt[5]{74}$ = _____	11. Round to tenths. $(13.8)^{0.9}$ = _____	12. Round to thousands. $(42)^{3.6}$ = _____

Continued on the following page.

SELF-TEST 23 - Continued

Evaluate each of the following. Round as directed.

13. Round to hundredths.	14. Round to thousandths.	15. Round to four decimal places.
$\left(\dfrac{29.4}{18.8}\right)^{3.5} =$ _____	$(2.5)^{-1.3} =$ _____	$(48)^{-0.7} =$ _____

---

16. Find "y". Round to three decimal places.

$$\boxed{y = x^n} \qquad x = 15 \\ n = -0.2$$

y = _____

17. Find "r". Round to tenths.

$$\boxed{r = \left(\frac{h}{p}\right)^{1.8}} \qquad h = 73.4 \\ p = 12.9$$

r = _____

---

18. Find M. Round to hundredths.

$$\boxed{M = KV^h} \qquad K = 5.3 \\ V = 0.86 \\ h = 3.4$$

M = _____

19. Find A. Round to the nearest whole number.

$$\boxed{A = P(1 + i)^n} \qquad P = 1,500 \\ i = 0.06 \\ n = 11$$

A = _____

ANSWERS:

1. 16,384	5. 3.22×10^{-14}	9. $8^{1.5}$	13. 4.78	17. r = 22.9
2. 167.9616	6. 12.1	10. $74^{0.2}$	14. 0.304	18. M = 3.17
3. 0.0078125	7. 1.70	11. 10.6	15. 0.0665	19. A = 2,847
4. 4.30×10^{16}	8. 0.559	12. 698,000	16. y = 0.582	

8-6   POWERS OF TEN WITH DECIMAL EXPONENTS

Any power of ten with a decimal exponent can be converted to an ordinary number, and any ordinary number can be converted to a power of ten. We will discuss the conversion methods in this section.

---

41.   We use the $\boxed{y^x}$ key to convert powers of ten with decimal exponents to ordinary numbers. Do these.

a) $10^{1.7892} =$ _____   (Round to tenths.)

b) $10^{-3.5066} =$ _____   (Round to millionths.)

a) 61.5

b) 0.000311

42. $10^1 = 10$ and $10^2 = 100$. Therefore, any power of ten with an exponent between 1 and 2 equals an ordinary number between 10 and 100 . To show that fact, do these. Round to tenths.

   a) $10^{1.2} =$ _____   b) $10^{1.7091} =$ _____

---

43. $10^3 = 1,000$ and $10^4 = 10,000$. Therefore, any power of ten with an exponent between 3 and 4 equals an ordinary number between 1,000 and 10,000 . To show that fact, do these. Round to the nearest whole number.

   a) $10^{3.25} =$ _____   b) $10^{3.9837} =$ _____

a) 15.8   b) 51.2

---

44. a) Since $10^{0.7518}$ lies between 10^0 and 10^1 , $10^{0.7518}$ equals a number between 1 and 10 .

   To the nearest hundredth, $10^{0.7518} =$ _____

   b) Since $10^{5.3799}$ lies between 10^5 and 10^6 , $10^{5.3799}$ equals a number between 100,000 and 1,000,000 .

   To the nearest thousand, $10^{5.3799} =$ _____

a) 1,778   b) 9,632

---

45. Any power of ten with a <u>negative</u> decimal exponent equals an ordinary number between 0 and 1. To show that fact, do these.

   a) $10^{-1.3} =$ _____   (Round to four decimal places.)

   b) $10^{-3.85} =$ _____   (Round to six decimal places.)

   c) $10^{-5.2075} =$ _____   (Round to seven decimal places.)

a) 5.65

b) 240,000

---

46. In $\boxed{y = 10^x}$, use the $\boxed{y^x}$ key to find the values of "y" corresponding to the following values of the exponent "x".

   a) If x = 1.25 , y = _____ (Round to tenths.)

   b) If x = 0.4 , y = _____ (Round to hundredths.)

   c) If x = -2.5 , y = _____ (Round to five decimal places.)

a) 0.0501

b) 0.000141

c) 0.0000062

---

a) 17.8

b) 2.51

c) 0.00316

47. To convert an ordinary number to a power of ten, we use the $\boxed{\log}$ key to find the exponent. We did so below for 475 and 0.0936.

Note: $\boxed{\log}$ is an abbreviation for "logarithm" which means exponent.

Enter	Press	Display
475	$\boxed{\log}$	2.6766936
0.0936	$\boxed{\log}$	-1.0287242

Rounding each exponent to four decimal places, we get:

$475 = 10^{2.6767}$ $0.0936 =$ _____

---

48. $100 = 10^2$ and $1,000 = 10^3$. Therefore, the exponent of the power-of-ten form of any number between 100 and 1,000 lies between 2 and 3. To show that fact, convert these powers of ten. Round each exponent to four decimal places.

a) $225 =$ _____   b) $879 =$ _____

$10^{-1.0287}$

---

49. $10,000 = 10^4$ and $100,000 = 10^5$. Therefore, the exponent of the power-of-ten form of any number between 10,000 and 100,000 lies between 4 and 5. To show that fact, convert these to powers of ten. Round each exponent to four decimal places.

a) $35,700 =$ _____   b) $69,699 =$ _____

a) $10^{2.3522}$ b) $10^{2.9440}$

---

50. a) Since 5.49 lies between 1 and 10, the exponent of its power-of-ten form lies between 0 and 1.

Rounding the exponent to four decimal places, $5.49 =$ _____

b) Since 7,825 lies between 1,000 and 10,000, the exponent of its power-of-ten form lies between 3 and 4.

Rounding the exponent to four decimal places, $7,825 =$ _____

a) $10^{4.5527}$ b) $10^{4.8432}$

---

51. The power-of-ten form of any number between 0 and 1 has a negative exponent. To show that fact, convert these to powers of ten. Round the exponents to four decimal places.

a) $0.679 =$ _____   b) $0.000175 =$ _____

a) $10^{0.7396}$

b) $10^{3.8935}$

---

52. In $\boxed{y = 10^x}$, use the $\boxed{\log}$ key to find the exponent "x" for the following values of "y". Round the exponent to four decimal places.

a) If $y = 9,210$, $x =$ _____   b) If $y = 0.008$, $x =$ _____

a) $10^{-0.1681}$

b) $10^{-3.7570}$

a) 3.9643   b) -2.0969

8-7   COMMON LOGARITHMS

When a number is written in power-of-ten form, the exponent of the 10 is called the <u>common</u> <u>logarithm</u> of the number. We will discuss common logarithms and logarithmic notation in this section.

---

53. The <u>exponent</u> of the power-of-ten form of a number is called the <u>common</u> <u>logarithm</u> of the number. Since $7.65 = 10^{0.8837}$, the logarithm of 7.65 is 0.8837 . Since $291 = 10^{2.4639}$, the logarithm of 291 is _____ .	
54. The logarithm of a number is <u>only the exponent</u> of the power of ten. It does not include the base "10". That is: Since $87.6 = 10^{1.9425}$: the logarithm of 87.6 is 1.9425 (the logarithm <u>is</u> <u>not</u> $10^{1.9425}$) $0.682 = 10^{-0.1662}$. Is -0.1662 or $10^{-0.1662}$ the logarithm of 0.682? _____	2.4639
55. To find the <u>logarithm</u> of a number on a calculator, enter the number and press the $\boxed{\text{log}}$ key. Do these. Round to four decimal places. a) The logarithm of 761,000 is _____ . b) The logarithm of 0.0829 is _____ .	-0.1662
56. The logarithm of 27.9 is 1.4456 . Therefore, $27.9 = 10^{1.4456}$ The logarithm of 0.55 is -0.2596 . Therefore, $0.55 = 10\boxed{}$	a) 5.8814 b) -1.0814
57. If the logarithm of a number is 2.7959 : a) In power-of-ten form, the number is _____ . b) Rounded to the nearest whole number, the number is _____ .	$10^{-0.2596}$
58. a) If the logarithm of a number is 0.8549 , the number is _____ . (Round to hundredths.) b) If the logarithm of a number is -2.0633 , the number is _____ . (Round to five decimal places.)	a) $10^{2.7959}$ b) 625
	a) 7.16 , from: $10^{0.8549}$ b) 0.00864 , from: $10^{-2.0633}$

59.   The phrase "logarithm of" is usually abbreviated to "log". For example:

log 1,643 means: the logarithm of 1,643

Using the ⌐log⌐ key, complete these. Round to four decimal places.

a) log 1,643 = _____   b) log 0.0005 = _____

---

60.   Negative numbers do not have logarithms. Therefore, finding the logarithm of a negative number is an IMPOSSIBLE operation. A calculator shows that fact with a flashing display, by printing out "Error", or in some other way. Try these on a calculator.

log(-155)   log(-2.81)

Note: When you are using a calculator for a "log" problem and the calculator display shows an IMPOSSIBLE operation, you know that you have made a mistake.

a) 3.2156   b) -3.3010

---

61.   Any basic power-of-ten equation can be written as a "log" equation. For example:

$56.2 = 10^{1.7497}$ can be written: log 56.2 = 1.7497

$10^{-0.6925} = 0.203$ can be written: -0.6925 = log 0.203

Write each equation below as a "log" equation.

a) $0.0097 = 10^{-2.0132}$ b) $10^{5.8136} = 651,000$

_____   _____

---

62.   When a basic power-of-ten equation contains a letter, it can also be written as a "log" equation. For example:

$y = 10^{1.2514}$ can be written: log y = 1.2514

$10^{x} = 0.714$ can be written: x = log 0.714

Write each equation as a "log" equation.

a) $11.8 = 10^{t}$ b) $10^{-0.6711} = m$

_____   _____

a) log 0.0097 = -2.0132

b) 5.8136 = log 651,000

---

a) log 11.8 = t

b) -0.6711 = log m

63.   Any basic "log" equation can be converted to a basic power-of-ten equation.
       For example:

$$\log 299 = 2.4757 \quad \text{can be written:} \quad 299 = 10^{2.4757}$$

$$-0.1871 = \log 0.65 \quad \text{can be written:} \quad 10^{-0.1871} = 0.65$$

Write each equation below as a power-of-ten equation.

   a) $\log 0.017 = -1.7696$ b) $4.5846 = \log 38,420$

   _____   _____

64.   When a basic "log" equation contains a letter, it can also be written as a
       power-of-ten equation. For example:

$$\log x = 3.6099 \quad \text{can be written:} \quad x = 10^{3.6099}$$

$$y = \log 0.074 \quad \text{can be written:} \quad 10^{y} = 0.074$$

Write each equation as a power-of-ten equation.

   a) $-1.5277 = \log d$ b) $\log 803 = a$

   _____   _____

a) $0.017 = 10^{-1.7696}$

b) $10^{4.5846} = 38,420$

a) $10^{-1.5277} = d$ b) $803 = 10^{a}$

8-8   SOLVING POWER-OF-TEN AND "LOG" EQUATIONS

In this section, we will discuss the methods for solving power-of-ten and "log" equations that contain a letter.

65.   In each power-of-ten equation below, the letter stands for <u>an ordinary
       number</u>. Use the $\boxed{y^{x}}$ key to solve each equation.

   a) Round to tenths.   b) Round to five decimal places.
       $10^{1.6294} = x$ $y = 10^{-2.1095}$

       $x = $ _____ $y = $ _____

66.   In each power-of-ten equation below, the letter stands for <u>an exponent</u>.
       Use the $\boxed{\log}$ key to solve each equation. Round each exponent to four
       decimal places.

   a) $0.016 = 10^{t}$ b) $10^{b} = 81,900$

       $t = $ _____ $b = $ _____

a) $x = 42.6$

b) $y = 0.00777$

a) $t = -1.7959$

b) $b = 4.9133$

67. When a "log" equation contains a letter, we can solve it by writing the equation in power-of-ten form. Let's use that method to solve the equation below.

$$\log V = 2.7427$$

a) Write the equation in power-of-ten form. _____

b) Therefore, rounded to the nearest whole number, $V =$ _____

---

68. Let's solve $\log 0.456 = y$.

a) Write the equation in power-of-ten form. _____

b) Therefore, rounded to four decimal places, $y =$ _____

a) $V = 10^{2.7427}$

b) $V = 553$

---

69. Let's solve $-1.7086 = \log x$.

a) Write the equation in power-of-ten form. _____

b) Therefore, rounded to four decimal places, $x =$ _____

a) $0.456 = 10^{y}$

b) $y = -0.3410$

---

70. Let's solve $m = \log 3,190$.

a) Write the equation in power-of-ten form. _____

b) Therefore, rounded to four decimal places, $m =$ _____

a) $10^{-1.7086} = x$

b) $x = 0.0196$

---

71. When the letter follows "log" in an equation, it stands for the ordinary number. For example:

In $\log x = -1.53$, "x" stands for the ordinary number since: $x = 10^{-1.53}$

In $2.25 = \log y$, "y" stands for the ordinary number since: $10^{2.25} = y$

In such cases, convert the "log" equation to a power-of-ten equation and use $\boxed{y^x}$. Solve these.

a) Round to tenths.   b) Round to thousandths.

$\log t = 1.7539$ $-0.49 = \log m$

$t =$ _____ $m =$ _____

a) $10^{m} = 3,190$

b) $m = 3.5038$

a) $t = 56.7$, from:
   $t = 10^{1.7539}$

b) $m = 0.324$, from:
   $10^{-0.49} = m$

72.   When the letter is on the opposite side of "log" in an equation, it stands for the logarithm (or exponent). For example:

In $\log 3.87 = d$, "d" stands for the logarithm since: $3.87 = 10^d$

In $h = \log 0.025$, "h" stands for the logarithm since: $10^h = 0.025$

In such cases, we can simply use the $\boxed{\log}$ key without converting to a power-of-ten equation. Solve these. Round to four decimal places.

a) $\log 0.051 = R$ b) $G = \log 995,000$

R = _____   G = _____

---

73.   Solve: a) Round to hundredths.   b) Round to thousandths.

$\log 1,290 = V$ $\log D = -0.207$

V = _____   D = _____

a) R = -1.2924 , from:

$0.051 = 10^R$

b) G = 5.9978 , from:

$10^G = 995,000$

---

74.   Solve: a) Round to thousands.   b) Round to hundredths.

$6.37 = \log a$ $k = \log 0.091$

a = _____   k = _____

a) V = 3.11

b) D = 0.621

a) a = 2,344,000   b) k = -1.04

8-9   EVALUATING "LOG" FORMULAS

In this section, we will show how a calculator can be used to evaluate formulas containing "log" expressions.

---

75.   The formula below contains a "log" expression.

$$\boxed{D = 10 \log R}$$

To show that there are two factors on the right side, we put them in parentheses below.

$$\boxed{D = (10)(\log R)}$$

To find D when R = 752 , we multiply 10 times log 752. The steps are shown. Notice that we entered 752 and then pressed $\boxed{\log}$ before pressing $\boxed{=}$. We rounded the answer to tenths.

Enter	Press	Display	
10	$\boxed{\text{x}}$	10.	
752	$\boxed{\log}\ \boxed{=}$	28.762178	(D = 28.8)

Continued on following page.

75.   Continued

Use a calculator for these. Round to tenths in (a) and to the nearest whole number in (b).

a) In $\boxed{D = 10 \log R}$, when R = 0.0386 , D = _____

b) In $\boxed{H = w \log T}$, when w = 100 and T = 32.5 , H = _____

---

76.   The steps for the evaluation below are shown. Notice again that we entered the value for Q and then pressed $\boxed{\log}$. We rounded the answer to tenths.

In $\boxed{P = A - K \log Q}$, find P when A = 59 , K = 19 , and Q = 0.215 .

Enter	Press	Display	
59	$\boxed{-}$	59.	
19	$\boxed{x}$	19.	
0.215	$\boxed{\log}$ $\boxed{=}$	71.683669	(P = 71.7)

Use a calculator for this one. Round to tenths.

In $\boxed{P = A - K \log Q}$, when A = 94 , K = 30 , and Q = 3.9 ,

P = _____

a) -14.1

b) 151

---

77.   In the formula below, $\log\left(\dfrac{P_2}{P_1}\right)$ is the log of a fraction or division. Two calculator methods are shown. We rounded the answer to hundredths.

In $\boxed{D = 10 \log\left(\dfrac{P_2}{P_1}\right)}$, find D when P_2 = 750 and P_1 = 4,875 .

1) If a calculator has parentheses symbols, we can use them to evaluate $\dfrac{P_2}{P_1}$ before pressing $\boxed{\log}$.

Enter	Press	Display	
10	$\boxed{x}$ $\boxed{(}$	10.	
750	$\boxed{\div}$	750.	
4,875	$\boxed{)}$ $\boxed{\log}$ $\boxed{=}$	-8.1291336	(D = -8.13)

76.3

Continued on following page.

77.   Continued

2) If a calculator does not have parentheses symbols, we must evaluate $\log\left(\dfrac{P_2}{P_1}\right)$ first and then multiply by 10. Notice that we pressed $\boxed{=}$ to complete the division before pressing $\boxed{\log}$.

Enter	Press	Display
750	$\boxed{\div}$	750.
4,875	$\boxed{=}$ $\boxed{\log}$ $\boxed{x}$	-.81291336
10	$\boxed{=}$	-8.1291336 (D = -8.13)

Use a calculator for this one. Round to hundredths.

In $\boxed{M = 2.5 \log\left(\dfrac{I_1}{I}\right)}$, when $I_1 = 79.3$ and $I = 16.4$, M = _____

---

78.   In the formula below, $\log(W + H)$ is the log of an addition. Two calculator methods are shown. We rounded the answer to hundredths.   1.71

In $\boxed{B = K \log(W + H)}$, find B when $K = 2.79$, $W = 18.3$, and
$H = 41.9$.

1) If a calculator has parentheses symbols, we can use them to evaluate $(W + H)$ before pressing $\boxed{\log}$.

Enter	Press	Display
2.79	$\boxed{x}$ $\boxed{(}$	2.79
18.3	$\boxed{+}$	18.3
41.9	$\boxed{)}$ $\boxed{\log}$ $\boxed{=}$	4.9650742 (B = 4.97)

2) If a calculator does not have parentheses symbols, we must evaluate $\log(W + H)$ first and then multiply by the value of K. Notice that we pressed $\boxed{=}$ to complete the addition before pressing $\boxed{\log}$.

Enter	Press	Display
18.3	$\boxed{+}$	18.3
41.9	$\boxed{=}$ $\boxed{\log}$ $\boxed{x}$	1.7795965
2.79	$\boxed{=}$	4.9650742 (B = 4.97)

Use a calculator for this one. Round to hundredths.

In $\boxed{G = 1.75 \log(a + b)}$, when $a = 3.25$ and $b = 9.87$,

G = _____

1.96

79. In $\boxed{P_H = -\log A_H}$, "$-\log A_H$" means "the <u>opposite</u> of log A_H". Therefore, after finding log A_H , we press $\boxed{+/-}$ to get its opposite. An example
is given. We rounded the answer to hundredths.

In $\boxed{P_H = -\log A_H}$, find P_H when $A_H = 0.000095$.

Enter	Press	Display
0.000095	$\boxed{\log}$ $\boxed{+/-}$	4.0222764 ($P_H = 4.02$)

Using the same steps, do this one. Round to hundredths.

In $\boxed{P_H = -\log A_H}$, when $A_H = 0.000061$, $P_H = $ _____

---

80. A two-step process is needed to find "R" in the evaluation below. We
rounded the answer to tenths.

In $\boxed{\log R = \dfrac{D}{10}}$, find R when $D = 17.5$.

1) First we find "log R" by substituting.

$$\log R = \frac{D}{10} = \frac{17.5}{10} = 1.75$$

2) Then we use $\boxed{y^x}$ to find R. The steps are:

Enter	Press	Display
10	$\boxed{y^x}$	10.
1.75	$\boxed{=}$	56.234133 (R = 56.2)

Use the same steps for this one. Round to the nearest whole number.

In $\boxed{\log R = \dfrac{D}{10}}$, when $D = 28.9$, $R = $ _____

4.21

776, since: log R = 2.89

SE<u>LF</u>-<u>TEST</u> <u>24</u> (<u>pages</u> <u>317-328</u>)

Convert each power of ten to an ordinary number.

1. Round to thousands.	2. Round to tenths.	3. Round to thousandths.
$10^{5.2354}$ = _____	$10^{1.62}$ = _____	$10^{-0.1387}$ = _____

Convert each number to a power of ten. Round each exponent to four decimal places.

4. 8.59 = _____	5. 596 = _____	6. 0.0204 = _____

Find the numerical value of each of the following. Round to four decimal places.

7. log 93,600 = _____	8. log 5.18 = _____	9. log 0.0072 = _____

Write each power-of-ten equation as a "log" equation.

10. $10^{2.8762}$ = 752 _____	11. 4.88 = 10^h _____

Write each "log" equation as a power-of-ten equation.

12. log 0.018 = -1.7447 _____	13. 1.9148 = log P _____

14. Find R. Round to tenths.	15. Find "y". Round to four decimal places.	16. Find F. Round to thousandths.
log R = 1.8063 R = _____	0.00723 = 10^y y = _____	-0.3916 = log F F = _____

17. Find D. Round to the nearest whole number.	18. Find B. Round to hundredths.
$$D = 20 \log\left(\frac{E_2}{E_1}\right)$$ $\quad E_2 = 9.65$ $\quad E_1 = 0.038$ D = _____	$\boxed{B = K \log(W + H)}$ $\quad K = 0.75$ $\quad W = 23.8$ $\quad H = 47.5$ B = _____

ANSWERS:

1. 172,000	5. $10^{2.7752}$	9. -2.1427	13. $10^{1.9148}$ = P	17. D = 48
2. 41.7	6. $10^{-1.6904}$	10. 2.8762 = log 752	14. R = 64.0	18. B = 1.39
3. 0.727	7. 4.9713	11. log 4.88 = h	15. y = -2.1409	
4. $10^{0.9340}$	8. 0.7143	12. 0.018 = $10^{-1.7447}$	16. F = 0.406	

8-10   POWERS OF "e"

Some formulas contain powers in which the base is a number called "e". The numerical value of "e" is 2.7182818... . In this section, we will convert powers of "e" to ordinary numbers and ordinary numbers to powers of "e".

81.   To convert $e^{1.5}$ to an ordinary number, one of the two methods below is used. We rounded the answer to two decimal places.

1) If your calculator has an $\boxed{e^x}$ key, enter 1.5 and press $\boxed{e^x}$.

Enter	Press	Display	
1.5	$\boxed{e^x}$	4.4816891	$(e^{1.5} = 4.48)$

2) If your calculator does not have an $\boxed{e^x}$ key, enter 1.5 and press $\boxed{INV}\,\boxed{\ln x}$.

Enter	Press	Display	
1.5	$\boxed{INV}\,\boxed{\ln x}$	4.4816891	$(e^{1.5} = 4.48)$

Convert to an ordinary number. Round to the indicated place.

a) $e^{2.19} =$ _____   (Round to hundredths)

b) $e^{-3.75} =$ _____   (Round to four decimal places)

82.   Since $e = e^1$, we can confirm the fact that $e = 2.7182818...$ by entering "1" and pressing either $\boxed{e^x}$ or $\boxed{INV}\,\boxed{\ln x}$. Do so.

Convert to an ordinary number. Round to the indicated place.

a) $e^{5.1} =$ _____   (Round to the nearest whole number)

b) $e^{-1} =$ _____   (Round to thousandths)

a) 8.94

b) 0.0235

83.   Just as $10^0 = 1$, $e^0 = 1$. To confirm that fact, enter "0" and press either $\boxed{e^x}$ or $\boxed{INV}\,\boxed{\ln x}$.

Any power of "e" with a positive exponent equals an ordinary number larger than "1". To confirm that fact, do these. Round to the indicated place.

a) $e^{0.5} =$ _____   (Round to hundredths)

b) $e^{3.68} =$ _____   (Round to tenths)

c) $e^{9.757} =$ _____   (Round to hundreds)

a) 164

b) 0.368

a) 1.65   b) 39.6   c) 17,300

84. Any power of "e" with a negative exponent equals an ordinary number between 0 and 1. To confirm that fact, do these. Round to the indicated place.

   a) $e^{-0.545}$ = _____ (Round to thousandths)

   b) $e^{-3.37}$ = _____ (Round to four decimal places)

   c) $e^{-8.5}$ = _____ (Round to millionths)

---

85. In $\boxed{y = e^x}$:

   a) when x = 6.2 , y = _____ (Round to the nearest whole number)

   b) when x = -0.25 , y = _____ (Round to thousandths)

a) 0.580

b) 0.0344

c) 0.000203

---

86. To convert an ordinary number to a power of "e", we use the $\boxed{\ln x}$ key to find the exponent. We did so below for 67.8 and 0.045 .

Enter	Press	Display
67.8	$\boxed{\ln x}$	4.2165622
0.045	$\boxed{\ln x}$	-3.1010928

Rounding each exponent to four decimal places, we get:

   $67.8 = e^{4.2166}$ $0.045 =$ _____

a) 493

b) 0.779

---

87. Using the $\boxed{\ln x}$ key to find the exponent, convert each ordinary number to a power of "e". Round each exponent to two decimal places.

   a) 7.95 = _____   b) 0.00635 = _____

$e^{-3.1011}$

---

88. When "1" is converted to a power of "e", the exponent is "0". That is: $1 = e^0$ To confirm that fact, enter "1" and press $\boxed{\ln x}$.

When a number larger than "1" is converted to a power of "e", the exponent is a positive number. To confirm that fact, convert these numbers to powers of "e". Round each exponent to three decimal places.

   a) 1.25 = _____   b) 287 = _____   c) 45,900 = _____

a) $e^{2.07}$

b) $e^{-5.06}$

---

89. When a number between "0" and "1" is converted to a power of "e", the exponent is a negative number. To confirm that fact, convert these numbers to powers of "e". Round each exponent to two decimal places.

   a) 0.979 = _____   b) 0.000408 = _____

a) $e^{0.223}$

b) $e^{5.659}$

c) $e^{10.734}$

---

a) $e^{-0.02}$ b) $e^{-7.80}$

90.   In $\boxed{y = e^x}$:

      a) when y = 99.5 ,   x = _____ (Round to four decimal places)

      b) when y = 0.0275 ,   x = _____ (Round to two decimal places)

---

a) 4.6002   b) -3.59

8-11   EVALUATING FORMULAS CONTAINING POWERS OF "e"

In this section, we will show how a calculator can be used to evaluate formulas containing powers of "e".

---

91.   We can use $\boxed{e^x}$ or $\boxed{INV}$ $\boxed{\ln x}$ for each evaluation below.

      a) In $\boxed{T = e^x}$, when x = 2.5 ,   T = _____ (Round to tenths)

      b) In $\boxed{R = e^p}$, when p = 9.15 ,   R = _____
                                         (Round to the nearest whole number)

---

92.   In the formula below, the "-x" means "the opposite of x". Therefore, we press $\boxed{+/-}$ after entering the value of "x" and then press either $\boxed{e^x}$ or $\boxed{INV}$ $\boxed{\ln x}$. We rounded the answer to thousandths.

      In $\boxed{S = e^{-x}}$, find S when x = 1.7 .

Enter	Press	Display	
1.7	$\boxed{+/-}$ $\boxed{e^x}$	.18268352	(S = 0.183)
	or $\boxed{INV}$ $\boxed{\ln x}$		

Do this one. Round to four decimal places.

      In $\boxed{V = e^{-p}}$, when p = 4.1 ,   V = _____

a) 12.2

b) 9,414

0.0166

93.   To perform the evaluation below, we press $\boxed{+/-}$ after entering 1.4 for "x" in e^{-x}. Notice that we press $\boxed{=}$ to complete the addition in the numerator before dividing. We rounded the answer to hundredths.

In $\boxed{H = \dfrac{e^x + e^{-x}}{2}}$, find H when x = 1.4 .

Enter	Press	Display
1.4	$\boxed{e^x}$ $\boxed{+}$	4.0552
	or $\boxed{INV}$ $\boxed{\ln x}$ $\boxed{+}$	
1.4	$\boxed{+/-}$ $\boxed{e^x}$ $\boxed{=}$ $\boxed{\div}$	4.3017969
	or $\boxed{+/-}$ $\boxed{INV}$ $\boxed{\ln x}$ $\boxed{=}$ $\boxed{\div}$	
2	$\boxed{=}$	2.1508985 (H = 2.15)

Using the same steps, do this one. Round to hundredths.

In $\boxed{H = \dfrac{e^x + e^{-x}}{2}}$, when x = 0.8 , H = _____

---

94.   In the formula below, "-ft" means "the <u>opposite</u> of ft". Therefore, we press $\boxed{+/-}$ after completing the multiplication "ft" by pressing $\boxed{=}$. We rounded the answer to five decimal places.

In $\boxed{R = e^{-ft}}$, find R when f = 0.5 and t = 10.6 .

Enter	Press	Display
0.5	$\boxed{x}$	0.5
10.6	$\boxed{=}$ $\boxed{+/-}$ $\boxed{e^x}$	.00499159 (R = 0.00499)
	or $\boxed{INV}$ $\boxed{\ln x}$	

Use the same steps for this one. Round to thousandths.

In $\boxed{R = e^{-ft}}$, when f = 0.1 and t = 24 , R = _____

1.34

0.091

95.   In the formula below, we must multiply a power of "e" by 14.7 . Two
methods are shown. We rounded the answer to tenths.

In $\boxed{P = 14.7e^{-0.2h}}$, find P when h = 1.8 .

1) If a calculator has parentheses symbols, we can use the parentheses
to evaluate "-0.2h" before pressing $\boxed{e^x}$ or $\boxed{\text{INV}}$ $\boxed{\ln x}$.

Enter	Press	Display
14.7	$\boxed{x}$ $\boxed{(}$	14.7
0.2	$\boxed{+/-}$ $\boxed{x}$	-0.2
1.8	$\boxed{)}$ $\boxed{e^x}$ $\boxed{=}$	10.255842 (P = 10.3)
	or $\boxed{\text{INV}}$ $\boxed{\ln x}$ $\boxed{=}$	

2) If a calculator does not have parentheses symbols, we have to
evaluate the power of "e" first before multiplying by 14.7 .

Enter	Press	Display
0.2	$\boxed{+/-}$ $\boxed{x}$	-0.2
1.8	$\boxed{=}$ $\boxed{e^x}$ $\boxed{x}$	.69767633
	or $\boxed{\text{INV}}$ $\boxed{\ln x}$ $\boxed{x}$	
14.7	$\boxed{=}$	10.255842 (P = 10.3)

Use a calculator for this one. Round to hundredths.

In $\boxed{P = 14.7e^{-0.2h}}$, when h = 2.1 , P = _____

---

96.   Two methods for the evaluation below are shown. We rounded the answer | 9.66
to thousandths.

In $\boxed{A = ke^{-ct}}$, find A when k = 4, c = 0.02, and t = 140 .

1) If a calculator has parentheses symbols, we can use them to evaluate
"-ct" before pressing $\boxed{e^x}$ or $\boxed{\text{INV}}$ $\boxed{\ln x}$. Notice that we press
$\boxed{+/-}$ after pressing $\boxed{)}$ to get "-ct".

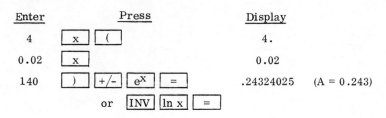

Enter	Press	Display
4	$\boxed{x}$ $\boxed{(}$	4.
0.02	$\boxed{x}$	0.02
140	$\boxed{)}$ $\boxed{+/-}$ $\boxed{e^x}$ $\boxed{=}$	.24324025 (A = 0.243)
	or $\boxed{\text{INV}}$ $\boxed{\ln x}$ $\boxed{=}$	

Continued on following page.

96.   Continued

2) If a calculator does not have parentheses symbols, we must evaluate the power of "e" first before multiplying by the value of "k". Notice that we press $\boxed{+/-}$ to get "-ct" before pressing $\boxed{e^x}$ or $\boxed{INV}$ $\boxed{\ln x}$.

Enter	Press	Display
0.02	$\boxed{x}$	0.02
140	$\boxed{=}$ $\boxed{+/-}$ $\boxed{e^x}$ $\boxed{x}$	.06081006
	or $\boxed{INV}$ $\boxed{\ln x}$ $\boxed{x}$	
4	$\boxed{=}$	.24324025 (A = 0.243)

Use a calculator for this one. Round to thousandths.

In $\boxed{A = ke^{-ct}}$, when k = 5, c = 0.04, and t = 55, A = _____

---

97.   Two methods for the evaluation below are also shown. We rounded the answer to tenths.

$$\boxed{i = I_o e^{-\frac{Rt}{L}}}$$, find "i" when $I_0 = 70$, R = 15.7 , t = 2.5 , and L = 21.9 .

1) If a calculator has parentheses symbols, we can use them to evaluate "$-\frac{Rt}{L}$" before pressing $\boxed{e^x}$ or $\boxed{INV}$ $\boxed{\ln x}$. Notice that we press $\boxed{+/-}$ after $\boxed{)}$ to get "$-\frac{Rt}{L}$" .

Enter	Press	Display
70	$\boxed{x}$ $\boxed{(}$	70.
15.7	$\boxed{x}$	15.7
2.5	$\boxed{\div}$	39.25
21.9	$\boxed{)}$ $\boxed{+/-}$ $\boxed{e^x}$ $\boxed{=}$	11.661092 (i = 11.7)
	or $\boxed{INV}$ $\boxed{\ln x}$ $\boxed{=}$	

0.554

Continued on following page.

97.   Continued

2) If a calculator does not have parentheses symbols, we must evaluate the power of "e" before multiplying by the value of I_0. Notice that we press $\boxed{+/-}$ to get $"-\dfrac{Rt}{L}"$ before pressing $\boxed{e^X}$ or $\boxed{INV}$ $\boxed{\ln x}$.

Enter	Press	Display
15.7	$\boxed{x}$	15.7
2.5	$\boxed{\div}$	39.25
21.9	$\boxed{=}$ $\boxed{+/-}$ $\boxed{e^X}$ $\boxed{x}$	.16658702
	or $\boxed{INV}$ $\boxed{\ln x}$ $\boxed{x}$	
70	$\boxed{=}$	11.661092 (i = 11.7)

Using the same method, do this one. Round to tenths.

In $\boxed{V = Ee^{-\frac{t}{RC}}}$, when E = 200 , t = 0.75 , R = 12 , and C = 0.06 , V = _____

70.6

8-12   NATURAL LOGARITHMS

When a number is written in power-of-"e" form, the exponent of the "e" is called the <u>natural</u> <u>logarithm</u> of the number. In this section, we will discuss natural logarithms and contrast them with common logarithms. We will also contrast natural logarithmic notation with common logarithmic notation.

98.   Any positive number can be written in either power-of-ten or power-of-"e" form. For example:

$$325 = 10^{2.5119} \qquad\qquad 325 = e^{5.7838}$$

a) The <u>common</u> logarithm of a number is the <u>exponent</u> when the number is written in <u>power-of-ten</u> form.

   The <u>common</u> logarithm of 325 is _____.

b) The <u>natural</u> logarithm of a number is the <u>exponent</u> when the number is written in <u>power-of-"e"</u> form.

   The <u>natural</u> logarithm of 325 is _____.

a) 2.5119

b) 5.7838

99.   Since $0.875 = e^{-0.1335}$ and $0.875 = 10^{-0.0580}$:

   a) The <u>common</u> logarithm of 0.875 is _____ .

   b) The <u>natural</u> logarithm of 0.875 is _____ .

---

100.   a) To find the <u>common</u> logarithm of a number, we enter the number and press $\boxed{\log}$.

   To four decimal places, the <u>common</u> logarithm of 16.7 is _____ .

   b) To find the <u>natural</u> logarithm of a number, we enter the number and press $\boxed{\ln x}$.

   To four decimal places, the <u>natural</u> logarithm of 16.7 is _____ .

a) -0.0580

b) -0.1335

---

101.   The <u>common</u> logarithm of 0.0189 is -1.7235 . The <u>natural</u> logarithm of 0.0189 is -3.9686 . Therefore:

$$0.0189 = 10^{-1.7235} \qquad 0.0189 = e^{\boxed{}}$$

a) 1.2227

b) 2.8154

---

102.   a) If the <u>common</u> logarithm of a number is 2.5639 , the number is $10^{2.5639}$ or _____ . (Round to the nearest whole number)

   b) If the <u>natural</u> logarithm of a number is 2.5639 , the number is $e^{2.5639}$ or _____ . (Round to the nearest tenth)

$e^{\boxed{-3.9686}}$

---

103.   a) If the <u>common</u> logarithm of a number is -1.5 , the number is _____ . (Round to four decimal places)

   b) If the <u>natural</u> logarithm of a number is -1.5 , the number is _____ . (Round to three decimal places)

a) 366

b) 13.0

---

104.   The abbreviations "<u>log</u>" and "<u>ln</u>" are used for logarithms.

   "<u>log</u> 265" means "the <u>common</u> logarithm of 265".

   "<u>ln</u> 265" means "the <u>natural</u> logarithm of 265".

   Using the $\boxed{\log}$ and $\boxed{\ln x}$ keys, complete these. Round to four decimal places.

   a) log 265 = _____   b) ln 265 = _____

a) 0.0316 , from:
   $10^{-1.5}$

b) 0.223 , from:
   $e^{-1.5}$

---

a) 2.4232

b) 5.5797

105. Any power-of-ten or power-of-"e" equation can be written as a logarithmic equation. For example:

$$27.5 = 10^{1.4393} \text{ can be written: } \log 27.5 = 1.4393$$

$$27.5 = e^{3.3142} \text{ can be written: } \ln 27.5 = 3.3142$$

Write each equation below as a logarithmic equation.

a) $10^{-1.5288} = 0.0296$ b) $e^{-2.78} = 0.062$

---

106. When a power-of-ten or power-of-"e" equation contains a letter, it can also be written as a logarithmic equation. For example:

$$t = 10^{-2.5684} \text{ can be written: } \log t = -2.5684$$

$$1.29 = e^{x} \text{ can be written: } \ln 1.29 = x$$

Write each equation below as a logarithmic equation.

a) $10^{y} = 6.59$ b) $e^{-2.75} = h$

a) $-1.5288 = \log 0.0296$

b) $-2.78 = \ln 0.062$

---

107. Any logarithmic equation can be written as a power-of-ten or power-of-"e" equation. For example:

$$\log 1.33 = 0.1239 \text{ can be written: } 1.33 = 10^{0.1239}$$

$$\ln 1.33 = 0.2852 \text{ can be written: } 1.33 = e^{0.2852}$$

Write each logarithmic equation as a power-of-ten or power-of-"e" equation.

a) $2.8209 = \log 662$ b) $4.1759 = \ln 65.1$

a) $y = \log 6.59$

b) $-2.75 = \ln h$

---

108. When a logarithmic equation contains a letter, it can also be written as a power-of-ten or power-of-"e" equation. For example:

$$\log 0.589 = d \text{ can be written: } 0.589 = 10^{d}$$

$$\ln t = -3.55 \text{ can be written: } t = e^{-3.55}$$

Write each logarithmic equation as a power-of-ten or power-of-"e" equation.

a) $-0.2066 = \log x$ b) $y = \ln 0.265$

a) $10^{2.8209} = 662$

b) $e^{4.1759} = 65.1$

---

a) $10^{-0.2066} = x$

b) $e^{y} = 0.265$

8-13   SOLVING POWER-OF-"e" AND "Ln" EQUATIONS

In this section, we will discuss the methods for solving power-of-"e" and "ln" equations that contain a letter.

---

109.   In each power-of-"e" equation below, the letter stands for <u>an ordinary</u> number. Use $\boxed{e^x}$ or $\boxed{INV}$ $\boxed{\ln x}$ to solve each equation.

    a) Round to tenths.

        $e^{4.25} = t$

        $t = $ _____

    b) Round to four decimal places.

        $y = e^{-2.685}$

        $y = $ _____

---

110.   In each power-of-"e" equation below, the letter stands for <u>an exponent</u>. Use the $\boxed{\ln x}$ key to solve each equation. Round each exponent to three decimal places.

    a) $2,750 = e^x$

        $x = $ _____

    b) $e^y = 0.0999$

        $y = $ _____

a) t = 70.1

b) y = 0.0682

---

111.   When a "ln" equation contains a letter, we can solve it by writing the equation in power-of-"e" form. Let's use that method to solve the equation below.

        $\ln R = 1.8576$

    a) Write the equation in power-of-"e" form. _____

    b) Therefore, rounded to hundredths, R = _____ .

a) x = 7.919

b) y = -2.304

---

112.   Let's solve $\ln 0.193 = y$.

    a) Write the equation in power-of-"e" form. _____

    b) Therefore, rounded to four decimal places, y = _____ .

a) $R = e^{1.8576}$

b) R = 6.41

---

113.   Let's solve $-2.56 = \ln V$.

    a) Write the equation in power-of-"e" form. _____

    b) Therefore, rounded to four decimal places, V = _____ .

a) $0.193 = e^y$

b) y = -1.6451

---

114.   Let's solve $t = \ln 6,295$.

    a) Write the equation in power-of-"e" form. _____

    b) Therefore, rounded to hundredths, t = _____ .

a) $e^{-2.56} = V$

b) V = 0.0773

115. When the letter follows "ln" in an equation, it stands for the ordinary number. For example:

In ln P = -1.75 , "P" stands for the ordinary number since: $P = e^{-1.75}$

In 3.69 = ln H , "H" stands for the ordinary number since: $e^{3.69} = H$

In such cases, convert the "ln" equation to a power-of-"e" equation and use $\boxed{e^x}$ or $\boxed{INV}$ $\boxed{ln\ x}$. Solve these.

a) Round to tenths.   b) Round to thousandths.

    ln B = 4.488   -0.36 = ln G

    B = _____   G = _____

a) $e^t = 6,295$	
b) $t = 8.75$	

116. When the letter is on the opposite side of "ln" in an equation, it stands for the logarithm (or exponent). For example:

In ln 10.8 = v , "v" stands for the logarithm since: $10.8 = e^v$

In y = ln 0.725 , "y" stands for the logarithm since: $e^y = 0.725$

In such cases, we can simply use the $\boxed{ln\ x}$ key without converting to a power-of-"e" equation. Solve these. Round to three decimal places.

a) ln 0.087 = F   b) Q = ln 13,500

    F = _____   Q = _____

a) B = 88.9 , from:
   $B = e^{4.488}$

b) G = 0.698 , from:
   $e^{-0.36} = G$

117. Solve: a) Round to hundredths.   b) Round to thousandths.

            ln 966 = m   ln C = -0.199

            m = _____   C = _____

a) F = -2.442
b) Q = 9.510

118. Solve: a) Round to the nearest   b) Round to hundredths.
                whole number.
                                          t = ln 0.056
            4.95 = ln b

            b = _____   t = _____

a) m = 6.87
b) C = 0.820

a) b = 141
b) t = -2.88

8-14   EVALUATING "Ln" FORMULAS

In this section, we will show how a calculator can be used to evaluate formulas containing "ln" expressions.

---

119.   The steps for the evaluation below are shown. Notice that we entered 50 and pressed $\boxed{\ln x}$ before pressing $\boxed{=}$. We rounded the answer to tenths.

In $\boxed{D = k \ln P}$, find D when $k = 10$ and $P = 50$.

Enter	Press	Display
10	$\boxed{x}$	10.
50	$\boxed{\ln x}$ $\boxed{=}$	39.12023 (D = 39.1)

Use a calculator for this one. Round to tenths.

In $\boxed{D = k \ln P}$, when $k = 7.5$ and $P = 18.5$, $D =$ _____

---

120.   The steps for the evaluation below are shown. Notice again that we entered the value for P and then pressed $\boxed{\ln x}$. We rounded the answer to thousandths.

In $\boxed{w = \dfrac{d}{\ln P}}$, find "w" when $d = 4.24$ and $P = 70$.

Enter	Press	Display
4.24	$\boxed{\div}$	4.24
70	$\boxed{\ln x}$ $\boxed{=}$	.99800041 (w = 0.998)

Use a calculator for this one. Round to hundredths.

In $\boxed{w = \dfrac{d}{\ln P}}$, when $d = 3.99$ and $P = 80$, $w =$ _____

21.9

---

121.   In the formula below, $\ln\left(\dfrac{A_o}{A}\right)$ is the natural log of a fraction or division. Notice that we press $\boxed{=}$ to complete that division before pressing $\boxed{\ln x}$, and then divide by the value of "k". We rounded the answer to hundredths.

In $\boxed{t = \dfrac{\ln\left(\dfrac{A_o}{A}\right)}{k}}$, find "t" when $A_o = 60$, $A = 20$, and $k = 0.31$.

Enter	Press	Display
60	$\boxed{\div}$	60.
20	$\boxed{=}$ $\boxed{\ln x}$ $\boxed{\div}$	1.0986123
0.31	$\boxed{=}$	3.5439106 (t = 3.54)

0.91

Continued on following page.

121.   Continued

Use a calculator for this one. Round to hundredths.

$$\boxed{k = \frac{\ln\left(\dfrac{A_o}{A}\right)}{t}}$$

In , when $A_o = 38.9$, $A = 9.45$, and $t = 1.29$,

$$k = \underline{\hspace{2cm}}$$

---

122.   Two calculator methods are shown for the evaluation below. We rounded the answer to hundredths.

In $\boxed{v = c \ln\left(\dfrac{M}{m}\right)}$, find "v" when $c = 1.88$, $M = 750$, and $m = 112$.

1) If a calculator has parentheses symbols, we can use them to evaluate $\left(\dfrac{M}{m}\right)$ before pressing $\boxed{\ln x}$.

Enter	Press	Display
1.88	$\boxed{x}$ $\boxed{(}$	1.88
750	$\boxed{\div}$	750.
112	$\boxed{)}$ $\boxed{\ln x}$ $\boxed{=}$	3.5749598 (v = 3.57)

2) If a calculator does not have parentheses symbols, we must evaluate $\ln\left(\dfrac{M}{m}\right)$ first and then multiply by 1.88 . Notice that we pressed $\boxed{=}$ to complete the division before pressing $\boxed{\ln x}$.

Enter	Press	Display
750	$\boxed{\div}$	750.
112	$\boxed{=}$ $\boxed{\ln x}$ $\boxed{x}$	1.9015743
1.88	$\boxed{=}$	3.5749598 (v = 3.57)

Use a calculator for this one. Round to hundredths.

In $\boxed{v = c \ln\left(\dfrac{M}{m}\right)}$, when $c = 1.86$, $M = 869$, and $m = 90$,

$$v = \underline{\hspace{2cm}}$$

1.10

4.22

123. In $\boxed{Q = -\ln B}$, "-ln B" means "the <u>opposite</u> of ln B". Therefore, after finding ln B , we press $\boxed{+/-}$ to get its opposite. An example is shown. We rounded the answer to hundredths.

In $\boxed{Q = -\ln B}$, find Q when B = 3.5 .

Enter	Press	Display	
3.5	$\boxed{\ln x}$ $\boxed{+/-}$	-1.252763	(Q = -1.25)

Using the same steps, do this one. Round to hundredths.

In $\boxed{Q = -\ln B}$, when B = 120 , Q = _____

---

124. A two-step process is needed to find "A" in the evaluation below. We rounded the answer to tenths.

In $\boxed{\ln A = \dfrac{h}{t}}$, find A when h = 80 and t = 20 .

1) First we find "ln A" by substituting:

$$\ln A \;=\; \frac{80}{20} \;=\; 4$$

2) Then we use $\boxed{e^x}$ or $\boxed{\text{INV}}\,\boxed{\ln x}$ to find A. The steps are:

Enter	Press	Display	
4	$\boxed{e^x}$	54.59815	(A = 54.6)
	or $\boxed{\text{INV}}$ $\boxed{\ln x}$		

Use the same steps for this one. Round to the nearest whole number.

In $\boxed{\ln A = \dfrac{h}{t}}$, when h = 150 and t = 30 , A = _____

-4.79

148

SELF-TEST 25 (pages 329-343)

Convert each power of "e" to an ordinary number.

1. $e^{4.25} =$ _____ (Round to tenths)

2. $e^{-1.36} =$ _____ (Round to thousandths)

Convert each number to a power of "e".

3. $915 =$ _____ (Round to hundredths)

4. $1.82 =$ _____ (Round to four decimal places)

---

5. Find F. Round to thousands.

$$\boxed{F = Ae^t} \quad \begin{array}{l} A = 18.7 \\ t = 9.42 \end{array}$$

F = _____

6. Find S when $x = 2.56$. Round to hundredths.

$$\boxed{S = \frac{e^x - e^{-x}}{2}}$$

S = _____

7. Find W. Round to tenths.

$$\boxed{W = ke^{-ct}} \quad \begin{array}{l} k = 51.7 \\ c = 0.065 \\ t = 12 \end{array}$$

W = _____

---

8. To four decimal places, the natural logarithm of 46.2 is _____ .

---

9. Write this power-of-"e" equation as a logarithmic equation.

$$314 = e^{5.75}$$ _____

10. Write this logarithmic equation as a power-of-"e" equation.

$$-1.016 = \ln 0.362$$ _____

---

11. Find G. Round to tenths.

$$\ln G = 2.5181$$

G = _____

12. Find "x". Round to hundredths.

$$17.4 = e^x$$

x = _____

13. Find B. Round to thousandths.

$$-1.93 = \ln B$$

B = _____

---

14. Find "r" when $T = 0.727$. Round to thousandths.

$$\boxed{r = -\ln T}$$

r = _____

15. Find W. Round to the nearest whole number.

$$\boxed{W = c \ln\left(\frac{a}{b}\right)} \quad \begin{array}{l} c = 290 \\ a = 7,300 \\ b = 510 \end{array}$$

W = _____

16. Find P. Round to hundredths.

$$\boxed{\ln P = \frac{b}{h}} \quad \begin{array}{l} b = 81.3 \\ h = 56.9 \end{array}$$

P = _____

ANSWERS:

1. 70.1	5. F = 231,000	9. ln 314 = 5.75	13. B = 0.145
2. 0.257	6. S = 6.43	10. $e^{-1.016} = 0.362$	14. r = 0.319
3. $e^{6.82}$	7. W = 23.7	11. G = 12.4	15. W = 772
4. $e^{0.5988}$	8. 3.8330	12. x = 2.86	16. P = 4.17

SUPPLEMENTARY PROBLEMS - CHAPTER 8

Assignment 23

Evaluate these powers. Record all digits shown on the display.

1. 3^{10} 2. 12^5 3. $(8.1)^4$ 4. $(0.2)^6$ 5. $(1.35)^3$

Evaluate these powers. Report each answer in scientific notation with the first factor rounded to two decimal places.

6. $(480)^8$ 7. $(3.79)^{19}$ 8. $(0.025)^7$ 9. $(0.146)^{18}$

Evaluate these roots. Round as directed.

10. Round to tenths.   11. Round to hundredths.   12. Round to hundredths.

$\sqrt[4]{936,000}$ $\sqrt[7]{350}$ $\sqrt[6]{8.92}$

13. Round to two decimal places.   14. Round to thousandths.   15. Round to four decimal places.

$\sqrt[5]{0.44}$ $\sqrt[3]{0.0617}$ $\sqrt[9]{0.001839}$

Convert each radical to a power with a decimal exponent. Do not evaluate.

16. $\sqrt[5]{18^3}$ 17. $\sqrt[8]{6^4}$ 18. $\sqrt{3^9}$ 19. $\sqrt[4]{542}$

Evaluate these powers. Round as directed.

20. Round to tenths.   21. Round to hundredths.   22. Round to nearest whole number.

$(263)^{0.75}$ $(3.94)^{1.5}$ $(18.6)^{2.3}$

23. Round to millionths.   24. Round to thousandths.   25. Round to hundredths.

6^{-3} $(2.74)^{-1.2}$ $(0.183)^{-0.5}$

26. In $\boxed{W = K^r}$, find W when $K = 3.9$ and $r = 2.7$. Round to tenths.

27. In $\boxed{R = PQ^a}$, find R when $P = 1.38$, $Q = 3.27$, and $a = 0.6$. Round to hundredths.

28. In $\boxed{D = K(t + 1)^n}$, find D when $K = 80$, $t = 3.9$, and $n = 1.3$. Round to the nearest whole number.

29. In $\boxed{H = \left(\dfrac{d}{t}\right)^{0.6}}$, find H when $d = 47.6$ and $t = 61.3$. Round to thousandths.

Assignment 24

Convert each power of ten to an ordinary number. Round as directed.

1. Round to hundredths. 2. Round to thousands. 3. Round to thousandths. 4. Round to millionths.

$10^{0.58}$ $10^{5.3948}$ $10^{-0.17}$ $10^{-3.0259}$

Convert each number to a power of ten. Round each exponent to four decimal places.

5. 8,250   6. 41.7   7. 0.063   8. 0.000216

Find the numerical value of each of the following. Round to four decimal places.

9. $\log 5,320,000$ 10. $\log 9.41$ 11. $\log 0.78$ 12. $\log 0.00115$

Write each power-of-ten equation as a "log" equation.

13. $185 = 10^{2.2672}$ 14. $10^{-0.7595} = 0.174$ 15. $10^X = 27.4$ 16. $G = 10^{-2.3814}$

Write each "log" equation as a power-of-ten equation.

17. $2.4166 = \log 261$ 18. $\log 0.02 = -1.6990$ 19. $\log R = 5.8037$ 20. $y = \log 7,190$

Find the numerical value of each letter. Round as directed.

21. Round to hundredths.   22. Round to millionths.   23. Round to four decimal places.

$\qquad 0.8673 = \log G \qquad\qquad \log T = -3.8673 \qquad\qquad 19,345 = 10^b$

24. In $\boxed{N = 10 \log K}$, find N when $K = 50$. Round to the nearest whole number.

25. In $\boxed{P = -\log A}$, find P when $A = 0.0000037$. Round to hundredths.

26. In $\boxed{M = 2.5 \log\left(\dfrac{I_1}{I}\right)}$, find M when $I_1 = 430$ and $I = 15$. Round to tenths.

27. In $\boxed{r = 0.95 \log(p + t)}$, find "r" when $p = 25.8$ and $t = 37.6$. Round to hundredths.

Assignment 25

Convert each power of "e" to an ordinary number.

1. Round to tenths.   2. Round to thousandths.   3. Round to hundreds.

$\qquad e^{3.98} \qquad\qquad\qquad e^{-1.72} \qquad\qquad\qquad e^{9.35}$

Convert each number to a power of "e".

4. Round to hundredths.   5. Round to thousandths.   6. Round to four decimal places.

$\qquad 116 \qquad\qquad\qquad 0.0384 \qquad\qquad\qquad 4.72$

7. In $\boxed{H = ae^p}$, find H when $a = 2.4$ and $p = 1.72$. Round to tenths.

8. In $\boxed{y = \dfrac{e^x + e^{-x}}{2}}$, find "y" when $x = 1.36$. Round to hundredths.

9. In $\boxed{B = ke^{-ct}}$, find B when $k = 485$, $c = 7.2$, and $t = 0.096$. Round to the nearest whole number.

Find these natural logarithms. Round each to four decimal places.

10. $\ln 29.8$ 11. $\ln 7,560$ 12. $\ln 2.04$ 13. $\ln 0.00192$

Convert each power-of-"e" equation to a logarithmic equation.

14. $e^{4.15} = 63.4$ 15. $0.0828 = e^{-2.49}$

Convert each logarithmic equation to a power-of-"e" equation.

16. $\ln 9.31 = 2.23$ 17. $-0.851 = \ln 0.427$

18. Find R.   19. Find "y".   20. Find P.
   Round to hundredths.   Round to tenths.   Round to four decimal places.

$\qquad 1.3206 = \ln R \qquad\qquad e^y = 3,500 \qquad\qquad \ln P = -3.725$

21. In $\boxed{r = \dfrac{d}{\ln A}}$, find "r" when $d = 836$ and $A = 11.5$. Round to the nearest whole number.

22. In $\boxed{t = k \ln\left(\dfrac{E}{G}\right)}$, find "t" when $k = 3.53$, $E = 984$, and $G = 127$. Round to hundredths.

23. In $\boxed{\ln N = \dfrac{a}{w}}$, find N when $a = 95.7$ and $w = 23.6$. Round to tenths.

Chapter 9

MEASUREMENT CONCEPTS

In this chapter, we will discuss the following properties of measurements: precision, lower and upper limits, absolute error, relative error, accuracy, and significant digits. We will also discuss the rules that are used to report answers when calculating with measurements.

9-1   THE APPROXIMATE NATURE OF MEASUREMENTS

Measurements of quantities like length, weight (or mass), time, temperature, voltage, and pressure are not exact. They are only approximate. We will discuss what is meant by the approximate nature of measurements in this section. The measurement of the length of a steel block will be used as an example.

---

1.   The measuring instrument below is called a 6-inch scale. The shaded figure represents a steel block.

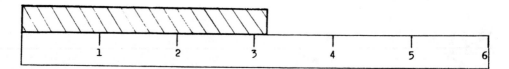

To the nearest <u>whole-number</u> inch, the length of the steel block is _____ .

---

2.   To measure the same steel block to the nearest <u>tenth</u> of an inch, we must use a scale subdivided into tenths of an inch.

3 inches (or 3")

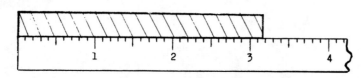

To the <u>nearest</u> tenth of an inch, the length of the steel block is _____ .

3.2 inches (or 3.2")

346

3.   To measure the same steel block to the nearest <u>hundredth</u> of an inch, we must use a scale subdivided into hundredths of an inch. An enlargement of such a scale at the right end of the steel block is shown below.

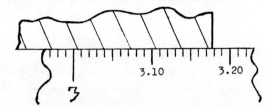

The end of the steel block lies between 3.10" and 3.20". To the nearest hundredth of an inch, the length of the steel block is _____.

---

4.   In the last three frames, we got the following length measurements for the same steel block. The more exact measurements were obtained by using a 6-inch scale with smaller subdivisions.

    3"   (to the nearest whole-number inch)

    3.2"   (to the nearest tenth of an inch)

    3.18"   (to the nearest hundredth of an inch)

The smallest possible subdivision of a 6-inch scale is <u>hundredths of an inch</u>. Therefore, using a 6-inch scale, the most exact measurement we can get of the length of the steel block is 3.18".

    a) Does 3.18" give the length to the nearest thousandth of an inch? _____

    b) Is 3.18" the <u>exact</u> length of the steel block? _____

Answer (frame 3): 3.18 inches (or 3.18")

---

5.   Using a different instrument called a <u>micrometer</u>, we can measure the length of the steel block to the nearest <u>thousandth</u> of an inch. The measurement is 3.177".

    a) Does 3.177" give the length to the nearest ten-thousandth of an inch? _____

    b) Is 3.177" the <u>exact</u> length of the steel block? _____

Answer (frame 4):
a) No
b) No

---

6.   Using a special micrometer, we can measure the length of the steel block to the nearest <u>ten-thousandth</u> of an inch. The measurement is 3.1768".

    a) Does 3.1768" give the length to the nearest hundred-thousandth of an inch? _____

    b) Is 3.1768" the <u>exact</u> length of the steel block? _____

Answer (frame 5):
a) No
b) No

Answer (frame 6):
a) No   b) No

7.   Using precision gage blocks, we can measure the length of the steel block
      to the nearest hundred-thousandth of an inch. The measurement is 3.17682".

      Using an optical method involving interference patterns of light waves, we
      can measure the length of the steel block to the nearest millionth of an inch.
      The measurement is 3.176819".

      a) Does either 3.17682" or 3.176819" give the length to the nearest
          ten-millionth of an inch? _____

      b) Is either 3.17682" or 3.176819" the exact length of the steel
          block? _____

---

8.   In measuring the steel block, we used scales with smaller and smaller
      subdivisions. The smallest subdivision used was millionths of an inch.
      But even 3.176819" was not exact because it does not give the length to
      the nearest ten-millionth of an inch. Therefore, we can make the
      following statement.

      | All length measurement is approximate. |

      By saying that all length measurement is approximate, we mean this:
      Even though we use a scale with a small subdivision, we could always,
      at least in principle, use a scale with a smaller subdivision. Therefore,
      we do not know the length to that smaller subdivision.

      A similar argument could be made to show that measurements of quantities
      like weight (or mass), time, temperature, voltage, and pressure are also
      NEVER EXACT. They are only approximate.

a) No

b) No

9-2   PRECISION OF MEASUREMENTS

Every measurement has a property called its precision. We will discuss the precision of measurements in
this section.

      Note: In this section, we will avoid precisions to the left of the "ones" place.
             Precisions of that type will be discussed in a later section.

---

9.   When a measurement is reported to the nearest calibration mark on the
      scale, the place of the final digit in the measurement indicates the smallest
      subdivision of the measuring scale. For example:

      In 9.65" ,   the "5" is in the hundredths place. Therefore, the
               smallest subdivision of the scale was hundredths.

      In 4.1839" ,   the "9" is in the ten-thousandths place. Therefore,
               the smallest subdivision of the scale was ten-
               thousandths.

Continued on following page.

9.   Continued

The smallest subdivision of the scale used to make a measurement is called the <u>precision</u> of the measurement. Therefore:

The <u>precision</u> of 9.65" is <u>hundredths</u> of an inch.

The <u>precision</u> of 4.1839" is _____ of an inch.

---

10.   State the precision of each measurement.

a) 5.7 liters _____

b) 0.128 gram _____

c) 0.004291 second _____

ten-thousandths

---

11.   Since measurements of the same quantity can have different precisions, they can be compared in terms of their precision. To do so, the following principle is used:

> The smaller the subdivision of the measuring scale, the greater is the precision of the measurement.

a) Which measurement below has the <u>greatest</u> precision? _____

    5,024.9 feet   68.075 feet   1.95 feet

b) Which measurement below has the <u>least</u> precision? _____

    7.5 seconds   127 seconds   42.66 seconds

a) tenths of a liter

b) thousandths of a gram

c) millionths of a second

---

12.   With ordinary numbers, we can insert or drop final 0's after the decimal point. That is:

$$7 = 7.0 = 7.00 = 7.000$$

With measurements like 7.0" or 7.00", we cannot insert or drop final 0's after the decimal point because they indicate the precision of the measurement.

a) 7.0" is precise to the nearest _____ of an inch.

b) 7.00" is precise to the nearest _____ of an inch.

c) Can we say that 7.0" = 7.00" ? _____

a) 68.075 feet

b) 127 seconds

---

a) tenth

b) hundredth

c) No, because they have different precisions.

13. a) The final "0" in 5.070 grams means that the measurement is precise to the nearest _____ of a gram.

     b) The two final 0's in 0.7500 second mean that the measurement is precise to the nearest _____ of a second.

---

14. An actual measurement of 12.5 liters should not be reported as 12.50 liters because:

     a) The actual measurement (12.5 liters) is precise to the nearest _____ of a liter.

     b) The reported measurement (12.50 liters) is precise to the nearest _____ of a liter.

     An actual measurement of 34.700 grams should not be reported as 34.7 grams because:

     c) The actual measurement (34.700 grams) is precise to the nearest _____ of a gram.

     d) The reported measurement (34.7 grams) is precise to the nearest _____ of a gram.

a) thousandth

b) ten-thousandth

---

15. a) Underline the measurement with the greatest precision.

         125.70 cm   9.0800 cm   85.007 cm

     b) Underline the measurement with the least precision.

         427.8 lb   64.000 lb   25.70 lb

a) tenth
b) hundredth

c) thousandth
d) tenth

a) 9.0800 cm   b) 427.8 lb

9-3   UPPER AND LOWER LIMITS OF MEASUREMENTS

Any measurement reported to the nearest calibration mark on the scale has an upper limit and a lower limit. Though the measurement is approximate, we know that its actual value lies between those two limits. We will discuss the limits of measurements in this section.

---

16. The scale below is precise to the nearest inch. Any point between M and N is read as 3".

     Since point M represents 2.5" and point N represents 3.5", any point on the scale between 2.5" and 3.5" is read as 3". We can state the same fact in a different way:

         If a length measurement is reported as 3", the actual length lies somewhere between _____ inches and _____ inches.

17. Though a measurement is not exact, it does tell us something about the actual value of the measurement. For example:

> If a weight measurement is reported as 27 grams, we know that the actual value lies between 26.5 grams and 27.5 grams.
>
> If a time measurement is reported as 141 seconds, we know that the actual value lies between _____ seconds and _____ seconds.

| 2.5 inches and 3.5 inches |

---

18. The scale below is precise to the nearest tenth of an inch. Any point between S and T is read as 43.8".

Since point S represents 43.75" and point T represents 43.85", any point on the scale between 43.75" and 43.85" is read as 43.8". We can state the same fact in a different way:

> If a length measurement is reported as 43.8", the actual length lies somewhere between _____ inches and _____ inches.

| 140.5 seconds and 141.5 seconds |

---

19. If a weight measurement is reported as 4 grams, we know that the actual weight lies between 3.5 grams and 4.5 grams.

> 3.5 grams is called the lower limit of the measurement.
>
> 4.5 grams is called the upper limit of the measurement.

If a volume measurement is reported as 7.6 liters, we know that the actual volume lies between 7.55 liters and 7.65 liters.

a) The lower limit of the 7.6 liter measurement is _____ liters.

b) The upper limit of the 7.6 liter measurement is _____ liters.

| 43.75 inches and 43.85 inches |

---

20. The final 5's in the lower and upper limits of a measurement do not change the precision of the measurement. For example:

> Though the lower and upper limits of a 9" measurement are 8.5" and 9.5", 9" is precise only to the nearest inch.
>
> Though the lower and upper limits of a 14.2" measurement are 14.15" and 14.25", 14.2" is precise only to the nearest _____ of an inch.

| a) 7.55 |
| b) 7.65 |

---

tenth

21.   To find the limits of a "whole-number" measurement like 13 seconds, we add 0.5 to get the upper limit and subtract 0.5 to get the lower limit. The process is shown below. Notice that we put a "0" after the decimal point in 13 to help with the addition and subtraction.

13 seconds $\begin{bmatrix} +0.5 = 13.5 & \text{(upper limit)} \\ 13.0 & \\ -0.5 = 12.5 & \text{(lower limit)} \end{bmatrix}$

Using the same method, find the limits of the measurement below.

27 grams $\begin{bmatrix} +0.5 = \underline{\hspace{1.5cm}} & \text{(upper limit)} \\ 27.0 & \\ -0.5 = \underline{\hspace{1.5cm}} & \text{(lower limit)} \end{bmatrix}$

---

22.   To find the limits of a "tenths" measurement like 7.2 meters, we add 0.05 to get the upper limit and subtract 0.05 to get the lower limit. Do so below. Notice that we put a "0" at the right end of 7.2 to help with the addition and subtraction.

7.2 meters $\begin{bmatrix} +0.05 = \underline{\hspace{1.5cm}} & \text{(upper limit)} \\ 7.20 & \\ -0.05 = \underline{\hspace{1.5cm}} & \text{(lower limit)} \end{bmatrix}$

27.5 (upper limit)

26.5 (lower limit)

---

23.   To find the limits of a "hundredths" measurement like 5.29 inches, we add 0.005 to get the upper limit and subtract 0.005 to get the lower limit. Do so below.

5.29 inches $\begin{bmatrix} +0.005 = \underline{\hspace{1.8cm}} & \text{(upper limit)} \\ 5.290 & \\ -0.005 = \underline{\hspace{1.8cm}} & \text{(lower limit)} \end{bmatrix}$

7.25 (upper limit)

7.15 (lower limit)

---

24.   When adding and subtracting a number to find the limits of a measurement, the only non-zero digit in the number is a "5". The "5" is always in the place immediately to the right of the right-most digit in the measurement. That is:

5.295 (upper limit)

5.285 (lower limit)

Measurement	Number Added And Subtracted	
25 grams	0.5	("5" in tenths place)
9.7 seconds	0.05	("5" in hundredths place)
1.84 liters	0.005	("5" in thousandths place)

Find the limits of each measurement below.

Measurement	Lower Limit	Upper Limit
a) 14.1 inches	_____	_____
b) 8.75 meters	_____	_____
c) 129 miles	_____	_____

25.   The same principle applies to measurements that are more precise than hundredths. That is, when adding and subtracting to find the limits, the "5" always appears in the place immediately to the right of the right-most digit in the measurement. That is:

Measurement	Number Added And Subtracted	
0.425 centimeter	0.0005	("5" in <u>ten-thousandths</u> place)
0.0926 volt	0.00005	("5" in <u>hundred-thousandths</u> place)
0.00282 second	0.000005	("5" in <u>millionths</u> place)

Find the limits of each measurement below.

Measurement	Lower Limit	Upper Limit
a) 4.574 meters		
b) 0.3085 gram		
c) 0.00901 centimeter		

a) 14.05   14.15

b) 8.745   8.755

c) 128.5   129.5

---

26.   The method we have been using is especially helpful in finding the limits of measurements ending with a "0" or 0's. An example is shown below.

$$\underline{3.00}\text{ grams}\quad \begin{bmatrix} +0.005 = 3.005 \text{ (upper limit)} \\ 3.000 \\ -0.005 = 2.995 \text{ (lower limit)} \end{bmatrix}$$

Find the limits of each measurement.

Measurement	Lower Limit	Upper Limit
a) 19.0 feet		
b) 0.700 ounce		
c) 0.0540 second		

a) 4.5735   4.5745

b) 0.30845   0.30855

c) 0.009005   0.009015

---

a) 18.95   19.05

b) 0.6995   0.7005

c) 0.05395   0.05405

9-4   ABSOLUTE ERROR OF MEASUREMENTS

When a measuring scale is read to the nearest calibration mark, there is some error involved in the reported measurement. The largest possible error that can occur by reading to the nearest calibration mark is called the <u>absolute</u> <u>error</u> of the measurement. We will discuss the <u>absolute</u> <u>error</u> of measurements in this section.

---

27. A reported measurement of 3" lies somewhere between 2.5" and 3.5". The largest possible error that can occur by reading to the nearest calibration mark is 0.5", since: 2.5" is 0.5" less than 3". 3.5" is 0.5" more than 3". The largest possible error that can occur by reading to the nearest calibration mark is called the <u>absolute</u> <u>error</u> of the measurement. Therefore, the <u>absolute</u> <u>error</u> of a 3" measurement is _____ .	
28. A reported measurement of 6.7 grams lies somewhere between 6.65 grams and 6.75 grams. The largest possible error that can occur by reading to the nearest calibration mark is 0.05 gram, since: 6.65 grams is 0.05 gram less than 6.7 grams. 6.75 grams is 0.05 gram more than 6.7 grams. Therefore, the <u>absolute</u> <u>error</u> of the measurement is _____ .	0.5"
29. A reported measurement of 1.21 seconds lies somewhere between 1.205 seconds and 1.215 seconds. 1.205 seconds is 0.005 second less than 1.21 seconds. 1.215 seconds is 0.005 second more than 1.21 seconds. Therefore, the <u>absolute</u> <u>error</u> of the measurement is _____ .	0.05 gram
30. You can see that the absolute error of a measurement is the same number added or subtracted to find the limits of a measurement. State the absolute error of each measurement below. a) 17.8 feet _____ c) 155 miles _____ b) 29 grams _____ d) 2.63 liters _____	0.005 second
	a) 0.05 foot b) 0.5 gram c) 0.5 mile d) 0.005 liter

31. Since 17.113 seconds has <u>three</u> digits after the decimal point, its absolute error has <u>three</u> 0's after the decimal point plus a "5".

The absolute error of 17.113 seconds is 0.0005 second.

State the absolute error of each measurement.

   a) 0.052 gram _____   b) 0.00270 second _____

---

32. Measurements are frequently reported with their absolute error attached as we have done below. The "±" symbol means that the absolute error is <u>both</u> <u>added</u> <u>and</u> <u>subtracted</u> to find the limits.

                 25 ± 0.5 inches

                 7.9 ± 0.05 grams

                 5.33 ± 0.005 meters

Insert the absolute error in each measurement below.

   a) 13 ± _____ ounces   c) 0.075 ± _____ mile

   b) 0.69 ± _____ liter   d) 0.00200 ± _____ second

a) 0.0005 gram

b) 0.000005 second

---

   a) 13 ± <u>0.5</u> ounces   c) 0.075 ± <u>0.0005</u> mile

   b) 0.69 ± <u>0.005</u> liter   d) 0.00200 ± <u>0.000005</u> second

---

9-5   ADDING AND SUBTRACTING MEASUREMENTS

In this section, we will discuss the addition and subtraction of measurements. Some rules are given for the precision that should be used when reporting sums and differences.

---

33. When adding or subtracting measurements, the measurements usually have <u>the</u> <u>same</u> <u>precision</u>. For additions and subtractions of that type, we use the following rule:

> WHEN ADDING (OR SUBTRACTING) MEASUREMENTS <u>WITH</u> <u>THE</u> <u>SAME</u> PRECISION, THE SUM (OR DIFFERENCE) SHOULD HAVE <u>THE</u> <u>SAME</u> <u>PRECISION</u> <u>AS</u> <u>THE</u> <u>MEASUREMENTS</u>.

Continued on following page.

33. Continued

On the diagram below, the two length measurements are both precise to "hundredths of an inch".

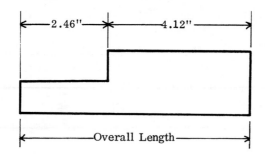

To find the overall length, we must add 2.46" and 4.12". According to the rule on the preceding page:

a) The sum should be reported to the nearest _____ of an inch.

b) The overall length is _____ .

---

34. Using the same rule, perform each addition and subtraction.

a) 2.65 grams + 7.66 grams + 0.28 gram = _____ grams

b) 13.1 seconds - 6.8 seconds = _____ seconds

a) hundredth

b) 6.58"

---

35. We wrote a "0" in the hundredths place of the sum below in order to report the proper precision.

2.45 liters + 4.35 liters = 6.80 liters

Complete these. Use 0's to report each answer to the proper precision.

a) 12.5 millimeters - 5.5 millimeters = _____ millimeters

b) 0.344 gram + 0.077 gram - 0.121 gram = _____ gram

a) 10.59 grams

b) 6.3 seconds

---

36. Though most additions and subtractions of measurements involve measurements with the same precision, occasionally measurements with different precisions are involved. For additions and subtractions of that type, we use the following rule:

> WHEN ADDING (OR SUBTRACTING) MEASUREMENTS WITH DIFFERENT PRECISIONS, THE PRECISION OF THE SUM (OR DIFFERENCE) SHOULD BE THE SAME AS THAT OF THE LEAST PRECISE MEASUREMENT.

a) 7.0 millimeters

b) 0.300 gram

Continued on following page.

36.   Continued

On the diagram below, the measurements have different precisions.

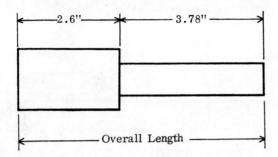

To find the overall length, we must add 2.6" and 3.78". We begin by adding the measurements as they stand.

$$2.6" + 3.78" = 6.38"$$

The <u>least</u> <u>precise</u> measurement is 2.6". Since 2.6" is precise to the nearest tenth of an inch, we should round the sum to the nearest tenth of an inch.

With proper precision, the overall length is _____ .

---

37.   Rounding a sum (or difference) to the precision of the least precise measurement makes sense for this reason:

Precision depends on the quality of the measuring instruments used. It cannot be improved by any calculation. Therefore, a sum (or difference) cannot be more precise than the least precise measurement involved in the calculation.

Let's use the same rule for the addition below.

0.2489 second + 1.24 seconds + 0.418 second

Adding the measurements as they stand, we get 1.9069 seconds.

a) The least precise measurement is _____ .

b) The sum should be rounded to the nearest _____ of a second.

c) With proper precision, the sum is _____ .

6.4"

---

a) 1.24 seconds

b) hundredth

c) 1.91 seconds

38.   Let's use the same rule for the subtraction below.

18 inches - 2.15 inches

Subtracting the measurements as they stand, we get 15.85 inches.

a) The least precise measurement is _____ .

b) The difference should be rounded to the nearest _____ inch.

c) With proper precision, the difference is _____ .

---

39.   Do these. Report each answer with the proper precision.

a) 7.5 liters + 6.23 liters = _____ liters

b) 0.289 second - 0.12 second = _____ second

a) 18 inches

b) whole-number

c) 16 inches

---

40.   Do these. Report each answer with the proper precision.

a) 2.5" + 4.17" + 3.595" = _____

b) 0.00230 gram + 1.255 grams - 0.0465 gram = _____ grams

a) 13.7 liters

b) 0.17 second

---

41.   When rounding a sum (or difference) to get the proper precision, we can get a "0" in the final place or places. Those 0's must be reported. For example, the sum below was rounded from 13.95 feet.

6.5 feet + 7.45 feet = 14.0 feet

Do these. Report each answer with the proper precision.

a) 0.2829 second - 0.18 second = _____ second

b) 0.031 gram + 2.95 grams + 0.0153 gram = _____ grams

a) 10.3"

b) 1.211 grams

---

a) 0.10 second

b) 3.00 grams

SELF-TEST 26 (pages 346-359)

State the precision of each measurement.

1.   0.000397 second

2.   38.60 gram

_____

_____

3. Which measurement has the greatest precision? _____

0.0870 cm   54.9 cm   203.64 cm   1.137 cm

4. Which measurement has the least precision? _____

137.2"   0.96"   5.80"   0.0144"   3.091"

5. Which measurements have the same precision? _____

115.34 sec   32.9 sec   0.712 sec   6.50 sec   93 sec

Find the upper and lower limits of each measurement.

6. 8.72"   _____ _____

7. 0.390 gram   _____ _____

State the absolute error of each measurement.

8. 23.4 feet   _____

9. 0.0250 meter   _____

Do the following additions and subtractions of measurements.

10. 4.85" + 2.37" = _____

11. 45.7 cm - 27.68 cm = _____

12. 0.0658 sec + 1.07 sec - 0.164 sec = _____

13. 0.2184 gram - 0.096 gram + 0.0075 gram = _____

Answers:
1. millionths of a second
2. hundredths of a gram
3. 0.0870 cm
4. 137.2"
5. 115.34 sec and 6.50 sec
6. 8.715 and 8.725
7. 0.3895 and 0.3905
8. 0.05 feet
9. 0.00005 meter
10. 7.22"
11. 18.0 cm
12. 0.97 sec
13. 0.130 gram

9-6   RELATIVE ERROR AND ACCURACY OF MEASUREMENTS

In this section we will define the <u>relative error</u> and <u>accuracy</u> of measurements and show how the two concepts are related. We will contrast accuracy with precision and discuss a common misuse of the word "accurate" in statements about measurements.

---

42.   The two lengths below were measured to the <u>nearest tenth of an inch</u>.

   A steel block. The reported measurement is 7.1".

   A driveway. The reported measurement is 1,080.6".

   Though both measurements have the same precision, it is more difficult to make a long measurement like 1,080.6" with only a 0.05" absolute error than to make a short measurement like 7.1" with a 0.05" absolute error. In that sense, the longer measurement is "better". <u>Relative error</u> and <u>accuracy</u> are measurement concepts used to describe that meaning of the word "better".

---

43.   The <u>relative error</u> of a measurement is a comparison of the absolute error of the measurement with the measurement itself. That is:

$$\boxed{\text{Relative Error} \; = \; \frac{\text{Absolute Error}}{\text{Measurement}}}$$

Since the absolute error of a measurement of 5.0" is 0.05",

   the <u>relative error</u> of 5.0" is $\dfrac{0.05"}{5.0"} = 0.01$.

Since the absolute error of a measurement of 50.0" is 0.05",

   the <u>relative error</u> of 50.0" is $\dfrac{0.05"}{50.0"} =$ _____ .

---

44.   Since relative error is a comparison of two quantities, it is a <u>ratio</u>. Like any ratio, <u>there are no units attached when the ratio is converted to a decimal number</u>.

   a) Find the relative error of 47 cm. Round to four decimal places.

   Relative Error = _____ = _____

   b) Find the relative error of 8.15 cm. Round to six decimal places.

   Relative Error = _____ = _____

0.001

a) $\dfrac{0.5 \text{ cm}}{47 \text{ cm}} = 0.0106$ b) $\dfrac{0.005 \text{ cm}}{8.15 \text{ cm}} = 0.000614$

45.   Relative error is sometimes stated as a percent. Convert the following relative errors to percents.

   The relative error of 5.0" is 0.01 or 1%.

   a) The relative error of 50.0" is 0.001 or _____ .

   b) The relative error of 47 cm is 0.0106 or _____ .

---

46.   We can compare two measurements by comparing their relative errors. The one <u>with</u> <u>the</u> <u>smaller</u> <u>relative</u> <u>error</u> is "better". By "better", we mean that its absolute error is a smaller part of the measurement itself. For example:

   The relative error of 5.0" is 0.01 (or 1%).

   The relative error of 50.0" is 0.001 (or 0.1%).

Which measurement is "better" because it has a smaller relative error? _____

| a) 0.1% |
| b) 1.06% |

---

47.   We found these relative errors in an earlier frame.

   The relative error of 47 cm is 0.0106 (or 1.06%).

   The relative error of 8.15 cm is 0.000614 (or 0.0614%).

Which measurement is "better" because it has a smaller relative error? _____

50.0"

---

48.   Instead of using the word "better" when comparing the relative errors of measurements, we use the word "<u>accurate</u>". Therefore:

   If one measurement has <u>a</u> <u>smaller</u> <u>relative</u> <u>error</u> than another, we say that it is <u>more</u> <u>accurate</u>.

   If one measurement has <u>a</u> <u>larger</u> <u>relative</u> <u>error</u> than another, we say that it is <u>less</u> <u>accurate</u>.

   a) Since the relative error (0.001) of 50.0" is <u>smaller</u> than the relative error (0.01) of 5.0", we say that 50.0" is _____ accurate than 5.0".

   b) Since the relative error (0.0106) of 47 cm is <u>larger</u> than the relative error (0.000614) of 8.15 cm, we say that 47 cm is _____ accurate than 8.15 cm.

8.15 cm

| a) more |
| b) less |

49. To decide whether one measurement is more or less accurate than another,
     we must compute their relative errors first.

        a) The relative error of 500.0 ft is _____ .

        b) The relative error of 5,000.0 ft is _____ .

        c) Which measurement is <u>more</u> accurate? _____

---

50. Let's compare the accuracy of 12.7 sec and 3.48 sec.

        a) To five decimal places, the relative error of 12.7 sec
            is _____ .

        b) To five decimal places, the relative error of 3.48 sec
            is _____ .

        c) Which measurement is <u>less</u> accurate? _____

a) 0.0001 (or 0.01%)

b) 0.00001 (or 0.001%)

c) 5,000.0 ft

---

51. Measurements with the same precision do not usually have the same
     accuracy. For example:

            The relative error of 3.14 ft is 0.001592 .

            The relative error of 45.72 ft is 0.000109 .

     Which measurement is <u>more</u> accurate? _____

a) 0.00394

b) 0.00144

c) 12.7 sec

---

52. If two measurements have different precisions, <u>the more precise is not
     always the more accurate</u>. For example:

            The relative error of 99.8 liters is 0.000501 .

            The relative error of 2.54 liters is 0.001969 .

        a) Which measurement is <u>more precise</u>? _____

        b) Which measurement is <u>more accurate</u>? _____

45.72 ft

---

53. Two measurements with different precisions can have the same accuracy.
     For example:

            The relative error of 2.6 meters is 0.0192 .

            The relative error of 0.26 meter is 0.0192 .

        a) Which measurement is <u>more precise</u>? _____

        b) Which measurement is <u>more accurate</u>? _____

a) 2.54 liters

b) 99.8 liters

54. The <u>precision</u> of a measurement can be stated either in terms of a place or in terms of the number of decimal places. For example:

2.47 seconds is precise to <u>hundredths</u>
or to <u>two</u> <u>decimal</u> <u>places</u>.

0.946 gram is precise to <u>thousandths</u>
or to <u>three</u> <u>decimal</u> <u>places</u>.

The word "accurate" is commonly misused in making statements about precision. For example, phrases like the following are used when describing precisions.

<u>accurate</u> to hundredths

<u>accurate</u> to three decimal places

That use of the word "accurate" is always incorrect because "accurate" refers to relative error. It does not refer to precision.

a) Instead of saying "accurate to hundredths",
we should say "_____ to hundredths".

b) Instead of saying "accurate to three decimal places",
we should say "_____ to three decimal places".

a) 0.26 meter

b) Neither. Both have the same accuracy.

---

55. a) A weight measurement of 14.5 ounces is _____ (accurate/precise) to the nearest tenth of an ounce.

b) A time measurement of 0.0725 second is _____ (accurate/precise) to four decimal places.

a) precise

b) precise

---

a) precise   b) precise

9-7   SIGNIFICANT DIGITS IN MEASUREMENTS

In this section, we will discuss the meaning of <u>significant</u> <u>digits</u> in measurements. Then in the next section, we will show the relationship between significant digits, relative error, and accuracy.

<u>Note</u>: Though measurements always include a measuring unit, we will not use measuring units in this section in order to simplify the discussion.

---

56. The word "significant" means "meaningful". Therefore, the <u>significant</u> <u>digits</u> in a measurement are the <u>meaningful</u> <u>digits</u> in the measurement. In a measurement context, the "meaningful" digits are those that appear in a name-form based on the precision of the measurement. For example:

Since 0.5 is precise to <u>tenths</u>, 0.5 = 5 <u>tenths</u>.

a) Since 1.7 is precise to <u>tenths</u>, 1.7 = _____ tenths.

b) Since 0.43 is precise to <u>hundredths</u>, 0.43 = _____ hundredths.

c) Since 25.96 is precise to <u>hundredths</u>, 25.96 = _____ hundredths.

57.   Write each measurement in a name-form based on its precision.

   a) 4.608 = _____

   b) 0.0095 = _____

   c) 0.00625 = _____

a) 17 tenths
b) 43 hundredths
c) 2,596 hundredths

---

58.   When a measurement is written in a name-form based on its precision, all non-zero digits appear in the name-form. For example:

$$7.4 = 74 \text{ tenths}$$

$$35.96 = 3,596 \text{ hundredths}$$

Therefore all non-zero digits in a measurement are significant. That is:

   a) 7.4 contains _____ significant digits.

   b) 35.96 contains _____ significant digits.

a) 4,608 thousandths
b) 95 ten-thousandths
c) 625 hundred-thousandths

---

59.   When a measurement is written in a name-form based on its precision, some 0's appear in the name-form and some do not. For example:

   5.09 = 509 hundredths   (The "0" appears)

   0.008 = 8 thousandths   (The 0's do not appear)

Therefore, some 0's in measurements are significant and some are not.

   a) Is the "0" in 5.09 significant? _____

   b) Are the 0's in 0.008 significant? _____

a) two
b) four

---

60.   Any "0" or 0's lying between non-zero digits in a measurement appear in the name-form. For example:

   20.6 = 206 tenths

   9.005 = 9,005 thousandths

Therefore all 0's of that type are significant. That is:

   a) 20.6 contains _____ significant digits.

   b) 9.005 contains _____ significant digits.

a) Yes
b) No

---

a) three
b) four

61. A "0" written in the "ones" place to emphasize the decimal point does not appear in the name form. For example:

$$0.7 = 7 \text{ tenths}$$

$$0.539 = 539 \text{ thousandths}$$

Therefore, a "0" of that type is not significant. That is:

a) 0.7 contains only _____ significant digit.

b) 0.539 contains only _____ significant digits.

---

62. A "0" or 0's after the decimal point but before the first non-zero digit in the measurement do not appear in the name-form. For example:

$$0.08 = 8 \text{ hundredths}$$

$$0.0025 = 25 \text{ ten-thousandths}$$

Therefore 0's of that type are also not significant. That is:

a) 0.08 contains only _____ significant digit.

b) 0.0025 contains only _____ significant digits.

| a) one |
| b) three |

---

63. When a measurement contains a final "0" or 0's after the decimal point, the "0" or 0's are included to indicate the precision of the measurement. Final 0's of that type appear in a name-form based on precision. For example:

$$3.0 = 30 \text{ tenths}$$

$$5.900 = 5,900 \text{ thousandths}$$

Therefore, final 0's of that type are significant digits. That is:

a) 3.0 contains _____ significant digits.

b) 5.900 contains _____ significant digits.

| a) one |
| b) two |

---

64. For each measurement, state whether the "0" or 0's with arrows over them are significant or not. (Write "Yes" or "No" in each blank.)

a) 0.645 _____   d) 20.747 _____

b) 0.7009 _____   e) 0.00888 _____

c) 0.020 _____   f) 4.00 _____

| a) two |
| b) four |

---

a) No	d) Yes
b) Yes	e) No
c) Yes	f) Yes

65.   State the number of significant digits in each measurement below.

a) 0.83 _____   d) 65.000 _____

b) 5.70 _____   e) 0.09003 _____

c) 0.002 _____   f) 200.001 _____

---

66.   For the measurement 600.0, the "0" in the tenths place tells us that the measurement is precise to tenths. Therefore it contains <u>four</u> significant digits.

For the measurement 600, we cannot tell whether it is precise to "ones", "tens", or "hundreds".

If precise to "<u>ones</u>", the measurement means "600 ones".
"600 ones" contains <u>three</u> significant digits.

If precise to "<u>tens</u>", the measurement means "60 tens".
"60 tens" contains <u>two</u> significant digits.

If precise to "<u>hundreds</u>", the measurement means "6 hundreds".
"6 hundreds" contains only _____ significant digit.

a) two   d) five
b) three   e) four
c) one   f) six

---

67.   If a measurement of 37,000 is reported, it could mean any of the following precisions:

37,000 ones
3,700 tens
370 hundreds
37 thousands

a) If 37,000 means "37,000 ones", it contains _____ significant digits.

b) If 37,000 means "37 thousands", it contains _____ significant digits.

c) A measurement of 37,000 could have as few as _____ significant digits and as many as _____ significant digits.

one

---

68.   If possible, state the <u>precision</u> of each measurement.

a) 27.00 _____   c) 400.0 _____

b) 9,000 _____   d) 620,000 _____

a) five
b) two
c) two ..... five

---

a) hundredths
b) not possible
c) tenths
d) not possible

69. If possible, state the number of significant digits in each measurement.

    a) 24,000 _____   c) 870.0 _____

    b) 30.000 _____   d) 910 _____

---

70. We can use scientific notation to indicate the number of significant digits in a measurement like 460,000.

$$460,000 \text{ can equal } 4.60000 \times 10^5$$

Since all four 0's are retained in 4.60000, all four 0's are significant. Therefore 4.60000×10^5 has <u>six</u> significant digits.

    a) If 460,000 is written as 4.600×10^5, it means that the measurement has _____ significant digits.

    b) If 460,000 is written as 4.6×10^5, it means that the measurement has _____ significant digits.

    a) not possible

    b) five

    c) four

    d) not possible

---

71. State the number of significant digits in each measurement.

    a) 2.03×10^4 _____   c) 8.1×10^3 _____

    b) 9.600×10^6 _____   d) 5.007×10^5 _____

    a) four

    b) two

---

72. Using scientific notation, write 7,200,000 to signify:

    a) five significant digits _____

    b) three significant digits _____

    a) three   c) two

    b) four   d) four

a) 7.2000×10^6 b) 7.20×10^6

SUMMARY OF 0'S AS SIGNIFICANT DIGITS IN MEASUREMENTS

1) 0's before the first non-zero digit in a measurement <u>are not</u> significant digits.

    <u>Example</u>:   0.0624 has <u>three</u> significant digits.

2) 0's lying between non-zero digits in a measurement <u>are</u> significant digits.

    <u>Example</u>:   50.008 has <u>five</u> significant digits.

3) Final 0's after the decimal point in a measurement <u>are</u> significant digits.

    <u>Example</u>:   9.100 has <u>four</u> significant digits.

Continued on following page.

SUMMARY OF 0'S AS SIGNIFICANT DIGITS IN MEASUREMENTS - Continued

4) Final 0's before the decimal point in a measurement with no digits after the decimal point may or may not be significant. We can only tell when the measurement is written in scientific notation.

Example: The number of significant digits in 7,000 could be either one, two, three, or four. However:

7×10^3 has one significant digit.

7.00×10^3 has three significant digits.

5) When a measurement larger than "1" has a "0" or 0's after the decimal point, all 0's immediately to the left of the decimal point are significant.

Example: 900.00 has five significant digits.

9-8   SIGNIFICANT DIGITS, RELATIVE ERROR, AND ACCURACY

The number of significant digits in measurements is roughly related to their relative errors. Therefore, to make statements about accuracy, we can compare the number of significant digits or the sequence of significant digits in the measurements. We will discuss the method in this section.

---

73. If a first measurement contains more significant digits than a second measurement, the first measurement is more accurate than the second. For example:

The relative error of 5.4" is 0.00926 .

The relative error of 14.8" is 0.00338 .

a) Which measurement is more accurate? _____

b) Which measurement contains more significant digits? _____

---

74. Given these two measurements: 3.9 sec and 4.68 sec

The relative error of 3.9 sec is 0.01282 .

The relative error of 4.68 sec is 0.00107 .

a) Which measurement is more accurate? _____

b) Which measurement contains more significant digits? _____

a) 14.8"

b) 14.8"

---

75. Using the fact that the measurement with more significant digits is always more accurate, underline the more accurate measurement in each pair below.

a) 250.75 cm or 19.22 cm   c) 0.0059 g or 4.05 g

b) 2.77 ft or 0.1825 ft   d) 0.700 volt or 0.0097 volt

a) 4.68 sec

b) 4.68 sec

76.   The two measurements 16.4 cm and 5.28 cm have the same number of significant digits.

        The relative error of 16.4 cm is 0.00305 .

        The relative error of 5.28 cm is 0.00095 .

When two measurements have the same number of significant digits, we must compare the "sequence" of significant digits in each to determine which is more accurate. The measurement whose "sequence" of significant digits is a larger number is more accurate.

        The sequence of significant digits in 16.4 cm is "164".

        The sequence of significant digits in 5.28 cm is "528".

    a) Which sequence of digits is larger? _____

    b) Therefore which measurement is more accurate? _____

    c) Is that fact confirmed by the relative errors above? _____

a) 250.75 cm

b) 0.1825 ft

c) 4.05 g

d) 0.700 volt

---

77.   Let's use the same principle to decide whether 0.28 gram or 0.0077 gram is more accurate.

    a) The sequence of significant digits in 0.28 gram is _____ .

    b) The sequence of significant digits in 0.0077 gram is _____ .

    c) Which sequence of digits is larger? _____

    d) Therefore which measurement is more accurate? _____

a) "528"

b) 5.28 cm

c) Yes

---

78.   Using the same principle, underline the more accurate measurement in each pair below.

    a) 508 ft or 2.95 ft   b) 0.309 sec or 0.0382 sec

a) "28"
b) "77"
c) "77"
d) 0.0077 gram

---

79.   When two measurements have the same number and sequence of significant digits, they have the same accuracy. To confirm that fact, let's compare the relative errors of 2.5" and 0.025".

    a) The relative error of 2.5" is $\dfrac{0.05''}{2.5''} =$ _____ .

    b) The relative error of 0.025" is $\dfrac{0.0005''}{0.025''} =$ _____ .

    c) Do the two measurements have the same accuracy? _____

a) 508 ft

b) 0.0382 sec

---

a) 0.02

b) 0.02

c) Yes

80.   Given these measurements: 159.3"   0.675"   2.03"   0.0069"

    a) The measurement with the <u>greatest</u> <u>accuracy</u> is _____ .

    b) The measurement with the <u>least</u> <u>accuracy</u> is _____ .

    c) The measurement with the <u>greatest</u> <u>precision</u> is _____ .

    d) The measurement with the <u>least</u> <u>precision</u> is _____ .

---

81.   0.003 cm   259.8 cm   49.62 cm   0.2066 cm   95.700 cm

Of the measurements above, list the one with the:

a) greatest accuracy _____   c) greatest precision _____

b) least accuracy _____   d) least precision _____

a) 159.3"

b) 0.0069"

c) 0.0069"

d) 159.3"

---

82.   Though the number of significant digits in a measurement is only roughly related to its accuracy, the phrase "<u>accurate</u> <u>to</u> <u>a</u> <u>certain</u> <u>number</u> <u>of</u> <u>significant</u> <u>digits</u>" is frequently used. For example:

    We say that 37.5 ft is accurate to <u>three</u> significant digits .

a) We say that 0.080 sec is accurate to _____ significant digits .

b) We say that 6.140 cm/hr is accurate to _____ significant digits .

a) 95.700 cm

b) 0.003 cm

c) 0.2066 cm

d) 259.8 cm

---

83.   a) To show that 8,100 meters is accurate to three significant digits, we write it as _____ .

    b) To show that 50,000 yd³ is accurate to two significant digits, we write it as _____ .

a) two

b) four

---

a) 8.10×10^3 meters

b) 5.0×10^4 yd^3

9-9   ROUNDING TO A DEFINITE NUMBER OF SIGNIFICANT DIGITS

When adding and subtracting measurements, the answers are reported to a certain precision. However, in other calculations with measurements, the answers are rounded to a certain number of significant digits. We will discuss rounding to a certain number of significant digits in this section, and then use that type of rounding for calculations in later sections.

Note: To simplify the instruction, we will not write measuring units in this section.

84.   To round a measurement to three significant digits, we use these steps:

1) Begin counting from the first non-zero digit at the left.

2) Count to the third digit (including 0's).

3) Round to the place of the third digit.

Two examples of rounding to three significant digits are shown below.

$$\downarrow$$
147.66 rounds to 148

$$\downarrow$$
0.98132 rounds to 0.981

When rounding to a place to the left of the "ones" place, we report the answer in scientific notation so that the number of significant digits is clearly shown. For example:

$$\downarrow$$
6,427,900 rounds to 6.43×10^6 (from 6,430,000)

Round to three significant digits. (The arrows indicate the place to be rounded to.)

a) $\downarrow$ 61,358 _____

b) $\downarrow$ 29.4166 _____

c) $\downarrow$ 701,029 _____

d) $\downarrow$ 0.0098472 _____

---

85.   Round to two significant digits. (The arrows indicate the place to be rounded to.)

a) $\downarrow$ 1,836.75 _____

b) $\downarrow$ 90,740,000 _____

c) $\downarrow$ 0.06513 _____

d) $\downarrow$ 0.001099 _____

a) 6.14×10^4
   (from 61,400)

b) 29.4

c) 7.01×10^5
   (from 701,000)

d) 0.00985

---

86.   Round to four significant digits. (The arrows indicate the place to be rounded to.)

a) $\downarrow$ 609,088 _____

b) $\downarrow$ 15.00625 _____

c) $\downarrow$ 0.298317 _____

d) $\downarrow$ 5,940.33 _____

a) 1.8×10^3
   (from 1,800)

b) 9.1×10^7
   (from 91,000,000)

c) 0.065

d) 0.0011

87. Round to <u>one</u> significant digit. (The arrows indicate the place to be rounded to.)

   a) $\downarrow$ 395,250 _____

   b) $\downarrow$ 62.004 _____

   c) $\downarrow$ 0.1259 _____

   d) 0.007881 $\downarrow$ _____

a) 6.091×10^5
   (from 609,100)

b) 15.01

c) 0.2983

d) 5.940×10^3
   (from 5,940)

---

88. We rounded each measurement below to <u>three</u> significant digits. Notice that we got a "0" in the final decimal place. That "0" <u>must</u> <u>be</u> <u>retained</u> so that the rounded measurement has three significant digits.

   $\downarrow$
   17.034 rounds to 17.0

   $\downarrow$
   0.95966 rounds to 0.960

   a) Round 0.06049 to <u>two</u> significant digits. _____

   b) Round 124.963 to <u>four</u> significant digits. _____

a) 4×10^5
   (from 400,000)

b) 6×10^1 (from 60)

c) 0.1

d) 0.008

---

89. We rounded each measurement below to <u>three</u> significant digits. Notice that both final 0's must be retained.

   $\downarrow$
   6.0034 rounds to 6.00

   $\downarrow$
   0.049975 rounds to 0.0500

   a) Round 1.99518 to <u>three</u> significant digits. _____

   b) Round 0.700035 to <u>four</u> significant digits. _____

a) 0.060

b) 125.0

---

90. When asking to round to a definite number of significant digits, we sometimes omit the word "significant". That is:

   Instead of: Round to three significant digits.

   We say: Round to three digits.

   Round to <u>three</u> digits.

   a) 2.6951 _____

   b) 0.300175 _____

a) 2.00

b) 0.7000

---

91. Round to <u>two</u> digits.

   a) 0.01967 _____

   b) 403.75 _____

a) 2.70

b) 0.300

---

92. Round to <u>four</u> digits.

   a) 300,047 _____

   b) 0.0599978 _____

a) 0.020

b) 4.0×10^2 (from 400)

93.   When rounding to a definite number of significant digits, we sometimes have to drop a final "0" to get the right number of significant digits. For example:

Rounding 0.099753 to <u>two</u> digits, we get 0.10 .

<u>Note</u>: We had to drop a final "0" from 0.100 to get two significant digits.

a)   Round 9.9984 to <u>three</u> digits. _____

b)   Round 0.97523 to <u>one</u> digit. _____

c)   Round 0.0999956 to <u>four</u> digits. _____

a) 3.000 x 10^5
    (from 300,000)

b) 0.06000

---

a) 10.0   b) 1 (not 1.0)   c) 0.1000

<u>SELF-TEST</u> 27 (pages <u>360-373</u>)

1. Find the relative error of 126.47 meters. Round to six decimal places. _____

2. Find the relative error of 0.85 gram. Round to four decimal places. _____

3. The measurements 45.8" and 3.62" have relative errors of 0.0011 and 0.0014, respectively. Which measurement is <u>more</u> accurate? _____

State the number of significant digits in each measurement.

4. 0.0918 gram _____ | 5. 360.00 ft _____ | 6. 2.070 x 10^6 cm _____

In each pair below, which measurement has the <u>greater</u> accuracy?

7. 0.38 sec and 18.6 sec _____ | 9. 1.03 grams and 0.0409 gram _____

8. 217.4" and 0.593" _____ | 10. 5.13 meters and 34.7 meters _____

| 0.256 cm   152.93 cm   3.025 cm   0.0084 cm   61.7 cm |

For the five measurements above, which measurement has the:

11. greatest precision? | 12. least precision? | 13. greatest accuracy? | 14. least accuracy?

_____ | _____ | _____ | _____

Round each of the following to <u>three</u> significant digits.

15. 23.9481 _____ | 16. 6.995724 _____ | 17. 0.0130492 _____

18. Round 750.17625 to <u>five</u> digits. | 19. Round 0.99713 to <u>two</u> digits.

_____ | _____

20. Round 8,709,615 to <u>four</u> digits. Report the answer in scientific notation. _____

<u>ANSWERS</u>:

1. 0.000040	6. four	11. 0.0084 cm	16. 7.00
2. 0.0059	7. 18.6 sec	12. 61.7 cm	17. 0.0130
3. 45.8"	8. 217.4"	13. 152.93 cm	18. 750.18
4. three	9. 0.0409 gram	14. 0.0084 cm	19. 1.0
5. five	10. 5.13 meters	15. 23.9	20. 8.710 x 10^6

9-10   MULTIPLYING AND DIVIDING MEASUREMENTS

In this section, we will discuss the multiplication and division of measurements. We will emphasize the rule for the number of digits that should be retained when reporting the products and quotients.

---

94.   The problem below involves a multiplication of two measurements.

Find the area of a rectangle whose measured length is 17.5" and whose measured width is 12.5".

$$A = LW = (17.5")(12.5") = 218.75 \text{ in}^2$$

Notice that the calculator product contains <u>five</u> significant digits, even though each factor contains only <u>three</u> significant digits. If we report the product with <u>five</u> significant digits, we imply that the area is more accurate than either the length or the width. However, <u>since accuracy depends on the quality of the measuring instrument and cannot be improved by a calculation</u>, the area cannot be more accurate than either the length or the width.

To make the product above as accurate as the factors, we must round it to <u>three</u> significant digits. Doing so, we get:

$$A = LW = (17.5")(12.5") = \underline{\hspace{2in}}$$

---

95.   In order to keep the product from appearing more accurate than the factors when multiplying measurements, the following rule is used to report products.

| WHEN MULTIPLYING MEASUREMENTS, THE PRODUCT SHOULD CONTAIN THE SAME NUMBER OF SIGNIFICANT DIGITS AS THERE ARE IN THE FACTOR <u>WITH THE LEAST NUMBER OF SIGNIFICANT DIGITS</u>. |

Let's apply the rule to this multiplication: 42.83 ft x 5.25 ft

a) Which factor contains the least number of significant digits? _____

b) How many significant digits does it contain? _____

c) How many significant digits should the product contain? _____

219 in²

---

96.   In this multiplication: 9.66 cm x 8.6 cm x 17.25 cm

a) Which factor contains the least number of significant digits? _____

b) How many significant digits does it contain? _____

c) How many significant digits should the product contain? _____

a) 5.25 ft

b) three

c) three

97.   If all of the factors contain the same number of significant digits, the product should contain that number of significant digits. Therefore:

    a) For 37.46 m x 29.55 m , the product should contain _____ significant digits.

    b) For 2.5 yd x 7.5 yd x 6.4 yd , the product should contain _____ significant digits.

a) 8.6 cm

b) two

c) two

---

98.   In each multiplication below, the factors are measurements. Following the example, state the number of significant digits that each product should contain.

Multiplication	Significant Digits In Product
0.95 x 17.6	two
a) 225.5 x 0.0900	_____
b) 0.20775 x 5.00366	_____
c) 7.13 x 9.00 x 84.33	_____

a) four

b) two

---

99.   In each multiplication below, the factors are measurements. Multiply and report each product with the correct number of significant digits.

    a) 5.66 x 9.17 = _____   c) 0.30 x 0.30 = _____

    b) 2.629 x 0.404 = _____   d) 8.7 x 9.035 = _____

a) three

b) five

c) three

---

100.   In each multiplication below, the factors are measurements. Report each product with the correct number of significant digits.

    a) 0.2640 x 7.449 = _____

    b) 1.79 x 64.8 x 0.925 = _____

    c) 346 x 7.0 x 15.88 = _____

a) 51.9   c) 0.090

b) 1.06   d) 79

---

101.   In order to show the correct number of significant digits in a product, we sometimes have to report the product in scientific notation. For example:

$$255 \text{ in} \times 325 \text{ in} = 8.29 \times 10^4 \text{ in}^2$$

Do these. Report the products in scientific notation to show the proper number of significant digits.

    a) 46.9 cm x 241 cm = _____

    b) 7.8 ft x 0.925 ft x 18.0 ft = _____

a) 1.967

b) 107

c) 3.8 x 10^4
    (from 38,000)

102.   In the multiplication below, one factor is written in scientific notation. Since 6.7 cm contains only <u>two</u> significant digits, the product contains only <u>two</u> significant digits. Notice that the product is written in scientific notation to show the correct number of significant digits.

$$(3.49 \times 10^4 \text{ cm}) \times 6.7 \text{ cm} = 2.3 \times 10^5 \text{ cm}^2$$

Use the $\boxed{\text{EE}}$ key to perform these multiplications. Report each product in scientific notation with the correct number of significant digits.

a) $(5.25 \times 10^3 \text{ in}) \times 29.45 \text{ in} = $ _____

b) $(1.29 \times 10^3 \text{ m}) \times (4.60 \times 10^4 \text{ m}) = $ _____

a) $1.13 \times 10^4 \text{ cm}^2$

b) $1.3 \times 10^2 \text{ ft}^3$

---

103.   When dividing two measurements, the same rule applies. That is:

> THE QUOTIENT SHOULD CONTAIN THE SAME NUMBER OF SIGNIFICANT DIGITS AS THERE ARE IN THE TERM <u>WITH</u> <u>THE</u> <u>LEAST</u> <u>NUMBER</u> <u>OF</u> <u>SIGNIFICANT</u> <u>DIGITS</u>.

For example, if both terms in the divisions below are measurements:

For $\dfrac{65.6}{2.48}$, the quotient should contain <u>three</u> significant digits.

a) For $\dfrac{7.045}{99}$, the quotient should contain _____ significant digits.

b) For $\dfrac{37.35}{456.91}$, the quotient should contain _____ significant digits.

c) For $\dfrac{9.80}{12.65}$, the quotient should contain _____ significant digits.

a) $1.55 \times 10^5 \text{ in}^2$

b) $5.93 \times 10^7 \text{ m}^2$

---

104.   In each division below, the terms are measurements. Divide and report each quotient with the correct number of significant digits.

a) $\dfrac{0.784}{2.49} = $ _____

c) $\dfrac{420.80}{13.40} = $ _____

b) $\dfrac{64.9}{7.6} = $ _____

d) $\dfrac{6.20}{9.00} = $ _____

a) two

b) four

c) three

---

105.   In each division below, the terms are measurements. Report each quotient in scientific notation to show the correct number of significant digits.

a) $\dfrac{483.7}{0.216} = $ _____

b) $\dfrac{9.96 \times 10^6}{1.040 \times 10^2} = $ _____

a) 0.315   c) 31.40

b) 8.5   d) 0.689

a) 2.24×10^3

b) 9.58×10^4

106. When the terms are measurements written in scientific notation, we can sometimes report the quotient as an ordinary number.

$$\frac{7.26 \times 10^3}{5.99 \times 10^4} = 1.21 \times 10^{-1} = 0.121$$

Do these. If necessary, report the quotient in scientific notation to show the correct number of significant digits.

a) $\dfrac{1.24 \times 10^4}{0.903}$ = _____   b) $\dfrac{725}{3.46 \times 10^3}$ = _____

a) 1.37×10^4 b) 0.210, from: 2.10×10^{-1}

9-11   SQUARES, SQUARE ROOTS, AND THE ARITHMETIC MEAN OF MEASUREMENTS

In this section, we will discuss the rule for the number of significant digits that should be reported in the square or square root of a measurement and in the arithmetic mean (or "average") of a set of measurements.

107. The square or square root of a measurement should contain the same number of significant digits as the original measurement. For example:

$(2.56 \text{ cm})^2 = 6.55 \text{ cm}^2$ (reported <u>three</u> significant digits)

$\sqrt{85 \text{ in}^2} = 9.2 \text{ in}$ (reported <u>two</u> significant digits)

Each number below is a measurement. Report each square or square root with the correct number of significant digits.

a) $(21.8)^2$ = _____   c) $\sqrt{457}$ = _____

b) $(0.095)^2$ = _____   d) $\sqrt{.01980}$ = _____

108. We sometimes have to report the square of a measurement in scientific notation to show the correct number of significant digits. Square these measurements.

a) $(899)^2$ = _____   c) $(6.5 \times 10^3)^2$ = _____

b) $(0.000055)^2$ = _____   d) $(1.906 \times 10^{-5})^2$ = _____

a) 475   c) 21.4

b) 0.0090   d) 0.1407

109. We sometimes have to report the square root of a measurement in scientific notation to show the correct number of significant digits. Find the square roots of these measurements.

a) $\sqrt{7.9 \times 10^6}$ = _____

b) $\sqrt{4.13 \times 10^{-12}}$ = _____

a) 8.08×10^5

b) 3.0×10^{-9}

c) 4.2×10^7

d) 3.633×10^{-10}

110.   When finding the square root of a measurement written in scientific notation, we can sometimes report the square root as an ordinary number. For example:

a) $\sqrt{5.24 \times 10^4} = 2.29 \times 10^2 = $ _____

b) $\sqrt{3.8 \times 10^{-6}} = 1.9 \times 10^{-3} = $ _____

a) 2.8×10^3

b) 2.03×10^{-6}

---

111.   To find the <u>arithmetic</u> <u>mean</u> (or "average") of a set of measurements, we add the measurements and divide by the number of measurements. The arithmetic mean is reported with the same number of significant digits as the original measurements. An example is shown.

Find the <u>mean</u> of 7.15", 6.92", and 2.67".

The <u>mean</u> is $\dfrac{7.15" + 6.92" + 2.67"}{3} = 5.58"$

a) Find the mean of 4.8 grams, 9.7 grams, and 5.8 grams.

The mean is _____ .

b) Find the mean of 141 cm, 212 cm, 193 cm, 156 cm, and 201 cm.

The mean is _____ .

a) 229

b) 0.0019

a) 6.8 grams   b) 181 cm

9-12   COMBINED OPERATIONS WITH MEASUREMENTS

In this section, we will discuss the rule for reporting the number of significant digits in the answers for combined operations involving measurements.

---

112.   In any combined operation with measurements that involves multiplying, dividing, squaring, or finding a square root, <u>the</u> <u>answer</u> <u>should</u> <u>contain</u> <u>as</u> <u>many</u> <u>significant</u> <u>digits</u> <u>as</u> <u>the</u> <u>measurement</u> <u>with</u> <u>the</u> <u>least</u> <u>number</u> <u>of</u> <u>significant</u> <u>digits</u>. That is:

For $\dfrac{(1.54)(6.03)}{0.917}$, the answer should contain three significant digits.

a) For $(14.56)^2(25.33)$, the answer should contain _____ significant digits.

b) For $\sqrt{\dfrac{0.413}{1.980}}$, the answer should contain _____ significant digits.

a) four

b) three

113. The same rule applies even when the combined operation contains an addition or subtraction, as long as it also contains a multiplication, division, squaring, or square root. That is:

For $\dfrac{16.5 + 71.6}{14.3}$, the answer should contain <u>three</u> significant digits.

For $\sqrt{17^2 + 48^2}$, the answer should contain _____ significant digits.

---

114. All of the numbers below are measurements. Report each answer with the correct number of significant digits.

  a) $\dfrac{4.70}{(86.3)(6.2)}$ = _____

  b) $\dfrac{(253.96)(9.014)}{(72.38)(0.040700)}$ = _____

  c) $\dfrac{17.8 + 21.6}{14.9}$ = _____

two

---

115. All of the numbers below are measurements. Report each answer with the correct number of significant digits.

  a) $(3.14)(8.77)^2$ = _____

  b) $\sqrt{\dfrac{1.480}{6.6}}$ = _____

  c) $\sqrt{(15.9)^2 + (27.1)^2}$ = _____

a) 0.0088
b) 777.1
c) 2.64

---

a) 242
b) 0.47
c) 31.4

9-13   ESTIMATING FINAL DIGITS WHEN READING SCALES

Up to this point we have discussed measurements reported to the nearest calibration mark on the scale. In practice, however, we usually estimate a final digit when reading scales if it is possible to do so. We will discuss estimating a final digit in this section.

116.   We want to use the scale below to measure the steel block.

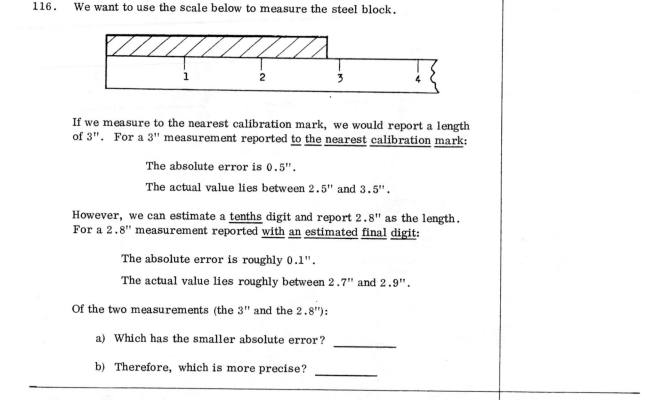

If we measure to the nearest calibration mark, we would report a length of 3". For a 3" measurement reported <u>to the nearest calibration mark</u>:

The absolute error is 0.5".

The actual value lies between 2.5" and 3.5".

However, we can estimate a <u>tenths</u> digit and report 2.8" as the length. For a 2.8" measurement reported <u>with an estimated final digit</u>:

The absolute error is roughly 0.1".

The actual value lies roughly between 2.7" and 2.9".

Of the two measurements (the 3" and the 2.8"):

a) Which has the smaller absolute error? _____

b) Therefore, which is more precise? _____

117.   Since the absolute error of the measurement with the estimated final digit is approximately 0.1", we can report it this way:

2.8" ± approximately 0.1"

Don't confuse a measurement of this type with a 2.8" measurement read <u>to the nearest calibration mark on a scale precise to tenths</u>. For a 2.8" measurement reported to the nearest calibration mark:

a) The absolute error is _____ .

b) The actual value lies between _____ and _____ .

a) 2.8"

b) 2.8"

a) 0.05"

b) 2.75" and 2.85"

118. The three measurements we have been discussing are given below with
their absolute errors.

    1) $3'' \pm 0.5''$ (read to the nearest calibration mark)

    2) $2.8'' \pm$ approximately $0.1''$ (an estimate of the tenths digit)

    3) $2.8'' \pm 0.05''$ (read to the nearest calibration mark)

When comparing measurements, the one with the smallest absolute error
is most precise.

    a) Which measurement has the largest absolute error? _____

    b) Which measurement has the smallest absolute error? _____

    c) Is (2) more precise than (1)? _____

    d) Is (2) more precise than (3)? _____

---

119. Here is a different block whose length is being measured by a 6-inch scale
divided into tenths of an inch:

To the nearest calibration mark, the length of the block is $1.9''$. For
this reported length of $1.9''$:

    The absolute error is $0.05''$.

    The actual value lies between $1.85''$ and $1.95''$.

If we estimate a final digit in the "hundredths" place, the reported
measurement is roughly $1.87''$. Since the measurement is clearly
between $1.85''$ and $1.89''$, we can report it as $1.87'' \pm 0.02''$.

    a) The absolute error of the $1.87''$ measurement is _____.

    b) Is the absolute error of the $1.87''$ measurement larger or smaller
than the absolute error of the $1.9''$ measurement? _____

    c) Which reported measurement is closer to the actual length of the
block, the $1.9''$ measurement or the $1.87''$ measurement? _____

    d) Is it worthwhile to carry the measurement farther by estimating a
final digit? _____

---

a) (1)

b) (3)

c) Yes

d) No

---

a) $0.02''$ c) $1.87''$

b) smaller   d) Yes

120.   When estimating a final digit, we must state that fact. For example:

2.83" (final digit estimated)

When a final digit is estimated, the absolute error of the measurement cannot be stated exactly. For the measurement above, the absolute error is probably 0.01" or greater. It is definitely not 0.005".

For which of the following reported measurements can the absolute error be stated exactly? _____

   a) 63.9 inches (to nearest calibration mark)

   b) 368.5 grams (final digit estimated)

   c) 4.57 pounds (final digit estimated)

   d) 18.29 centimeters (to nearest calibration mark)

---

121.   When a final digit is estimated, the reported measurement is not precise to that place. For example:

If 2.83" (final digit estimated) is reported, the measurement is not precise to hundredths.

However, when adding or subtracting measurements in which a final digit is estimated, the common practice is to treat them as if they were precise to the place of the estimated digit. For example, even though the final digit is estimated in each of the following measurements, we report the sum precise to hundredths:

2.83" + 1.79" = 4.62"

Each number below is a measurement whose final digit is estimated. Report each answer to the proper place.

   a) 17.9 + 5.86 = _____   b) 289 - 48.7 = _____

Only (a) and (d)

---

122.   When a final digit is estimated, the absolute error of the reported measurement is not known exactly. Therefore we cannot make an exact statement about its relative error and accuracy.

However, when multiplying or dividing measurements in which a final digit is estimated, the common practice is to treat them as if the estimated digit were a significant digit. For example, even though the final digit is estimated in each of the following measurements, we report three significant digits in the product.

14.9" x 7.85" = 117 in^2

Each number below is a measurement whose final digit is estimated. Report each answer with the correct number of significant digits.

   a) 6.53 x 7.14 = _____   b) $\dfrac{0.1875}{0.9653}$ = _____

a) 23.8   b) 240

a) 46.6   b) 0.1942

9-14   INSTRUMENT ERROR

One source of error in measurement is "absolute error". Absolute error is related to the nature of measuring scales. Another source of error in measurement is "instrument error". Instrument error is related to the construction of measuring instruments. We will discuss "instrument error" in this section.

123. There is another source of error in measurements called "instrument error". "Instrument error" occurs because of imperfections in measuring instruments. That is, it is a result of the fact that the measuring scale of the instrument does not perfectly coincide with the "standard" measuring unit. (Note: Most "standard" measuring units are kept in the National Bureau of Standards, Washington, D.C.) The amount of "instrument error" in measuring instruments differs for all instruments. It depends on the care with which the instrument is made. Though no measuring instrument is completely free from this type of error, more expensive instruments usually have less instrument error. Sometimes the "instrument error" is stated in percent form in the manufacturer's specifications. For example: A voltmeter is specified by the manufacturer to have "5% error". This means that, when measuring a voltage, the actual value of the voltage is within ±5% of the observed value. Therefore, if an observed value is 100 volts, the actual value lies between 95 volts (100 volts – 5%) and 105 volts (100 volts + 5%). If other measurements are made on the same voltmeter: a) An observed value of 400 volts means that the actual value lies between _____ volts and _____ volts. b) An observed value of 20 volts means that the actual value lies between _____ volts and _____ volts.	
124. If a voltmeter of "5% error" shows a reading of 80 volts, it is likely that the actual voltage lies between 76 volts and 84 volts (80 ± 5%). If the same voltmeter shows a reading of 30 volts, it is likely that the actual voltage lies between _____ volts and _____ volts.	a) 380 and 420 b) 19 and 21
125. A precision thermometer for an industrial furnace has a 1% error. For a thermometer reading of 2300° F, it is likely that the actual temperature of the furnace lies between _____ and _____ .	28.5 and 31.5 (from 30 ± 5%)
126. A car's speedometer has 10% error at the high end of its dial. For a speedometer reading of 80 miles per hour, it is likely that the actual speed of the car lies between _____ and _____ miles per hour.	2277° and 2323° F

127.   Instrument error is sometimes given in decimal form rather than percent form. For example, a vernier micrometer which measures lengths up to 2 inches, may be described as having an error of 0.0002 inch. Then for an instrument reading of 1.2047", it is likely that the actual length lies between 1.2045" and 1.2049" (1.2047" ± 0.0002").

For this same instrument, a reading of 0.4680" means that it is likely that the actual length lies between _____ and _____ .

	72 and 88

Note: Besides "absolute error" and "instrument error", a third source of error in measurement is possible. The third source of error is "observer error". "Observer error" is the result of improper handling of the measuring instrument or faulty reading of its scale.

0.4678" and 0.4682"

SELF-TEST 28 (pages 374-384)

The numbers in problems 1-12 are measurements. Report each answer with the correct number of significant digits.

1. 4.710 x 0.0836
    = _____

2. 53.9 x 0.0062 x 8.140
    = _____

3. $\dfrac{71.83}{0.9206}$ = _____

4. $\dfrac{4.15}{80.47}$ = _____

5. $(6.527)^2$ = _____

6. $\sqrt{0.34}$ = _____

7. $\sqrt{(228)^2 + (457)^2}$
    = _____

8. $\dfrac{(0.504)(89)}{731.2}$ = _____

9. $(72.19)(3.460)^2$
    = _____

Report each answer in scientific notation with the correct number of significant digits.

10. $(7.2 \times 10^4) \times (9.4 \times 10^2)$
     = _____

11. $\sqrt{1.83 \times 10^{11}}$
     = _____

12. $\dfrac{59.24}{0.0008380}$ = _____

13. Find the arithmetic mean of the following set of measurements.   _____

87.6 meters   65.9 meters   91.2 meters   76.8 meters   84.3 meters

An enlarged portion of a 20-centimeter scale is shown at the right.

14. Read point P to the nearest calibration mark. _____

15. Read point P, estimating a final digit. _____

16. If a final digit is estimated at point P, the approximate absolute error is _____ (greater/less) than 0.05 cm.

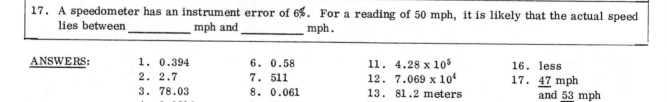

17. A speedometer has an instrument error of 6%. For a reading of 50 mph, it is likely that the actual speed lies between _____ mph and _____ mph.

ANSWERS:
1. 0.394	6. 0.58	11. 4.28×10^5	16. less
2. 2.7	7. 511	12. 7.069×10^4	17. 47 mph
3. 78.03	8. 0.061	13. 81.2 meters	and 53 mph
4. 0.0516	9. 864.2	14. 4.4 cm	
5. 42.60	10. 6.8×10^7	15. 4.42 cm	

SUPPLEMENTARY PROBLEMS - CHAPTER 9

Assignment 26

State the precision of each measurement.

    1. 21.8 liters   2. 0.000637 sec   3. 3.450 grams   4. 118.00 meters

In each pair, which measurement has the greater precision.

    5. 2.730" and 0.0195"   6. 382 km and 29.6 km   7. 0.00420 g and 0.0753 g

For these measurements: | 0.0960 cm   5.90 cm   73.8 cm   321.64 cm   0.705 cm |

    8. Which measurement has the greatest precision?

    9. Which measurement has the least precision?

   10. Which measurements have the same precision?

Find the upper and lower limits of each measurement.

   11. 30.2"   12. 7.01 cm   13. 0.480 sec   14. 119 ft

State the absolute error of each measurement.

   15. 65.0 cm   16. 268 grams   17. 319.40 ft   18. 0.0715 meter

Do these additions of measurements.

   19. 93.7" + 58.4"   21. 5.86 grams + 0.034 gram + 12.1 grams

   20. 2.686 cm + 3.71 cm   22. 0.712 sec + 0.85 sec + 0.0564 sec

Do these subtractions of measurements.

   23. 12.36" - 7.86"   24. 0.602 sec - 0.0414 sec   25. 343 ft - 77.82 ft

Do these combined additions and subtractions of measurements.

   26. 0.0887 g + 15.59 g - 5.604 g   27. 0.974 cm - 0.08863 cm + 1.5350 cm

Assignment 27

Find the relative error of each measurement.

    1. 20.0 cm   2. 0.125 gram   3. 400 ft   4. 2.5 sec

    5. 382.47 meters (Round to millionths)   6. 1.296" (Round to millionths)

7. The measurements 0.0042 second and 0.217 second have relative errors of 0.0119 and 0.0023, respectively. Which measurement is more accurate?

8. The measurements 2.74 centimeters and 5.18 centimeters have relative errors of 0.0018 and 0.0010, respectively. Which measurement is more accurate?

State the number of significant digits in each measurement.

    9. 28.5"   10. 0.073 sec   11. 4.610 cm   12. 225.39 ft

   13. 0.00800 gram   14. 3.4×10^4 meters   15. 5.20×10^3 km

In each pair, which measurement has the greater accuracy?

   16. 436 cm and 8.5 cm   17. 51.72" and 2.680"   18. 0.340 sec and 0.085 sec

For the set of time measurements below, which measurement has the:

| 3.724 sec | 0.016 sec | 44.8 sec | 2.3170 sec | 115 sec |

19. greatest accuracy?   20. greatest precision?   21. least accuracy?   22. least precision?

For the set of weight measurements below, which measurement has the:

| 864.2 grams | 0.0510 gram | 33.57 grams | 15.285 grams | 7.14 grams |

23. greatest precision?   24. least precision?   25. greatest accuracy?   26. least accuracy?

Round to three significant digits: 27. 36.4487   28. 0.0589637

Round to four significant digits: 29. 214.95376   30. 8.099628

Round to two digits. Report each answer in scientific notation.

31. 483,927   32. 695.30847   33. 1,462.554

Assignment 28

NOTE: The numbers in the following problems are measurements.

Multiply and report each answer with the correct number of significant digits.

1. 6.193 x 8.70   2. 0.3059 x 0.72817   3. 1.38 x 93.16 x 0.029

Divide and report each answer with the correct number of significant digits.

4. $\dfrac{75.38}{24.26}$ 5. $\dfrac{0.0090}{18.7}$ 6. $\dfrac{5.73}{0.0165}$ 7. $\dfrac{3,250.7}{81.436}$

Report each square and square root with the correct number of significant digits.

8. $(21.44)^2$ 9. $(0.0680)^2$ 10. $\sqrt{0.053}$ 11. $\sqrt{7,349}$

Report each answer with the correct number of significant digits.

12. $\dfrac{(409.5)(0.376)}{32.8}$ 13. $\dfrac{5.819}{(0.474)(8.6)}$ 14. $\dfrac{(41.63)(0.05290)}{(0.3874)(7.0358)}$

15. $\sqrt{(1.85)^2 + (3.26)^2}$ 16. $(87)(0.414)^2$ 17. $\dfrac{122.85 + 362.74}{21.419}$

Report each answer in scientific notation with the correct number of significant digits.

18. $(6.8)(9.7 \times 10^3)$ 19. $\dfrac{491.6}{0.003870}$ 20. $\sqrt{8.18 \times 10^7}$

21. $\dfrac{(792)(628.4)}{13.6}$ 22. $(5.21 \times 10^3) \times (8.9 \times 10^3)$ 23. $(41.70 \times 10^3)^2$

Find the arithmetic mean of each set of measurements.

24.   193 grams   208 grams   168 grams   176 grams

25.   5.42 cm   3.93 cm   4.27 cm   6.04 cm   4.77 cm

26. A wattmeter has an instrument error of 4%. For a reading of 250 watts, it is likely that the actual value lies between _____ watts and _____ watts.

Chapter 10

ESTIMATION

There are times when approximate answers or "estimates" are useful. To get approximate answers or estimates, we use a process called "estimation". In this chapter, we will discuss methods for estimating sums, differences, products, quotients, squares, square roots, and the answers for combined operations. Some uses of estimation are also discussed.

10-1   ESTIMATING SUMS AND DIFFERENCES

In this section, we will discuss the method for estimating sums and differences.

---

1.   To get an approximate answer or "estimate" for the addition at the left below, we rounded each number to thousands. Notice that 190,000 (the approximate sum or "estimate") is reasonably close to 189,425 (the exact sum). Estimate the other sum by rounding to thousands.

$$\begin{array}{rcl} 43,599 & \rightarrow & 44,000 \\ 65,127 & \rightarrow & 65,000 \\ +\,80,699 & \rightarrow & +\,81,000 \\ \hline 189,425 & & 190,000 \end{array}$$

$$\begin{array}{rcl} 7,825 & \rightarrow & \\ 1,250 & \rightarrow & \\ +\,6,535 & \rightarrow & +\,\underline{\hspace{2cm}} \end{array}$$

---

2.   When rounding to estimate a sum, the type of rounding needed depends on the actual problem. In this chapter, we will tell you the type of rounding to use for each problem.

a) Estimate by rounding to millions.

$$\begin{array}{rl} 7,764,255 & \rightarrow \\ 5,044,634 & \rightarrow \\ +\,2,587,000 & \rightarrow +\,\underline{\hspace{2cm}} \end{array}$$

b) Estimate by rounding to the nearest whole number.

$$\begin{array}{rl} 4.27 & \rightarrow \\ 6.51 & \rightarrow \\ +\,3.44 & \rightarrow +\,\underline{\hspace{2cm}} \end{array}$$

$$\begin{array}{r} 8,000 \\ 1,000 \\ +\,7,000 \\ \hline 16,000 \end{array}$$

a)
$$\begin{array}{r} 8,000,000 \\ 5,000,000 \\ +\,3,000,000 \\ \hline 16,000,000 \end{array}$$

b)
$$\begin{array}{r} 4 \\ 7 \\ +\,3 \\ \hline 14 \end{array}$$

3. We rounded to <u>tens</u> for the estimation at the left below. Round to <u>thousands</u> for the other estimation.

568 →	570	12,900 →	
37 →	40	3,465 →	
233 →	230	41,115 →	
+ 82 →	+ 80	+ 9,775 →	+ _____
	920		

4. We estimated the sum below by rounding to <u>hundreds</u>. Estimate the other sum by rounding to <u>hundredths</u>.

276 + 419 + 765   0.1859 + 0.0606 + 0.0755

```
   300
   400
 + 800
 1,500
```

13,000
3,000
41,000
+ 10,000
67,000

5. The same procedure is used to estimate a difference. For example, we estimated the difference at the left below by rounding to <u>hundreds</u>. Estimate the other difference by rounding to <u>thousands</u>.

387 →	400	78,150	
– 114 →	– 100	– 36,766 →	– _____
	300		

0.19
0.06
+ 0.08
0.33

6. a) Estimate by rounding to <u>hundreds</u>.

3,056 →
– 824 → – _____

b) Estimate by rounding to the <u>nearest</u> <u>whole</u> <u>number</u>.

26.88 →
– 5.23 → – _____

78,000
– 37,000
41,000

7. To estimate the difference at the left below, we rounded to <u>millions</u>. Estimate the other difference by rounding to <u>tenths</u>.

89,125,000 – 57,645,000   0.765 – 0.193

```
  89,000,000
– 58,000,000
  31,000,000
```

a) 3,100   b)   27
     – 800   – 5
    2,300   22

0.8
– 0.2
0.6

8. An estimate can be used to check the sensibleness of a calculator answer. That is, it can be used to detect a large error. For example:

    Suppose you get 393.17 as a calculator answer for this addition:

    $$27.95 + 64.03 + 49.64$$

    a) Estimate the answer by rounding to <u>tens</u>.
    The estimate is _____.

    b) Does 393.17 make sense as an answer? _____

---

9. Suppose you get 60,875 as a calculator answer for: 94,650 - 33,775

    a) Estimate the answer by rounding to <u>thousands</u>.
    The estimate is _____.

    b) Does 60,875 make sense as an answer? _____

a) 140

b) No

---

10. In some applied problems, an estimate of the answer is sufficient. Do these.

    a) In an electrical circuit, the total resistance is found by adding the separate resistances. If the separate resistances are 1.75 ohms, 3.66 ohms, 4.15 ohms, and 6.37 ohms, estimate the total resistance by rounding <u>to the nearest whole number</u>.

    b) In 1970, the population of the United States was 203,185,000 and the population of Japan was 98,275,000. Estimate the difference in the two populations <u>to the nearest million</u>.

a) 61,000

b) Yes

---

a) 16 ohms   b) 105,000,000 people

10-2   ESTIMATING PRODUCTS

In this section, we will discuss a method for estimating products.

---

11. When estimating products, we usually round the factors <u>to one digit</u>. For example, we rounded 3,265 to 3,000 below. Use the same method to estimate the other product.

    ```
    3,265  →  3,000        7,647,100  →
      x 7  →    x 7              x 5  →              x 5
              21,000
    ```

12. To estimate the product below, we rounded both factors <u>to</u> <u>one</u> <u>digit</u> and then multiplied. Use the same method to estimate the other product.

$$
\begin{array}{r}
67.9 \rightarrow 70 \\
\times\ 3.14 \rightarrow \times\ 3 \\
\hline
210
\end{array}
\qquad
\begin{array}{r}
9.66 \rightarrow \\
\times\ 7.83 \rightarrow \quad \times \rule{2cm}{0.4pt} \\
\end{array}
$$

$$
\begin{array}{r}
8{,}000{,}000 \\
\times\ 5 \\
\hline
40{,}000{,}000
\end{array}
$$

13. Following the example, estimate the other product by rounding both factors <u>to</u> <u>one</u> <u>digit</u>.

$$
\begin{array}{r}
0.0737 \rightarrow 0.07 \\
\times\ 0.825 \rightarrow \times\ 0.8 \\
\hline
0.056
\end{array}
\qquad
\begin{array}{r}
67{,}500 \rightarrow \\
\times\ 0.091 \rightarrow \quad \times \rule{2cm}{0.4pt}
\end{array}
$$

$$
\begin{array}{r}
10 \\
\times\ 8 \\
\hline
80
\end{array}
$$

14. Following the example, estimate the other product by rounding to one digit.

$$
\begin{array}{ccc}
4 \times 287 \\
\downarrow \quad \downarrow \\
4 \times 300 = 1{,}200
\end{array}
\qquad
\begin{array}{ccc}
6 \times 710{,}609 \\
\downarrow \qquad \downarrow \\
\underline{} \times \underline{} = \underline{}
\end{array}
$$

$$
\begin{array}{r}
70{,}000 \\
\times\ 0.09 \\
\hline
6300.00 \ \text{or}\ 6{,}300
\end{array}
$$

15. Estimate each product by rounding both factors to one digit.

a) 2.35 x 790
    ↓   ↓
    ___ x ___ = _____

b) 0.0075 x 6.94
    ↓   ↓
    ___ x ___ = _____

6 x 700,000 = 4,200,000

16. To estimate some products, we have to perform multiplications like the one at the left below. A two-step shortcut is described.

a) 2 x 800 = 1,600

b) 0.008 x 7 = 0.056

$$
\begin{array}{r}
400 \\
\times\ 70 \\
\end{array}
$$

	Step 1		Step 2
	400	(<u>two</u> 0's)	400
	x 70	(<u>one</u> "0")	x 70
	000	(<u>three</u> 0's)	28,000

<u>Step 1</u>: To get the number of 0's in the product, we add the total number of 0's in the factors. There are <u>three</u>, since 2 + 1 = 3 .

<u>Step 2</u>: To get the remaining digits of the product, we multiply "4" and "7".

Using the same shortcut, find each product.

a)   80
     x 60

b)   200
     x 30

c) 5,000
    x 700

a) 4,800   b) 6,000   c) 3,500,000

17. When using the shortcut, we can get another "0" in the product in addition to the total number of 0's in the factors. An example is shown below. Notice that we got a <u>fourth</u> "0" when multiplying "5" and "4". Use the shortcut for the other multiplications.

$$\begin{array}{r} 500 \\ \times\ 40 \\ \hline 20,000 \end{array}$$

a)
$$\begin{array}{r} 20 \\ \times\ 50 \\ \hline \end{array}$$

b)
$$\begin{array}{r} 500 \\ \times\ 800 \\ \hline \end{array}$$

---

18. Use the same shortcut for these.

a) 20 x 40 = _____

b) 10 x 900 = _____

c) 900 x 60 = _____

d) 400 x 500 = _____

a) 1,000

b) 400,000

---

19. To estimate the product below, we rounded each factor <u>to one digit</u>. Use the same method to estimate the other product.

$$\begin{array}{rcr} 387 & \rightarrow & 400 \\ \times\ 34 & \rightarrow & \times\ 30 \\ \hline & & 12,000 \end{array}$$

$$\begin{array}{rcl} 71,250 & \rightarrow & \\ \times\ 12 & \rightarrow & \times\ \underline{\hspace{1.5cm}} \end{array}$$

a) 800

b) 9,000

c) 54,000

d) 200,000

---

20. To estimate the product below, we rounded each factor to one digit. Use the same method to estimate the other product.

$$\begin{array}{rcr} 26.8 & \rightarrow & 30 \\ \times\ 71.9 & \rightarrow & \times\ 70 \\ \hline & & 2,100 \end{array}$$

$$\begin{array}{rcl} 88.7 & \rightarrow & \\ \times\ 66.6 & \rightarrow & \times\ \underline{\hspace{1.5cm}} \end{array}$$

$$\begin{array}{r} 70,000 \\ \times\ 10 \\ \hline 700,000 \end{array}$$

---

21. Estimate each product by rounding both factors to one digit.

a)
$$\begin{array}{rcl} 425 & \rightarrow & \\ \times\ 87.7 & \rightarrow & \times\ \underline{\hspace{1cm}} \end{array}$$

b)
$$\begin{array}{rcl} 51.9 & \rightarrow & \\ \times\ 767 & \rightarrow & \times\ \underline{\hspace{1cm}} \end{array}$$

$$\begin{array}{r} 90 \\ \times\ 70 \\ \hline 6,300 \end{array}$$

---

22. Following the example, estimate the other product.

67 x 615
↓   ↓
70 x 600 = 42,000

18.5 x 91.3
↓   ↓
___ x ___ = _____

a)
$$\begin{array}{r} 400 \\ \times\ 90 \\ \hline 36,000 \end{array}$$

b)
$$\begin{array}{r} 50 \\ \times\ 800 \\ \hline 40,000 \end{array}$$

---

23. Estimate each product by rounding both factors to one digit.

a) 567 x 71.6
   ↓   ↓
   ___ x ___ = _____

b) 14.3 x 9,775
   ↓   ↓
   ___ x ___ = _____

20 x 90 = 1,800

---

a) 600 x 70 = 42,000   b) 10 x 10,000 = 100,000

24.   To estimate the product below, we rounded each factor to one digit.

$$5.77 \times 12.8 \times 21.9$$
$$\downarrow \qquad \downarrow \qquad \downarrow$$
$$6 \ \times \ 10 \ \times \ 20 \ = 60 \times 20 = 1,200$$

Using the same method, estimate each product below.

a) 2.36 x 8.75 x 6.03
    ↓   ↓   ↓

    ____ x ____ x ____ = ____ x ____ = _____

b) 41.9 x 18.7 x 73.6
    ↓   ↓   ↓

    ____ x ____ x ____ = ____ x ____ = _____

---

25.   Suppose you get 3,555.1 as an answer for: 7.3 x 487

    a) Estimate the answer by rounding to one digit.
        The estimate is _____.

    b) Does 3,555.1 make sense as an answer? _____

a) 2 x 9 x 6 =
      18 x 6 = 108

b) 40 x 20 x 70 =
      800 x 70 = 56,000

---

26.   Suppose you get 17,214.3 as an answer for: 27.9 x 61.7

    a) Estimate the answer by rounding to one digit.
        The estimate is _____.

    b) Does 17,214.3 make sense as an answer? _____

a) 3,500

b) Yes

---

27.   Find each estimate by rounding to one digit.

    a) Estimate the area of a rectangular plot of ground whose length
        and width are 287 ft and 185 ft.

        _____

    b) Estimate the total rent paid in one year for an industrial building
        if the monthly rent is $925.

        _____

    c) Estimate the volume of a room whose length, width, and height
        are 19.5 ft, 11.5 ft, and 8.5 ft.

        _____

a) 1,800

b) No

---

a) 60,000 ft^2

b) $9,000

c) 1,800 ft^3

10-3   ESTIMATING SQUARES

In this section, we will discuss a method for estimating squares.

---

28.   Instead of the phrase "is approximately equal to", an equal sign <u>with a dot over it</u> can be used. That is:

> $\doteq$ means "is approximately equal to"

Therefore: $28.9 + 53.7 \doteq 30 + 50$ or 80

$17.9 \times 81.8 \doteq 20 \times 80$ or _____

---

29.   We can estimate a square by rounding the number to one digit. For example:

$$(6.83)^2 \doteq (7)^2 \text{ or } (7)(7) \text{ or } 49$$
$$(417)^2 \doteq (400)^2 \text{ or } (400)(400) \text{ or } 160,000$$
$$(0.0829)^2 \doteq (0.08)^2 \text{ or } (0.08)(0.08) \text{ or } 0.0064$$

Using the same method, estimate each square.

a) $(3.14)^2 \doteq$ _____   c) $(59.7)^2 \doteq$ _____

b) $(9.66)^2 \doteq$ _____   d) $(2,175)^2 \doteq$ _____

1,600

---

30.   Use the same method to estimate each square.

a) $(0.169)^2 \doteq$ _____   b) $(0.087)^2 \doteq$ _____

a) 9   c) 3,600
b) 100   d) 4,000,000

---

31.   Suppose you get $2,641.96$ as an answer for: $(51.4)^2$

a) Estimate the square by rounding to one digit.
The estimate is _____ .

b) Does $2,641.96$ make sense as an answer? _____

a) 0.04
b) 0.0081

---

32.   Suppose you get 0.0441 as an answer for: $(0.021)^2$

a) Estimate the square by rounding to one digit.
The estimate is _____ .

b) Does 0.0441 make sense as an answer? _____

a) 2,500
b) Yes

---

a) 0.0004
b) No

33.   Find each estimate by rounding to one digit.

a) Estimate the area of a square room with a 10.5 ft side.

_____

b) Estimate the area of a square piece of metal with a 72.6 cm side.

_____

| a) 100 ft^2 | b) 4,900 cm^2 |

10-4   THE SCIENTIFIC-NOTATION METHOD FOR ESTIMATING PRODUCTS AND SQUARES

The scientific-notation method is a general method for estimating products and squares.. However, it is most useful when multiplying or squaring quite large or quite small numbers. We will discuss the scientific-notation method in this section.

Note: If you have difficulty converting ordinary numbers to scientific notation, review the procedure in Chapter 3 - Powers of Ten and Scientific Notation.

34.   We can estimate the product for 3,780 x 61,200 by converting each factor to scientific notation. The steps are:

1) Write each factor in scientific notation.

2) Rearrange the factors so that the ordinary numbers and the powers of ten are together.

3) Round each ordinary number to one digit.

4) Perform the multiplication in each set of parentheses.

$3,780 \times 61,200 = (3.78 \times 10^3) \times (6.12 \times 10^4)$

$= (3.78 \times 6.12) \times (10^3 \times 10^4)$

$\doteq (4 \times 6) \times (10^3 \times 10^4)$

$\doteq 24 \times 10^7$

Therefore, $3,780 \times 61,200 \doteq 24 \times 10^7$ or what ordinary number? _____

35.   Use the same steps to complete this estimation.

$0.000671 \times 233,000 = (6.71 \times 10^{-4}) \times (2.33 \times 10^5)$

$= (6.71 \times 2.33) \times (10^{-4} \times 10^5)$

$\doteq (7 \times 2) \times (10^{-4} \times 10^5)$

a) $\doteq$ _____ x _____

b) $\doteq$ _____ (What ordinary number?)

240,000,000

a) 14×10^1

b) 140

36. Use the same steps to complete this estimation.

$$0.00078 \times 0.00912 = (7.8 \times 10^{-4}) \times (9.12 \times 10^{-3})$$
$$= (7.8 \times 9.12) \times (10^{-4} \times 10^{-3})$$
$$\doteq (8 \times 9) \times (10^{-4} \times 10^{-3})$$

a) $\doteq$ _____ x _____

b) $\doteq$ _____ (What ordinary number?)

a) 72×10^{-7}

b) 0.0000072

---

37. We used the scientific-notation method to estimate the product below.

$$0.0513 \times 69,800 = (5.13 \times 10^{-2}) \times (6.98 \times 10^{4})$$
$$= (5.13 \times 6.98) \times (10^{-2} \times 10^{4})$$
$$\doteq (5 \times 7) \times (10^{-2} \times 10^{4})$$
$$\doteq 35 \times 10^{2} \text{ or } 3,500$$

There is a shortcut for the scientific-notation method. We used the shortcut for the same estimation below.

$$0.0513 \times 69,800 \doteq 35 \times 10^{2} \text{ or } 3,500$$

(-2)   (+4)

5   x   7

Note: 1) The (-2) and (+4) above the numbers are the exponents of the 10 from scientific-notation form.

2) The 5 and 7 are the rounded first factors from scientific-notation form.

3) We got 35×10^{2} by multiplying the 5 and 7 and by adding the (-2) and (+4).

Use the shortcut to estimate the product below.

285 x 0.0000915 $\doteq$ _____ x _____

_____ x _____

---

38. Use the shortcut to estimate each product below.

a) 6,750,000 x 8,220,000 $\doteq$ _____ x _____

_____ x _____

b) 0.000591 x 0.0000128 $\doteq$ _____ x _____

_____ x _____

(+2) (-5)

$\doteq 27 \times 10^{-3}$

3 x 9   (or 0.027)

39.   Use the shortcut to estimate each product below.

a) 235,000 x 0.000086 $\doteq$ _____ x _____ or _____

_____ x _____

b) 0.00000395 x 6,280 $\doteq$ _____ x _____ or _____

_____ x _____

a) $\overset{+6}{\bigcirc}\ \overset{+6}{\bigcirc}$ $\doteq$ 56 x 10^{12}
       7 x 8

b) $\overset{-4}{\bigcirc}\ \overset{-5}{\bigcirc}$ $\doteq$ 6 x 10^{-9}
       6 x 1

---

40.   We used the scientific-notation method to estimate the product for the three-factor multiplication below.

$$1,820 \times 0.00416 \times 785 = (1.82 \times 10^3) \times (4.16 \times 10^{-3}) \times (7.85 \times 10^2)$$
$$= (1.82 \times 4.16 \times 7.85) \times (10^3 \times 10^{-3} \times 10^2)$$
$$\doteq (2 \times 4 \times 8) \times (10^3 \times 10^{-3} \times 10^2)$$
$$\doteq 64 \times 10^2 \text{ or } 6,400$$

The same type of shortcut can be used to estimate the product above by the scientific-notation method. That is:

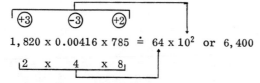

$$1,820 \times 0.00416 \times 785 \doteq 64 \times 10^2 \text{ or } 6,400$$

$$2\ \ \text{x}\ \ 4\ \ \text{x}\ \ 8$$

Use the shortcut to estimate this product.

325 x 0.00214 x 0.775 $\doteq$ _____ x _____ or _____

_____ x _____ x _____

a) $\overset{+5}{\bigcirc}\ \overset{-5}{\bigcirc}$ $\doteq$ 18 x 10^0
       2 x 9   (or 18)

b) $\overset{-6}{\bigcirc}\ \overset{+3}{\bigcirc}$ $\doteq$ 24 x 10^{-3}
       4 x 6   (or 0.024)

---

41.   Use the shortcut to estimate each product below.

a) 525 x 71,200 x 0.00195 $\doteq$ _____ x _____ or _____

_____ x _____ x _____

b) 0.0243 x 0.000058 x 7,550,000 $\doteq$ _____ x _____ or _____

_____ x _____ x _____

$\overset{+2}{\bigcirc}\ \overset{-3}{\bigcirc}\ \overset{-1}{\bigcirc}$ $\doteq$ 48 x 10^{-2}
3 x 2 x 8   (or 0.48)

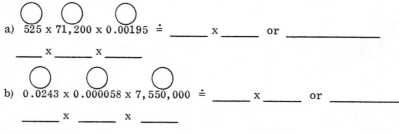

a) $\overset{+2}{\bigcirc}\ \overset{+4}{\bigcirc}\ \overset{-3}{\bigcirc}$ $\doteq$ 70 x 10^3
       5 x 7 x 2   (or 70,000)

b) $\overset{-2}{\bigcirc}\ \overset{-5}{\bigcirc}\ \overset{+6}{\bigcirc}$ $\doteq$ 96 x 10^{-1}
       2 x 6 x 8   (or 9.6)

42. Squaring a number in scientific notation is the same as squaring each of the factors. For example:

$$(5 \times 10^3)^2 = (5 \times 10^3) \times (5 \times 10^3)$$
$$= (5 \times 5) \times (10^3 \times 10^3)$$
$$= (5)^2 \times (10^3)^2$$

The principle above can be written in the following way:

$$(a \times 10^b)^2 = (a)^2 \times (10^b)^2$$

Using the principle, complete these:

a) $(2.49 \times 10^5)^2 = ($ $)^2 \times ($ $)^2$

b) $(7.13 \times 10^{-4})^2 = ($ $)^2 \times ($ $)^2$

---

43. We used the scientific-notation method to estimate the square below.

$$(61,500)^2 = (6.15 \times 10^4)^2$$
$$= (6.15)^2 \times (10^4)^2$$
$$\doteq (6)^2 \times (10^4)^2$$
$$\doteq 36 \times 10^8$$

Using the same steps, complete this estimation.

$$(0.000875)^2 = (8.75 \times 10^{-4})^2$$
$$= (8.75)^2 \times (10^{-4})^2$$
$$\doteq (9)^2 \times (10^{-4})^2$$
$$\doteq \underline{\quad\quad} \times \underline{\quad\quad}$$

a) $(2.49)^2 \times (10^5)^2$

b) $(7.13)^2 \times (10^{-4})^2$

---

44. We used a shortcut to estimate the square below by the scientific-notation method.

$$(237,000)^2 \doteq (2)^2 \times (10^5)^2 \text{ or } 4 \times 10^{10}$$

Using the same shortcut, estimate estimate each square below.

a) $(6,890,000)^2 \doteq ($ $)^2 \times ($ $)^2 \text{ or } \underline{\quad} \times \underline{\quad}$

b) $(0.0000412)^2 \doteq ($ $)^2 \times ($ $)^2 \text{ or } \underline{\quad} \times \underline{\quad}$

81×10^{-8}

---

45. Using the same shortcut, estimate each square below.

a) $(51,700,000)^2 \doteq$ _____

b) $(0.00788)^2 \doteq$ _____

a) $(7)^2 \times (10^6)^2$ or 49×10^{12}

b) $(4)^2 \times (10^{-5})^2$ or 16×10^{-10}

---

a) 25×10^{14} b) 64×10^{-6}

SELF-TEST 29 (pages 387-398)

Estimate each sum by rounding as directed.

1. Round to thousands.	2. Round to the nearest whole number.	3. Round to hundredths.
21,392 7,643 54,188 + 19,736	8.27 + 3.84 + 6.79	0.1682 + 0.0419 + 0.3957

Estimate each difference by rounding as directed.

4. Round to millions.	5. Round to tenths.	6. Round to hundreds.
29,815,394 - 13,278,106	3.0827 - 0.6195	2,418 - 794

Estimate each product by rounding each factor to one digit.

7. 718 x 37	8. 63.5 x 47.9	9. 78.2 x 3.36	10. 0.038 x 52.7

11. 8.73 x 0.216 x 32.5	12. 11.4 x 5.31 x 768

Estimate each square by rounding the number to one digit.	Estimate each square by using the scientific-notation method.
13. $(96.8)^2$ 14. $(0.0427)^2$	15. $(48,200)^2$ 16. $(0.0000817)^2$

Estimate each product by using the scientific-notation method.

17. 91,470 x 286,200	18. 0.000528 x 765,000	19. 0.00324 x 1,740 x 0.0861

ANSWERS:

1. 103,000	5. 2.5	9. 240	13. 10,000	17. 27×10^9
2. 19	6. 1,600	10. 2	14. 0.0016	18. 40×10^1 or 400
3. 0.61	7. 28,000	11. 54	15. 25×10^8	19. 54×10^{-2} or 0.54
4. 17,000,000	8. 3,000	12. 40,000	16. 64×10^{-10}	

10-5   ESTIMATING QUOTIENTS

In this section, we will discuss a method for estimating quotients when the divisor (or denominator) is a number between 1 and 10.

---

46. To estimate the quotient below, we divided to get one non-zero digit and then put 0's in the remaining places up to the decimal point. We began by writing the estimate "200" above the numerator to help get the correct number of 0's. Estimate the other quotient.

$$\overset{200}{\frac{717}{3}} \doteq 200 \qquad\qquad \frac{25,940}{6} \doteq \underline{\hspace{2cm}}$$

---

47. To estimate the quotient below, we ignored the digits to the right of the decimal point. Estimate the other quotient.

$$\overset{40.}{\frac{87.56}{2}} \doteq 40 \qquad\qquad \frac{76.95}{8} \doteq \underline{\hspace{2cm}}$$

4,000

---

48. To estimate the quotient below, we simply divided to get one non-zero digit. Estimate the other quotient.

$$\overset{.006}{\frac{0.0429}{7}} \doteq 0.006 \qquad\qquad \frac{0.000895}{4} \doteq \underline{\hspace{2cm}}$$

9

---

49. Estimate each quotient.

a) $\dfrac{1,536}{5} \doteq$ \underline{\hspace{2cm}}   c) $\dfrac{12.87}{9} \doteq$ \underline{\hspace{2cm}}

b) $\dfrac{25.66}{3} \doteq$ \underline{\hspace{2cm}}   d) $\dfrac{0.976}{4} \doteq$ \underline{\hspace{2cm}}

0.0002

---

50. To estimate the quotient below, we rounded 4.75 to 5 and then found the estimate in the usual way. Notice that we crossed out 4.75 and wrote 5 below it. Estimate the other quotient.

$$\overset{1,000}{\frac{5,140}{\underset{5}{\cancel{4.75}}}} \doteq 1,000 \qquad\qquad \frac{903}{1.95} \doteq \underline{\hspace{2cm}}$$

a) 300   c) 1
b) 8   d) 0.2

400, from: $\dfrac{903}{\underset{2}{\cancel{1.95}}}$ ← (with 400 above 903)

51.   We rounded 7.4 to 7 to estimate the quotient below. Estimate the other quotient.

$$\frac{\overset{10.}{86.5}}{\underset{7}{7.4}} \doteq 10 \qquad\qquad \frac{65.7}{2.8} \doteq \underline{\hspace{2cm}}$$

---

52.   We rounded 4.66 to 5 to estimate the quotient below. Estimate the other quotient.

$$\frac{\overset{.0004}{0.00225}}{\underset{5}{4.66}} \doteq 0.0004 \qquad\qquad \frac{0.04375}{6.791} \doteq \underline{\hspace{2cm}}$$

20, from: $\dfrac{\overset{20.}{65.7}}{\underset{3}{2.8}}$

---

53.   Estimate each quotient.

a) $\dfrac{429}{2.3} \doteq \underline{\hspace{2cm}}$ c) $\dfrac{27.7}{5.03} \doteq \underline{\hspace{2cm}}$

b) $\dfrac{94.6}{3.14} \doteq \underline{\hspace{2cm}}$ d) $\dfrac{1.42}{6.75} \doteq \underline{\hspace{2cm}}$

0.006, from: $\dfrac{\overset{.006}{0.04375}}{\underset{7}{6.791}}$

---

54.   Estimate each quotient.

a) $\dfrac{7,546,000}{9.4} \doteq \underline{\hspace{2cm}}$ c) $\dfrac{0.947}{1.88} \doteq \underline{\hspace{2cm}}$

b) $\dfrac{0.000069}{7.75} \doteq \underline{\hspace{2cm}}$ d) $\dfrac{0.246}{3.3} \doteq \underline{\hspace{2cm}}$

a) 200   c) 5
b) 30   d) 0.2

---

a) 800,000   b) 0.000008   c) 0.4   d) 0.08

10-6   USING DECIMAL-POINT-SHIFTS TO ESTIMATE QUOTIENTS

To estimate a quotient when the divisor (or denominator) is not a number between 1 and 10, we can shift the decimal points in both numbers to get an equivalent division in which the division (or denominator) is a number between 1 and 10. We will discuss that method in this section.

---

55.   Though the two divisions below look different, they are equivalent.
       Notice that the one on the right has a denominator between 1 and 10.

$$\frac{0.0295}{0.00722} = \frac{29.5}{7.22}$$

Continued on following page.

55.   Continued

To show that the two divisions are equivalent, we converted the one on the left to the one on the right below. We did so by multiplying by $\frac{10^3}{10^3}$ or "1".

$$\frac{0.0295}{0.00722} = \frac{0.0295}{0.00722} \times 1$$

$$= \frac{0.0295}{0.00722} \times \frac{10^3}{10^3}$$

$$= \frac{0.0295 \times 10^3}{0.00722 \times 10^3}$$

$$= \frac{29.5}{7.22}$$

The estimated quotient for $\frac{29.5}{7.22}$ is 4. Therefore, the estimated quotient for $\frac{0.0295}{0.00722}$ must also be _____ .

---

56.   Here is a diagram of what happened when we multiplied by $\frac{10^3}{10^3}$ in the last frame.

$$\frac{0.0295}{0.00722} = \frac{0.029{\,}5}{0.007{\,}22} = \frac{29.5}{7.22}$$

As you can see, multiplying by $\frac{10^3}{10^3}$ is the same as shifting the decimal point _____ places to the right in both terms.

| 4 |

---

57.   Though the two divisions below also look different, they are equivalent. Notice that the one on the right has a denominator between 1 and 10.

$$\frac{465,000}{237} = \frac{4,650}{2.37}$$

To show that the two divisions are equivalent, we converted the one on the left to the one on the right below. We did so by multiplying by $\frac{10^{-2}}{10^{-2}}$ or "1".

$$\frac{465,000}{237} = \frac{465,000}{237} \times 1$$

$$= \frac{465,000}{237} \times \frac{10^{-2}}{10^{-2}}$$

$$= \frac{465,000 \times 10^{-2}}{237 \times 10^{-2}}$$

$$= \frac{4,650}{2.37}$$

The estimated quotient for $\frac{4,650}{2.37}$ is 2,000. Therefore, the estimated quotient for $\frac{465,000}{237}$ must also be _____ .

| 3 |

58. Here is a diagram of what happened when we multiplied by $\frac{10^{-2}}{10^{-2}}$ in the last frame.

$$\frac{465,000}{237} = \frac{465,000.}{237.} = \frac{4,650}{2.37}$$

As you can see, multiplying by $\frac{10^{-2}}{10^{-2}}$ is the same as shifting the decimal point _____ places to the left in both terms.

| 2,000 |

59. To convert any division to an equivalent one in which the divisor (or denominator) is a number between 1 and 10, we can shift the decimal point the same number of places in the same direction in both terms. Two more examples are shown below.

$$\frac{4.58}{687} = \frac{04.58}{687.} = \frac{0.0458}{6.87}$$

$$\frac{215}{0.349} = \frac{215.0}{0.349} = \frac{2,150}{3.49}$$

Shift the decimal points in these to get an equivalent division whose denominator lies between 1 and 10.

a) $\frac{1,740,000}{27,500} = $ _____

b) $\frac{6.07}{9,470} = $ _____

c) $\frac{0.555}{0.000189} = $ _____

d) $\frac{0.00818}{0.391} = $ _____

| 2 |

60. The reason for shifting the decimal points to get an equivalent division whose denominator lies between 1 and 10 is to get a division whose quotient is easier to estimate. For example:

$$\frac{0.635}{0.0209} = \frac{0.635}{0.0209} = \frac{63.5}{2.09} \doteq \frac{30.}{2} \doteq 30$$

Using the same steps, estimate each quotient.

a) $\frac{43,900}{5,860}$ _____

b) $\frac{0.0162}{0.000224}$ _____

a) $\frac{174}{2.75}$ c) $\frac{5,550}{1.89}$

b) $\frac{0.00607}{9.47}$ d) $\frac{0.0818}{3.91}$

61. Use the decimal-point-shift method to estimate each quotient.

a) $\frac{8.69}{357}$ _____

b) $\frac{4,820}{6,190,000}$ _____

c) $\frac{0.000658}{0.0241}$ _____

d) $\frac{275,300}{875}$ _____

a) 7, from: $\frac{43.9}{5.86}$

b) 80, from: $\frac{162}{2.24}$

62. Suppose you get 55.1 as the rounded answer for $\frac{256}{4.65}$.

    a) Estimate the answer. The estimate is _____ .

    b) Does 55.1 make sense as an answer? _____

a) 0.02, from: $\frac{0.0869}{3.57}$

b) 0.0008, from: $\frac{0.00482}{6.19}$

c) 0.03, from: $\frac{0.0658}{2.41}$

d) 300, from: $\frac{2,753}{8.75}$

---

63. Suppose you get 41.6 as the rounded answer for $\frac{17.5}{0.0421}$.

    a) Estimate the answer. The estimate is _____ .

    b) Does 41.6 make sense as an answer? _____

a) 50

b) Yes

---

64. a) If an airplane travels 3,025 miles in 5.75 hours, estimate its average speed in "miles per hour".

                                            _____ miles per hour

    b) 1 kilometer is equal to 0.62 mile. Estimate the number of kilometers in 200 miles.

                                            _____ kilometers

a) 400

b) No

---

a) 500 miles per hour   b) 300 kilometers

10-7   THE SCIENTIFIC-NOTATION METHOD FOR ESTIMATING QUOTIENTS

The scientific-notation method is a useful way of estimating quotients when the division involves very large or very small numbers. We will discuss that method in this section.

---

65. We used the scientific-notation method to estimate the quotient below. The steps are described.

    1) Convert each term to scientific notation. $\frac{647,000}{314} = \frac{6.47 \times 10^5}{3.14 \times 10^2}$

    2) Combine the power-of-ten factors. $= \frac{6.47}{3.14} \times \frac{10^5}{10^2}$

                                                    $= \frac{6.47}{3.14} \times 10^3$

    3) Estimate the quotient for the simple division. $\doteq 2 \times 10^3$

    Therefore, the estimated quotient is 2×10^3 or _____ .

2,000

66.   We also used the scientific-notation method to estimate the quotient below.

$$\frac{0.000849}{0.0207} = \frac{8.49 \times 10^{-4}}{2.07 \times 10^{-2}}$$

$$= \frac{8.49}{2.07} \times \frac{10^{-4}}{10^{-2}}$$

$$\doteq 4 \times 10^{-2}$$

Therefore, the estimated quotient is 4×10^{-2} or _____ .

---

67.   We used the scientific-notation method to estimate the quotient below.

$$\frac{0.0887}{4,220} = \frac{8.87}{4.22} \times \frac{10^{-2}}{10^3} \doteq 2 \times 10^{-5} \text{ or } 0.00002$$

Using the same steps, estimate these quotients.

a) $\dfrac{947,000}{3,180} \doteq$ _____

b) $\dfrac{897}{0.000244} \doteq$ _____

0.04

---

68.   When using the scientific-notation method, we can get an estimate that is not in scientific notation. For example, the estimate below is not in scientific notation because 0.4 is not a number between 1 and 10.

$$\frac{2,090}{512,000} = \frac{2.09}{5.12} \times \frac{10^3}{10^5} \doteq 0.4 \times 10^{-2} \text{ or } 0.004$$

Using the same steps, estimate each quotient.

a) $\dfrac{0.0726}{0.000925} \doteq$ _____

b) $\dfrac{0.0365}{71,400} \doteq$ _____

a) 3×10^2 or 300

b) 4×10^6 or 4,000,000

---

69.   Use the scientific-notation method to estimate each quotient.

a) $\dfrac{13,270,000}{0.063} \doteq$ _____

b) $\dfrac{0.000651}{927,000} \doteq$ _____

a) 0.8×10^2 or 80

b) 0.5×10^{-6}
           or 0.0000005

---

a) 0.2×10^9 or 200,000,000

b) 0.7×10^{-9} or 0.0000000007

10-8   ESTIMATING SQUARE ROOTS

In this section, we will discuss methods for estimating the square roots of numbers.

70.   The square roots of the perfect squares from 1 to 100 are given below.

$$\sqrt{1} = 1 \qquad\qquad \sqrt{36} = 6$$
$$\sqrt{4} = 2 \qquad\qquad \sqrt{49} = 7$$
$$\sqrt{9} = 3 \qquad\qquad \sqrt{64} = 8$$
$$\sqrt{16} = 4 \qquad\qquad \sqrt{81} = 9$$
$$\sqrt{25} = 5 \qquad\qquad \sqrt{100} = 10$$

We can estimate the square roots of the other numbers from 1 to 100 by using the square root of the closest perfect square above. For example:

$$\sqrt{33} \doteq \sqrt{36} \text{ or } 6 \qquad\qquad \sqrt{67.5} \doteq \sqrt{64} \text{ or } 8$$

Using the same method, estimate each square root.

a) $\sqrt{8} \doteq$ _____   c) $\sqrt{17.9} \doteq$ _____

b) $\sqrt{52} \doteq$ _____   d) $\sqrt{97.6} \doteq$ _____

---

71.   To estimate the square root of a number larger than 100, we use these steps.

   1) Group the number into pairs of digits starting from the decimal point. For example:

   $$\sqrt{6,710} = \sqrt{6{,}7\,1\,0.}$$

   $$\sqrt{5,820,000} = \sqrt{5{,}8\,2\,0{,}0\,0\,0.}$$

   2) Estimate the square root of the left-most pair. Write that estimate above the left-most pair and write one "0" above each of the remaining groups to the decimal point. For example:

   $$\sqrt{6,710} \doteq \sqrt{\overset{8\ \ 0\ .}{6{,}7\,1\,0.}} \text{ or } 80$$

   $$\sqrt{5,820,000} \doteq \sqrt{\overset{2\ \ 0\ \ 0\ \ 0\ .}{5{,}8\,2\,0{,}0\,0\,0.}} \text{ or } 2,000$$

Complete the following estimations.

a) $\sqrt{856,000} \doteq \sqrt{\overset{9\ \ 0\ \ 0\ .}{8\,5\,6{,}0\,0\,0.}}$ or _____

b) $\sqrt{46,500} \doteq \sqrt{\overset{2\ \ 0\ \ 0\ .}{4\,6{,}5\,0\,0.}}$ or _____

a) 3, from: $\sqrt{9}$

b) 7, from: $\sqrt{49}$

c) 4, from: $\sqrt{16}$

d) 10, from: $\sqrt{100}$

72.   Using the same method, estimate each square root.

a) $\sqrt{6\,0\,8,7\,0\,0}$ ≐ _____   c) $\sqrt{9\,9,2\,0\,0}$ ≐ _____

b) $\sqrt{2,3\,8\,0}$ ≐ _____   d) $\sqrt{1,1\,5\,0,0\,0\,0}$ ≐ _____

a) 900

b) 200

---

73.   When estimating the square root of a number larger than 100, we ignore the digits to the right of the decimal point. For example:

$$\overset{2\ 0\ .}{\sqrt{5\,3\,5.7\,5}} \doteq 20 \qquad \overset{7\ 0\ .}{\sqrt{4,8\,0\,6.9\,9}} \doteq 70$$

Estimate each square root.

a) $\sqrt{8,0\,7\,5.6\,6}$ ≐ _____   b) $\sqrt{9\,5\,9.7\,5}$ ≐ _____

a) 800   c) 300

b) 50   d) 1,000

---

74.   To estimate the square root of a number smaller than "1", we use these steps.

1) Group the number into pairs of digits starting from the decimal point. For example:

$$\sqrt{0.0925} = \sqrt{0.0\,9\,2\,5}$$
$$\sqrt{0.00276} = \sqrt{0.0\,0\,2\,7\,6}$$

2) Bring up the decimal point. Write a "0" above any "00" group, and estimate the square root of the first non-zero group. For example:

$$\sqrt{0.0925} \doteq \overset{.\ \ 3}{\sqrt{0.0\,9\,2\,5}} \text{ or } 0.3$$

$$\sqrt{0.00276} \doteq \overset{.\ 0\ \ 5}{\sqrt{0.0\,0\,2\,7\,6}} \text{ or } 0.05$$

Complete the following estimations.

a) $\sqrt{0.000487} \doteq \overset{.\ 0\ \ 2}{\sqrt{0.0\,0\,0\,4\,8\,7}}$ or _____

b) $\sqrt{0.0000837} \doteq \overset{.\ 0\ \ 0\ \ 9}{\sqrt{0.0\,0\,0\,0\,8\,3\,7}}$ or _____

a) 90   b) 30

a) 0.02   b) 0.009

75.   Estimate each square root.

a) $\sqrt{0.00358} \doteq$ _____   c) $\sqrt{0.0126} \doteq$ _____

b) $\sqrt{0.515} \doteq$ _____   d) $\sqrt{0.0000472} \doteq$ _____

---

76.   Estimate each square root.

a) $\sqrt{113,000} \doteq$ _____   c) $\sqrt{17,500,000} \doteq$ _____

b) $\sqrt{0.636} \doteq$ _____   d) $\sqrt{0.00000083} \doteq$ _____

a) 0.06	c) 0.1
b) 0.7	d) 0.007

---

77.   Suppose you get 93,410 as the rounded answer for $\sqrt{87,250,000}$.

a) Estimate the answer. The answer is _____ .

b) Does 93,410 make sense as an answer? _____

a) 300	c) 4,000
b) 0.8	d) 0.0009

---

78.   Estimate the length of the side of a square

a) whose area is 1,647 yd². _____

b) whose area is 65.75 cm². _____

a) 9,000

b) No

---

a) 40 yd   b) 8 cm

10-9   ESTIMATING ANSWERS FOR COMBINED OPERATIONS

In this section, we will discuss methods for estimating answers to some simple combined operations.

79.   To estimate the answer below, we began by rounding each number to one digit. Estimate the other answer.

$$\frac{(2.17)(91.5)}{39.6} \doteq \frac{(2)(90)}{40} \text{ or } \frac{180}{40} \text{ or } 4$$

$$\frac{(31.5)(68.3)}{72.4} \doteq$$ _____

30, from: $\dfrac{2,100}{70}$

80.   Following the example, estimate the other answer.

$$\frac{52.6}{(10.9)(61.6)} \doteq \frac{50}{(10)(60)} \text{ or } \frac{50}{600} \text{ or } 0.08$$

$$\frac{926}{(2.03)(39.5)} \doteq \underline{\hspace{4cm}}$$

---

81.   Following the example, estimate the other answer.

$$\frac{27.5 + 52.3}{18.9} \doteq \frac{30 + 50}{20} \text{ or } \frac{80}{20} \text{ or } 4$$

$$\frac{819 + 388}{41.3} \doteq \underline{\hspace{4cm}}$$

10, from: $\frac{900}{80}$

---

82.   To estimate the answer below, we began by rounding both numbers to one digit. Estimate the other answer.

$$(21.5)^2(88.6) \doteq (20)^2(90) \text{ or } (400)(90) \text{ or } 36,000$$

$$(5.79)^2(12.3) \doteq \underline{\hspace{4cm}}$$

30, from: $\frac{1,200}{40}$

---

83.   Following the example, estimate the other answer.

$$\sqrt{(6.45)(8.09)} \doteq \sqrt{(6)(8)} \text{ or } \sqrt{48} \text{ or } 7$$

$$\sqrt{(29.1)(52.5)} \doteq \underline{\hspace{4cm}}$$

360, from: $(36)(10)$

---

84.   Following the example, estimate the other answer.

$$\sqrt{\frac{79.6}{2.14}} \doteq \sqrt{\frac{80}{2}} \text{ or } \sqrt{40} \text{ or } 6$$

$$\sqrt{\frac{616}{10.9}} \doteq \underline{\hspace{4cm}}$$

40, from: $\sqrt{1,500}$

---

85.   Following the example, estimate the other answer.

$$\sqrt{(6.8)^2 + (5.2)^2} \doteq \sqrt{7^2 + 5^2} \text{ or } \sqrt{49 + 25} \text{ or } \sqrt{74} \text{ or } 9$$

$$\sqrt{(81.2)^2 - (27.5)^2} \doteq \underline{\hspace{4cm}}$$

8, from: $\sqrt{60}$

70, from: $\sqrt{5,500}$

SELF-TEST 30 (pages 399-409)

Estimate each quotient.

1.

$$\frac{225.7}{3.2}$$

_____

2.

$$\frac{0.0946}{78.3}$$

_____

3.

$$\frac{553,000}{623}$$

_____

4.

$$\frac{0.00715}{0.206}$$

_____

5.

$$\frac{6,740}{0.00826}$$

_____

6.

$$\frac{0.338}{49,150}$$

_____

7. Four possible answers for $\frac{6.24}{192}$ are: 0.000325   0.00325   0.0325   0.325

Estimate the quotient. Which answer makes sense? _____

Estimate each square root.

8.

$\sqrt{37.2}$

_____

9.

$\sqrt{41,300}$

_____

10.

$\sqrt{0.476}$

_____

11.

$\sqrt{0.00000822}$

_____

Estimate each answer.

12.

$$\frac{415}{(22.3)(39.6)}$$

_____

13.

$$\frac{5.34 + 2.86}{3.92}$$

_____

14.

$\sqrt{(596)(4.27)}$

_____

ANSWERS:

1. 70

2. 0.001

3. 900

4. 0.03

5. 8×10^5
   or 800,000

6. 6×10^{-6}
   or 0.000006

7. 0.0325

8. 6

9. 200

10. 0.7

11. 0.003

12. 0.5

13. 2

14. 50

SUPPLEMENTARY PROBLEMS - CHAPTER 10

Assignment 29

Estimate each sum.

1. Round to hundreds.

738
592
225
+ 916

2. Round to millions.

18,217,596
5,803,147
+ 7,368,524

3. Round to thousands.

9,628
35,702
+ 6,437

4. Round to thousandths.

0.0547 + 0.0084

5. Round to one digit.

38 + 81 + 53

6. Round to hundredths.

0.063 + 0.147 + 0.019

Estimate each difference.

7. Round to thousands.

31,247
- 6,185

8. Round to the nearest whole number.

14.83
- 5.19

9. Round to millions.

43,271,800
- 15,904,200

10. Round to tenths.

2.315 - 0.58

11. Round to hundreds.

609 - 183

12. Round to thousandths.

0.04592 - 0.0373

13. A calculator answer of 169.3 was reported for: 57.4 + 82.3 + 79.6
    Using estimation, does 169.3 make sense as an answer?

14. In a physics experiment, the initial volume of a gas was 1.182 liters and the final volume was 0.710 liter. By rounding to tenths, estimate the change in volume.

Estimate each product by rounding each factor <u>to one digit</u>.

15.   41.6
      x 2.83

16.   7,130
      x 8.67

17.   294
      x 622

18.   93.5
      x 0.195

19. 49.4 x 31.8

20. 536 x 18

21. 0.0432 x 8.849

22. 2.78 x 8.16 x 19.4

23. 1.15 x 4.14 x 1.76

24. 65.7 x 12.6 x 178

25. If 18 kilograms of copper can be obtained from one ton of ore, estimate the amount of copper that can be obtained from 625 tons of ore.

Estimate each square by rounding the number <u>to one digit</u>.

26. $(7.14)^2$

27. $(48.6)^2$

28. $(0.78)^2$

29. $(0.0197)^2$

Estimate each answer by using the scientific-notation method.

30. 37,500 x 8,410

31. 0.00228 x 0.0364

32. $(71,300)^2$

Assignment 30

Estimate each quotient.

1. $\dfrac{22,600}{7}$ 2. $\dfrac{0.00704}{3}$ 3. $\dfrac{3.156}{4.92}$ 4. $\dfrac{571.92}{8.35}$

5. $\dfrac{0.736}{9.22}$ 6. $\dfrac{4,660}{5.37}$ 7. $\dfrac{811,300}{19.4}$ 8. $\dfrac{0.376}{68.5}$

9. $\dfrac{823}{0.185}$ 10. $\dfrac{0.216}{0.521}$ 11. $\dfrac{50.9}{0.0786}$ 12. $\dfrac{0.000732}{0.00827}$

13. $\dfrac{43,400}{0.00586}$ 14. $\dfrac{0.00905}{87,900}$ 15. $\dfrac{163}{2,852,000}$ 16. $\dfrac{31.7}{0.000485}$

17. Four possible answers for $\dfrac{5.56}{0.00613}$ are: 9,070   907   90.7   9.07

Estimate the quotient. Which answer makes sense?

18. Four possible answers for $\dfrac{283}{7,140}$ are: 0.00396   0.0396   0.396   3.96

Estimate the quotient. Which answer makes sense?

Estimate each square root.

19. $\sqrt{68}$ 20. $\sqrt{10.6}$ 21. $\sqrt{0.47}$ 22. $\sqrt{0.082}$

23. $\sqrt{52,800}$ 24. $\sqrt{0.000114}$ 25. $\sqrt{0.0000661}$ 26. $\sqrt{79,140,000}$

27. Estimate the length of the side of a square whose area is $3,470 \text{ cm}^2$.

28. Estimate the length of the side of a square whose area is 0.262 in^2.

Estimate each answer.

29. $\dfrac{(67.9)(3.24)}{5.15}$ 30. $\dfrac{(526)(11.4)}{6.77}$ 31. $\dfrac{81.6}{(1.79)(23.2)}$ 32. $\dfrac{6.26}{(29.5)(7.18)}$

33. $\dfrac{922 + 578}{2.93}$ 34. $\dfrac{12.6 + 48.2}{19.3}$ 35. $(1.85)^2(4.79)$ 36. $(50.8)^2(23.7)$

37. $\sqrt{(38.5)(91.6)}$ 38. $\sqrt{\dfrac{315}{5.04}}$ 39. $\sqrt{(37.4)^2 + (82.7)^2}$ 40. $\sqrt{(9.29)^2 - (5.88)^2}$

ANSWERS FOR SUPPLEMENTARY PROBLEMS

<u>CHAPTER 1</u> - <u>ELEMENTARY CALCULATOR OPERATIONS</u>

Assignment 1
1. hundreds   2. ten-thousands   3. billions   4. ones   5. ten-thousandths   6. hundredths
7. tenths   8. millionths   9. 7   10. 6   11. 9   12. 3   13. .23   14. .0080
15. .009   16. .000460   17. 48.03   18. 4.00739   19. 256.7   20. 18.031   21. 20,071,000
22. 403,620   23. 8,100,060,000   24. 4,208.2   25. .0032   26. 3.009   27. 15.70
28. .01200   29. .00107   30. 1,005.93   31. 49,810   32. 6.2   33. .001   34. 1,013
35. 310.05   36. .051

Assignment 2
1. 2.161   2. 843.4   3. 2,392   4. .08268   5. 2,159,000   6. 17.5   7. .0288   8. 6.6
9. 725.76   10. .00015   11. 96,100   12. 28.7   13. .26   14. .6724   15. .75   16. .018
17. 37,500   18. Not possible   19. 0   20. 82.32   21. -59   22. -760   23. 264
24. -135.3   25. 6.29   26. -.44   27. .625   28. -8,750   29. 45,369   30. Not possible
31. .0025   32. .08   33. 5,000   34. .92   35. .15625   36. 1.6

Assignment 3
1. 380,000   2. 85   3. 4.021   4. .000170   5. 30,400,000   6. 17,900   7. 523.10
8. .618   9. 109,000,000   10. 4,000,000,000   11. 5.87   12. .0240   13. 21,700
14. 20.26   15. 60,000,000   16. 310,000   17. 78.0   18. 240   19. .0147   20. 2,100
21. 4,600,000   22. 193,000   23. .000347   24. .756   25. .4684   26. .00380
27. .338   28. 13.5   29. 2.24   30. .006273   31. 874,000   32. 13,700

<u>CHAPTER 2</u> - <u>PERCENT, RATIO, PROPORTION</u>

Assignment 4
1. $\frac{3}{4}$ 2. $\frac{1}{5}$ 3. $\frac{1}{3}$ 4. $\frac{9}{10}$ 5. $\frac{1}{4}$ 6. .19   7. .039   8. .0065   9. 1.47   10. 3
11. 81.4%   12. 4.25%   13. .6%   14. 170%   15. 500%   16. 50%   17. 80%   18. 30%
19. $66\frac{2}{3}$%   20. 59%   21. 35%   22. 17%   23. 131%   24. 4.15%   25. .05%   26. 42.02%
27. 40%   28. 300   29. 50   30. $33\frac{1}{3}$%   31. $21.25   32. 700   33. 2   34. 400   35. 4%

Assignment 5
1. 17.5 pounds   2. 83%   3. 16 kilograms   4. 60.67%   5. $273.50   6. 667 grams
7. 85.8%   8. $25,000   9. .038%   10. 7,300 pounds   11. n = 760   12. x = 32
13. y = 55.8   14. t = 1.71   15. w = 50.7   16. R = 4.54   17. x = 3,400

Assignment 6
1. 3.14   2. $\frac{4}{3}$ 3. 644 miles   4. 2.2 pounds   5. 218 grams   6. 8.3 centimeters
7. 8 pounds   8. 214 washers   9. 40 teeth   10. 21.7 centimeters   11. 58.7 kilowatts

CHAPTER 3 - POWERS OF TEN AND SCIENTIFIC NOTATION

Assignment 7
1. $1,000,000$ 2. $.01$ 3. $10,000$ 4. $.000001$ 5. 10 6. 10^3 7. 10^{-3} 8. 10^8
9. 10^{-1} 10. 10^{-7} 11. 10^5 12. 10^1 13. 10^0 14. 10^3 15. 10^{-6} 16. 10^6
17. 10^{-3} 18. hundreds   19. billions   20. tenths   21. ten-thousandths   22. 10^{-2}
23. 10^2 24. 10^4 25. 10^{-6} 26. 10^4 27. 10^{-2} 28. 10^0 29. 10^3 30. 10^{-1}
31. 10^{-6} 32. 10^6 33. 10^2 34. 10^0 35. 10^2 36. 10^{-3} 37. 10^0 38. 10^{-14}

Assignment 8
1. 49.3 2. 760 3. $.0000592$ 4. $2,000,000$ 5. 21.5 6. $36,200$ 7. 80 8. $.00094$
9. $8,200$ 10. $100,000$ 11. $.005$ 12. $.0000046$ 13. $.6108$ 14. 2.92 15. $3,070,000$
16. 581.4 17. 2×10^{-2} 18. 7.6×10^{-4} 19. 4.9×10^4 20. 1×10^7 21. 9.15×10^2
22. 1.832×10^5 23. 5.604×10^{-3} 24. 2.18×10^{-1} 25. 5×10^5 26. 1.27×10^6
27. 3.9×10^{-3} 28. 6×10^{-8} 29. 4.21×10^0 30. 9.8×10^{-4} 31. 8.8×10^5
32. 2.075×10^3 33. 5.40×10^{12} 34. 4.89×10^{-13} 35. 6.17×10^{13} 36. 6.02×10^{23}
37. $12,756$ kilometers   38. $.000000782$ centimeter   39. 1.1574×10^{-5} day
40. $45,000,000,000$ years

Assignment 9
1. 4.2×10^{11} 2. 5.34×10^{-8} 3. 1.40×10^4 4. 4.29×10^{-6} 5. 5.2×10^8 6. 3.58×10^{-5}
7. 5.31×10^2 8. 4.03×10^{-2} 9. 5.24×10^0 10. 6.63×10^8 11. 2.68×10^{-5}
12. 9.04×10^6 13. 2.21×10^{-6} 14. 1.93×10^6 15. 2.32×10^{-3} 16. 7.84×10^2
17. 2.30×10^{-7} 18. 5.27×10^{-2} 19. 1.41×10^{-11} 20. 6.45×10^{-4} 21. 6.03×10^4
22. 1.26×10^{-2} 23. 1.99×10^4 24. 5.00×10^7 25. 4.29×10^{-13} 26. $N = 3.74 \times 10^8$
27. $x = 1.32 \times 10^0$ 28. $y = 4.67 \times 10^{-4}$ 29. 1.27 meters (from 1.27×10^0)
30. $18.5°C$ (from 1.85×10^1)

CHAPTER 4 - FORMULA EVALUATION

Assignment 10
1. $R = 332.9$ 2. $V = 6,400$ 3. $Q = 54.7$ 4. $L = 162$ 5. $t = 0.00174$ 6. $v = 6.77$
7. $X = 29.9$ 8. $P = 0.718$ 9. $C = -40$ 10. $T = 301$ 11. $H = 1.87$ 12. $m = 0.0872$
13. $E = 31,000,000$ 14. $F = 14.5$

Assignment 11
1. $P = 806$ 2. $A = 0.697$ 3. $R = 1.90$ 4. $s = 224$ 5. $a = 12.7$ 6. $p = 9.16$ 7. $E = 120$
8. $b = 480$ 9. $w = 77$ 10. $F = 86.6$ 11. $t = 0.054$ 12. $C = 0.424$ 13. $V = 710,000$
14. $s = 6.38$

Assignment 12
1. $v_1 = 180$ 2. $e = 32.5$ 3. $I_a = 0.534$ 4. $F = 2.26$ 5. $d = 1.30$ 6. $V_1 = 1,650$
7. $F_2 = 837$ 8. $P_1 = 75$ 9. $Q = 7.96 \times 10^{-7}$ 10. $t = 5.32 \times 10^{-10}$ 11. $Z = 6.96 \times 10^7$
12. $y = -54.5$ 13. $y = 16$ 14. $y = 381.25$

Assignment 13
1. $M = 4,900$ 2. $A = 1.71$ 3. $r = 24.6$ 4. $d = 5,600$ 5. $E = 3,100,000$ 6. $A = 93$
7. $\triangle L = 6.24 \times 10^{-2}$ 8. $I = 0.017$ 9. $R = 4.14$ 10. $v = 25.5$ 11. $m = 14,100$ 12. $R = 68.1$

CHAPTER 5 - MEASUREMENT SYSTEMS

Assignment 14
1. 102 in   2. $15,840$ ft   3. 4 yd   4. 7 mi   5. 127 cm   6. 9.14 m   7. 4.6 m
8. 80.5 km   9. 7.87 in   10. 124 mi   11. 328 ft   12. 3.2 ft   13. $4,200$ mm   14. 6.5 km
15. $.84$ hm   16. 225 cm   17. 59 cm   18. 60 dm   19. $.337$ km   20. 70 mm   21. $8,140$ m
22. 30 m   23. 4.8 m   24. $.5$ m   25. 6.33 m   26. $.075$ m   27. $1,000$ m   28. 26 m
29. 150 mm   30. 6.3 cm   31. 240 mm   32. $.9$ dm   33. 82 hm   34. 1 km   35. $.003$ km
36. $7,500$ dam

CHAPTER 5 - MEASUREMENT SYSTEMS (continued)

Assignment 15
1. 44 oz   2. 25 lb   3. 3.5 T   4. 1,000 lb   5. 354.4 g   6. 1.76 oz   7. 18 lb
8. 63.6 kg   9. 3,200 g   10. .57g   11. 430 mg   12. .914 kg   13. 40 mg   14. 2.8 cg
15. 610 dag   16. 1,900 dg   17. 390 sec   18. .8 min   19. 2.25 hr   20. 4,500 sec
21. .07 sec   22. 3,860 msec   23. 3.5 gal   24. 8.25 qt   25. 3.19 ℓ 26. 800 mℓ
27. 22ℓ 28. 10.3 qt   29. 5 mi/sec   30. 282 gal/min   31. 62.1 mi/hr   32. 14 cm/sec
33. 32.4 km/hr   34. 4 gal/sec   35. 48.3 km   36. 354.2 km   37. 644 km

Assignment 16
1. 20°C   2. 1,000°C   3. 5°F   4. 212°F   5. -220°F   6. -5°C   7. 27°C   8. 63°K
9. 10^6 10. 10^{-12} 11. 10^{12} 12. 10^{-3} 13. micro-   14. kilo-   15. nano-   16. gigi-
17. 18×10^{-3} sec   18. 5.6×10^3 m   19. 240×10^{-6} g   20. 5.2 Mm   21. 19 mg   22. 376 psec
23. 8.7 kg   24. 4.5 mm   25. 29.4 μsec   26. 34 mm   27. .6 μg   28. 750 μm
29. 1.2 msec   30. 5,100 km   31. 4 m   32. 82 mg   33. 7,070 psec   34. 93.8 hm

CHAPTER 6 - AREAS AND VOLUMES

Assignment 17
1. A = 2,020 cm^2, P = 185.8 cm   2. A = 16,000 ft^2, P = 508 ft   3. A = 38.8 m^2, P = 24.92 m
4. 4.5 ft^2 5. 29,300 cm^2 6. 20 in^2 7. 7.18 yd^2 8. 72.9 hectares   9. 184 cm^2
10. A = 520 cm^2 11. A = 5.08 in^2 12. A = 118,000 ft^2 13. A = 27.7 m^2 14. A = 388 yd^2
15. A = 2,996,000 ft^2 16. A = 1,500 cm^2 17. A = 900 in^2 18. A = 44.6 m^2

Assignment 18
1. C = 24.9 cm   2. C = 33,200 ft   3. A = 19.17 m^2 4. A = 522,000 yd^2 5. A = 1,350 cm^2
6. A = 4.3 m^2 7. A = 7.78 in^2 8. r = 1.12 cm   9. d = 17.8 ft   10. T = 1.00   11. M = 1,040
12. f = 496   13. V = 135,000 cm^3 14. V = 1,190 in^3 15. V = 3.4 ft^3 16. s = 3.52 cm
17. 11,800 in^3 18. 950 cm^3 19. 26.2 yd^3

Assignment 19
1. 4,100 ft^3 2. 4,710 cm^3 3. 18.6 m^3 4. 1,400 ft^2 5. 1,360 cm^2 6. 42.1 m^2
7. 391 cm^3 8. 6,900 in^3 9. 777 cm^3 10. 8.81 m^3 11. 900 cm^3 12. 140 in^3
13. 14.7 lb   14. 679 g   15. 25 kg   16. 415 lb

CHAPTER 7 - RIGHT-TRIANGLE TRIGONOMETRY

Assignment 20
1. F = 78°   2. P = 28°   3. A = 65°   4. t = 5,600 m   5. a = 32.8 cm   6. p = 5.52"
7. d = 26.2 ft   8. t = 630 in   9. w = 0.463 cm   10. G = 37°, H = 106°   11. A = 30°, B = 30°
12. R = 45°, S = 45°   13. A = 14,500 m^2 14. A = 254 cm^2 15. A = 20.1 in^2

Assignment 21
1. 0.2126   2. 0.9986   3. 0.3256   4. 1.   5. 0.5   6. 0.8660   7. 2.1445   8. 0.0175
9. A = 14°   10. P = 73°   11. H = 89°   12. Q = 72°   13. T = 7°   14. F = 82°   15. A = 45°
16. B = 12°   17. 29.1   18. 11.9   19. 77.6   20. 41.2   21. 347   22. 275   23. 110
24. 516   25. A = 23°   26. G = 59°   27. R = 71°   28. E = 9°   29. $\sin R = \dfrac{r}{s}$ 30. $\cos R = \dfrac{t}{s}$

31. $\tan R = \dfrac{r}{t}$ 32. $\sin T = \dfrac{t}{s}$ 33. $\cos T = \dfrac{r}{s}$ 34. $\tan T = \dfrac{t}{r}$ 35. $\cos B = \dfrac{a}{c}$ 36. $\sin A = \dfrac{a}{c}$

37. $\tan B = \dfrac{b}{a}$ 38. $\tan A = \dfrac{a}{b}$ 39. $\cos A = \dfrac{b}{c}$ 40. $\sin B = \dfrac{b}{c}$

Assignment 22
1. h = 3.63 cm   2. w = 3.31"   3. p = 7.22 m   4. d = 438 ft   5. t = 105 cm   6. b = 268"
7. F = 37°, G = 53°   8. A = 61°, B = 29°   9. R = 41°, S = 49°   10. h = 6.70", v = 5.62"
11. A = 34°   12. w = 11.2 cm

CHAPTER 8 - POWERS, ROOTS, LOGARITHMS

Assignment 23
1. 59,049   2. 248,832   3. 4,304.6721   4. 0.000064   5. 2.460375   6. 2.82×10^{21}
7. 9.87×10^{10} 8. 6.10×10^{-12} 9. 9.09×10^{-16} 10. 31.1   11. 2.31   12. 1.44
13. 0.85   14. 0.395   15. 0.4967   16. $18^{0.6}$ 17. $6^{0.5}$ 18. $3^{4.5}$ 19. $542^{0.25}$ 20. 65.3
21. 7.82   22. 832   23. 0.004630   24. 0.298   25. 2.34   26. $W = 39.4$ 27. $R = 2.81$
28. $D = 631$ 29. $H = 0.859$

Assignment 24
1. 3.80   2. 248,000   3. 0.676   4. 0.000942   5. $10^{3.9165}$ 6. $10^{1.6201}$ 7. $10^{-1.2007}$
8. $10^{-3.6655}$ 9. 6.7259   10. 0.9736   11. -0.1079   12. -2.9393   13. $\log 185 = 2.2672$
14. $-0.7595 = \log 0.174$ 15. $x = \log 27.4$ 16. $\log G = -2.3814$ 17. $10^{2.4166} = 261$
18. $0.02 = 10^{-1.6990}$ 19. $R = 10^{5.8037}$ 20. $10^y = 7,190$ 21. $G = 7.37$ 22. $T = 0.000136$
23. $b = 4.2866$ 24. $N = 17$ 25. $P = 5.43$ 26. $M = 3.6$ 27. $r = 1.71$

Assignment 25
1. 53.5   2. 0.179   3. 11,500   4. $e^{4.75}$ 5. $e^{-3.260}$ 6. $e^{1.5518}$ 7. $H = 13.4$
8. $y = 2.08$ 9. $B = 243$ 10. 3.3945   11. 8.9306   12. 0.7129   13. -6.2554
14. $4.15 = \ln 63.4$ 15. $\ln 0.0828 = -2.49$ 16. $9.31 = e^{2.23}$ 17. $e^{-0.851} = 0.427$ 18. $R = 3.75$
19. $y = 8.2$ 20. $P = 0.0241$ 21. $r = 342$ 22. $t = 7.23$ 23. $N = 57.7$

CHAPTER 9 - MEASUREMENT CONCEPTS

Assignment 26
1. tenths of a liter   2. millionths of a second   3. thousandths of a gram   4. hundredths of a meter
5. 0.0195"   6. 29.6 km   7. 0.00420 g   8. 0.0960 cm   9. 73.8 cm   10. 5.90 cm and 321.64 cm
11. 30.25" and 30.15"   12. 7.015 cm and 7.005 cm   13. 0.4805 sec and 0.4795 sec
14. 119.5 ft and 118.5 ft   15. 0.05 cm   16. 0.5 gram   17. 0.005 ft   18. 0.00005 meter
19. 152.1"   20. 6.40 cm   21. 18.0 grams   22. 1.62 sec   23. 4.50"   24. 0.561 sec
25. 265 ft   26. 10.07 g   27. 2.420 cm

Assignment 27
1. 0.0025   2. 0.004   3. 0.00125   4. 0.02   5. 0.000013   6. 0.000386   7. 0.217 sec
8. 5.18 cm   9. three   10. two   11. four   12. five   13. three   14. two   15. three
16. 436 cm   17. 51.72"   18. 0.340 sec   19. 2.3170 sec   20. 2.3170 sec   21. 0.016 sec
22. 115 sec   23. 0.0510 gram   24. 864.2 grams   25. 15.285 grams   26. 0.0510 grams
27. 36.4   28. 0.0590   29. 215.0   30. 8.100   31. 4.8×10^5 32. 7.0×10^2 33. 1.5×10^3

Assignment 28
1. 53.9   2. 0.2227   3. 3.7   4. 3.107   5. 0.00048   6. 347   7. 39.917   8. 459.7
9. 0.00462   10. 0.23   11. 85.73   12. 4.69   13. 1.4   14. 0.8080   15. 3.75   16. 15
17. 22.671   18. 6.6×10^4 19. 1.270×10^5 20. 9.04×10^3 21. 3.66×10^4 22. 4.6×10^7
23. 1.739×10^9 24. 186 grams   25. 4.89 cm   26. 240 watts and 260 watts

CHAPTER 10 - ESTIMATION

Assignment 29
1. 2,400   2. 31,000,000   3. 52,000   4. 0.063   5. 170   6. 0.23   7. 25,000   8. 10
9. 27,000,000   10. 1.7   11. 400   12. 0.009   13. No, since the estimated answer is 220.
14. 0.5 liter   15. 120   16. 63,000   17. 180,000   18. 18   19. 1,500   20. 10,000
21. 0.36   22. 480   23. 8   24. 140,000   25. 12,000 tons   26. 49   27. 2,500   28. 0.64
29. 0.0004   30. 32×10^7 (or 320,000,000)   31. 8×10^{-5} (or 0.00008)   32. 49×10^8 (or 4,900,000,000)

Assignment 30
1. 3,000   2. 0.002   3. 0.6   4. 70   5. 0.08   6. 900   7. 40,000   8. 0.005   9. 4,000
10. 0.4   11. 600   12. 0.09   13. 7×10^6 (or 7,000,000)   14. 1×10^{-7} (or 0.0000001)
15. 5×10^{-5} 16. 6×10^4 (or 60,000)   17. 907   18. 0.0396   19. 8   20. 3   21. 0.7
22. 0.3   23. 200   24. 0.01   25. 0.008   26. 9,000   27. 60 cm   28. 0.5 in   29. 40
30. 700   31. 2   32. 0.03   33. 500   34. 3   35. 20   36. 50,000   37. 60   38. 8
39. 90   40. 7

INDEX